Creating an RTS Game in Unity 2023

A comprehensive guide to creating your own strategy game from scratch using C#

Bruno Cicanci

Creating an RTS Game in Unity 2023

Group Product Manager: Rohit Rajkumar

Publishing Product Manager: Nitin Nainani

Senior Editor: Hayden Edwards

Technical Editor: K Bimala Singha

Copy Editor: Safis Editing

Project Coordinator: Aishwarya Mohan

Proofreader: Safis Editing

Indexer: Hemangini Bari

Production Designer: Vijay Kamble

Marketing Coordinators: Namita Velgekar and Nivedita Pandey

First published: October 2023

Production reference: 3070824

Published by Packt Publishing Ltd.

Grosvenor House

11 St Paul's Square

Birmingham

B3 1RB

ISBN 978-1-80461-324-5

www.packtpub.com

Contributors

About the author

Bruno Cicanci is a software engineer and game developer, with a BS computer science degree and professional experience in different technologies and platforms. Since 2009, he has worked at prestigious game studios such as Glu Mobile, Electronic Arts, and Aquiris, and has helped develop and publish many games, mainly using Unity for mobile devices.

Currently, Bruno develops games and reusable technologies at Ubisoft. Plus, he writes about game development on his personal blog, which has led him to present many talks at universities and events.

Originally from São Paulo, Brazil, Bruno currently resides in London, UK, with his wife and two adorable cats.

I want to thank the amazing team at Packt who helped me write and publish my first book.

About the reviewer

Prajwal G is a game developer who has been enchanted by the world of gaming since childhood. He started learning Unity for fun while at university, and since then, he has created a diverse range of 2D and 3D games, including educational odysseys, captivating kids' adventures, epic RPGs, and action-packed open worlds. Game development is his passion, and every line of code is an expression of dedication to the art of gaming.

Table of Contents

Preface   XV

Part 1: Foundations of RTS Games

1

Introducing Real-Time Strategy Games   3

What is a real-time strategy game?   3   Game outline   6
Classic examples   4   World   6
Modern games   4   Characters   7
                                              Enemies and bosses   7
Understanding the game and   Gameplay overview   8
level design of an RTS game   4   Game experience   9
Character   5   Mechanics and modes   10
Camera   5
Control   5   Summary   10
                                              Further reading   11
Creating a game design document   6

2

Setting Up Unity and the Dragoncraft Project   13

Technical requirements   13   Importing and organizing assets
Downloading and configuring   from the Unity Asset Store 20
the Unity Editor   14   Summary   22
Using Visual Studio Code as
the default IDE   16

3

Getting Started with Our Level Design   23

Technical requirements	24	Creating ScriptableObjects for game configuration	37
Setting up our base scene	24		
Scenes	24	Creating ScriptableObjects for level configuration	40
Lights	25		
Camera	26	Adding a custom editor for the level configuration	42
Ground foundation	27		
Level scene	28	Creating a map editor to speed up map creation	49
Creating the first map layout using Prefabs	30	Controlling the camera	55
Creating a custom group of Prefabs	35	Summary	61
		Further reading	61
Using ScriptableObjects to configure the map	36		

4

Creating the User Interface and HUD   63

Technical requirements	64	Creating the details panel using a custom camera	85
Using Canvas for a responsive UI	64		
Setting up the UI and HUD using Prefabs	70	Rendering the minimap	91
Creating the MenuButton GameObject and popup	72	Loading the UI scene additively	94
Displaying the resources counter	79	Summary	97
Adding the action buttons	82	Further reading	98

Part 2: The Combat Units

5

Spawning an Army of Units   101

Technical requirements	101	Creating the message interface	112
Configuring the unit ScriptableObject	102	Implementing the warrior unit spawner	113
		Creating the resource type	115
Spawning units using the Object Pooling pattern	104	Updating the UI	116
Implementing the Object Pooling pattern	105	**Creating a debugging tool for the Editor**	**118**
Creating a BaseSpawner class	107	Creating the Object Pool for the Warrior unit	118
Updating the UI using the Message Queue pattern	**108**	Creating the debug script	123
		Summary	**126**
Implementing the Message Queue pattern	109	**Further reading**	**126**

6

Commanding an Army of Units   127

Technical requirements	127	**Selecting the units**	**138**
Preparing the Prefabs and UI	**128**	Setting a custom color for the selected units	139
Adding more debug options	128	Defining the unit selector component	143
Preparing the Prefabs and UI	130	**Moving the units**	**147**
Drawing the selected area in the UI	132	**Summary**	**150**
Preparing the level scene	136	**Further reading**	**151**

7

Attacking and Defending Units   153

Technical requirements	153	Setting up the selected unit actions	163
Updating the UI with the selected units	**154**	**Attacking and defending with units**	**176**
		Setting up layers and collisions	179
Setting up the selected unit details	154	Updating the UI	181

Attacking and playing other animations   184

**Adding the new Mage unit
and ranged attack   191**

Creating the data, object pool, and spawner   191

Setting up ranged attacks with fireballs   196

Summary   204

Further reading   205

8

Implementing the Pathfinder   207

Technical requirements   207

Understanding the pathfinder   208

The Greedy Best-First and A* algorithms   208

The NavMesh   210

The Unity AI Navigation package   211

**Implementing the pathfinder
using the NavMesh   213**

The NavMesh component   213

The NavMesh Agent component   217

The NavMesh Obstacle component   220

Debugging the NavMesh   226

Summary   230

Further reading   231

Part 3: The Battlefield

9

Adding Enemies   235

Technical requirements   235

Configuring the enemies   236

Spawning the enemies   242

Creating the enemy messages   243

Creating the enemy component   244

Creating the enemy spawner   247

Configuring the enemy spawner   250

Testing the enemy spawner   253

Damage feedback in the UI   255

Managing the damage feedback   256

Fading the text over time   258

Setting up the Prefabs for
the damage feedback   260

Testing the damage feedback   263

Summary   267

Further reading   267

10

Creating an AI to Attack the Player   269

Technical requirements	269	Creating the collision component	290
Updating the physics settings	270	Creating the Enemy NavMesh	297
Calculating the damage	274	Updating Unit's NavMesh	301
Adding a base class for the data	274	**Implementing the character life cycle**	**304**
Adding a base class for the characters	277	**Summary**	**308**
Updating EnemyComponent	281	**Further reading**	**308**
Updating UnitComponent	284		
Adding damage to ranged attack	288		
Managing the collision and chase behavior	**289**		

11

Adding Enemies to the Map   311

Technical requirements	311	Creating the material and Prefab for the fog	329
Creating spawn points	312	Updating the Level Manager	332
Adding fog on the map	321	**Patrolling the area**	**336**
Adding a new layer and updating cameras	322	**Summary**	**340**
Creating the fog component	323	**Further reading**	**340**

Part 4: The Gameplay

12

Balancing the Game's Difficulty   345

Technical requirements	345	Adding more test scripts	356
Writing unit tests	346	**Using unit tests to simulate battles**	**359**
Setting up the UTF	346	Simulating a battle against a single enemy	362
Creating the first test script	350	Simulating battles against multiple enemies	369

Balancing the battle difficulty   374   Further reading   380
Summary   380

13

Producing and Gathering Resources   381

Technical requirements   382   Producing resources manually   401
Generating resources   382   Testing the resources   406
Managing the player's inventory   385   Summary   410
Producing resources automatically   390   Further reading   410
Gathering resources   398
Adding the new Resource layer   399

14

Crafting Buildings and Defense Towers   411

Technical requirements   412   Creating the Unit Store UI   449
Crafting and upgrading buildings   412   Testing the Unit Store   456
Dragging and dropping buildings   412   Defending with towers   457
Configuring the Resource Store   420   Adding the Tower model   458
Creating the Resource Store UI   426   Adding the new Tower layer   459
Creating the Store UI   433   Adding the Tower as a unit   460
Training units   445   Testing the Defense Tower   468
Configuring the Unit Store   446   Summary   469

15

Tracking Progression and Objectives   471

Technical requirements   471   Winning or losing the game   489
Setting up objectives   472   Adding the Game Over popup   492
Tracking objectives   476   Summary   496
Creating the Objective component   478   Further reading   497
Creating the Objectives UI   486
Pausing the game   488

16

Exporting and Expanding Your Game   499

Technical requirements	499	**Expanding Dragoncraft**	510
Exporting Dragoncraft for desktop	500	Adding music and sound effects	510
Preparing the Editor scripts	500	Adding more content	512
Exporting the desktop build manually	501	**Summary**	514
Creating a build system	503	**Further reading**	514
Adding more platforms	506		

Index   515

Other Books You May Enjoy   524

Preface

Real-time strategy (**RTS**) is a very challenging and competitive game genre, one of the first esports before the term even existed, and the base for many other genres and subgenres, showing the importance of RTS to the game industry.

My first contact with PC gaming was in the late 1990s, and I have unforgettable memories of *StarCraft* and, especially, *Warcraft III* – two amazing RTS games that got me into PC gaming. In the last couple of decades, the RTS genre has expanded into other subgenres and was also the foundation for **multiplayer online battle arena** (**MOBA**) hits such as *Dota* and *League of Legends*.

With the Unity Engine, one of the most used game engines in recent years, we have everything that we need to build the most amazing games. In this book, we will get the most out of the built-in features in the Unity Engine, allowing us to focus on creating an RTS game that has mechanics and gameplay inspired by the classics of this genre.

You could say that we are living in a golden age of game development, where we are no longer limited by technology but by our own creativity!

Who this book is for

If you are a beginner game developer who wants to learn the skills to develop a professional RTS game, a programmer thinking of getting into the games industry and need to develop a portfolio, or even an indie game developer who is looking for inspiration to release your own games, this book is for you.

Professional game developers and programmers with experience in C# or game designers and artists that seek a practical guide to bring game ideas to life will also benefit from this book.

What this book covers

In *Chapter 1, Introducing Real-Time Strategy Games*, we will learn what an RTS game is, the level and game design characteristics of this genre, and see the game design document of *Dragoncraft*, the game you will make in this book.

In *Chapter 2, Setting Up Unity and the Dragoncraft Project*, we will install the Unity Engine and our IDE, as well as setting up our project and importing the packages we will use from the Unity Asset Store.

In *Chapter 3, Getting Started with Our Level Design*, we start developing the first features of our RTS game by creating a level scene and the map using Prefabs, and then we will create a custom editor tool to speed up the level design and implement the camera movement.

In *Chapter 4, Creating the User Interface and HUD*, we will create our game UI using Unity's Canvas and UI elements, and we will also add more cameras to render 3D objects in the UI.

In *Chapter 5, Spawning an Army of Units*, you will learn how to create a flexible system to configure units using ScriptableObjects, spawn the new Warrior unit using the Object Pooling pattern, and update the UI using the Message Queue pattern.

In *Chapter 6, Commanding an Army of Units*, we will see how to select the units on the map and give them a command to move to any position in a scene.

In *Chapter 7, Attacking and Defending Units*, you will learn how to implement the attack and defense for the units using the Command pattern, and we will add a new Mage unit and create a ranged attack by throwing a fireball.

In *Chapter 8, Implementing the Pathfinder*, you will learn the different strategies to implement a Pathfinder algorithm, and then we will see how we can use the unit's NavMesh system with our units.

In *Chapter 9, Adding Enemies*, it is time to start adding enemies to our game by creating a configuration and spawning each of the three different types – an Orc, a Golem, and a Dragon.

In *Chapter 10, Creating an AI to Attack the Player*, you will learn how to set up Unity's physics settings, detect collision and calculate damage, and make enemies chase the units.

In *Chapter 11, Adding Enemies to the Map*, we will create spawn points on the map to add enemies in the level configuration and make them patrol a specific area, and we will also create fog to cover the unexplored parts of the map.

In *Chapter 12, Balancing the Game's Difficulty*, you will learn how to use unit tests to simulate battles between units and enemies, as well as how to analyze the results and iterate on a character's configuration to balance the difficulty.

In *Chapter 13, Producing and Gathering Resources*, you will learn how to produce resources automatically and command units to gather resources on the map.

In *Chapter 14, Crafting Buildings and Defense Towers*, we will add buildings to upgrade the resource generation, learn how to upgrade units to make them stronger, and add a new tower defense unit to protect the settlement against enemies that are within its attack range.

In *Chapter 15, Tracking Progression and Objectives*, you will learn how to create and track objectives that will help in the game progression to determine whether the player won or lost the game.

In *Chapter 16, Exporting and Expanding Your Game*, you will learn how to export your game for desktop platforms manually and then automate the process using a build system, as well as see how you can expand the game by adding more content.

To get the most out of this book

To get the most out of this book, a basic understanding of programming concepts and the C# programming language is required. Some experience with any game development engine or framework would be an advantage.

It is recommended that you familiarize yourself with the Unity Engine features and the interface. The Unity website has a lot of beginner-friendly content that can help you get a good understanding of the engine quickly.

Since we will develop an RTS game, it is also recommended that you play a few RTS games or watch gameplay videos on YouTube, especially the games *StarCraft* and *Warcraft III*, which are the ones we will use as a reference and inspiration to develop the features of our *Dragoncraft* game.

Software covered in the book	Operating system requirements
Unity Engine version 2023.1 or above	Windows or macOS
Visual Studio Code or Visual Studio 2022	Windows or macOS
C#	Windows or macOS

You will need an internet connection to download Unity, IDEs, and the assets required for the project. The disk space required for Unity Engine 2023.1 and Visual Studio Code or Visual Studio 2022 may vary, but make sure you have at least 30 GB of disk space before installing the software. The project itself should require at least 4 GB. Although it is not required, a dedicated graphics card is recommended.

You can find more information about the Unity Engine requirements in their documentation at `https://docs.unity3d.com/2023.1/Documentation/Manual/system-requirements.html`.

Download the example code files

You can download the example code files for this book from GitHub at `https://github.com/PacktPublishing/Creating-an-RTS-game-in-Unity-2023`. Sometimes code blocks are truncated to focus on the specific code/topic being discussed, so it is advised that you use the code on the GitHub repository to follow along with the project.

If there's an update to the code, it will be updated in the GutHub repository.

We also have other code bundles from our rich catalog of books and videos available at `https://github.com/PacktPublishing/`. Check them out!

Conventions used

There are a number of text conventions used throughout this book.

`Code in text`: Indicates code words in text, folder names, filenames, and pathnames. Here is an example: "Add the following new `Start` method to the `LevelManager` class."

A block of code is set as follows:

```
using UnityEngine;
namespace Dragoncraft
{
    public class LevelManager : MonoBehaviour
    {
        private void Start()
        {
        }
    }
}
```

When we wish to draw your attention to a particular part of a code block, the relevant lines or items are set in bold:

```
private void Start()
{
    _initialPosition = transform.position;
    _camera = GetComponent<Camera>();
}
```

Bold: Indicates a new term, an important word, or words that you see on screen. For instance, words in menus or dialog boxes appear in **bold**. Here is an example: "In the **Inspector** view, click on the **Add Component** button and search for the **LevelComponent** script."

> **Tips or important notes**
> Appear like this.

Get in touch

Feedback from our readers is always welcome.

General feedback: If you have questions about any aspect of this book, email us at `customercare@packtpub.com` and mention the book title in the subject of your message.

Errata: Although we have taken every care to ensure the accuracy of our content, mistakes do happen. If you have found a mistake in this book, we would be grateful if you would report this to us. Please visit www.packtpub.com/support/errata and fill in the form.

Piracy: If you come across any illegal copies of our works in any form on the internet, we would be grateful if you would provide us with the location address or website name. Please contact us at copyright@packtpub.com with a link to the material.

If you are interested in becoming an author: If there is a topic that you have expertise in and you are interested in either writing or contributing to a book, please visit authors.packtpub.com.

Share Your Thoughts

Once you've read *Creating an RTS Game in Unity 2023*, we'd love to hear your thoughts! Scan the QR code below to go straight to the Amazon review page for this book and share your feedback.

https://packt.link/r/1-804-61324-X

Your review is important to us and the tech community and will help us make sure we're delivering excellent quality content.

Download a free PDF copy of this book

Thanks for purchasing this book!

Do you like to read on the go but are unable to carry your print books everywhere?

Is your eBook purchase not compatible with the device of your choice?

Don't worry, now with every Packt book you get a DRM-free PDF version of that book at no cost.

Read anywhere, any place, on any device. Search, copy, and paste code from your favorite technical books directly into your application.

The perks don't stop there, you can get exclusive access to discounts, newsletters, and great free content in your inbox daily

Follow these simple steps to get the benefits:

1. Scan the QR code or visit the link below

https://packt.link/free-ebook/9781804613245

2. Submit your proof of purchase
3. That's it! We'll send your free PDF and other benefits to your email directly

Part 1:
Foundations of RTS Games

In this first part of the book, you will learn what a real-time strategy game is and see a short game design document of *Dragoncraft*, the RTS game we are going to develop throughout this book. Then, you will learn how to install the Unity Engine and set up a new project, as well as how to import packages from the Unity Asset Store that we will use to build our game.

After having the Unity project set up and all packages imported, you will learn how to create a level, build a custom editor tool to configure the GameObjects on the map, implement camera movement, and everything else that is needed for our level design. You will also learn how to create a flexible UI and HUD on Unity, and how to use multiple cameras to render 3D GameObjects in the UI.

This part includes the following chapters:

- *Chapter 1, Introducing Real-Time Strategy Games*
- *Chapter 2, Setting Up Unity and the Dragoncraft Project*
- *Chapter 3, Getting Started with Our Level Design*
- *Chapter 4, Creating the User Interface and HUD*

1

Introducing Real-Time Strategy Games

Video games are highly complex graphic software, and, at the same time, an art form used to create interactive and immersive experiences. They are not easy to develop and most of them are difficult to master. **Real-time strategy** (**RTS**) games require the player to think ahead regarding each possible movement that they are going to perform, as well as what the opponent might do in response. If you then add a real-time constraint to this, you get one of the most challenging and competitive video games possible.

In this first chapter, you will be introduced to the definition of an RTS game and analyze a few classic examples of this sub-genre so that you understand what the main game mechanics and features are. This chapter will also cover the game and level design that defines the game project that is going to be developed throughout this book.

So, in this chapter, we will cover the following topics:

- What is a real-time strategy game?
- Understanding the game and level design of an RTS game
- Creating a game design document

What is a real-time strategy game?

An **RTS** game, as its name suggests, is a subgenre of strategy game where the player plays in "real time" without needing to wait a turn. This becomes especially challenging when the match starts to build up and the player needs to handle multiple situations, make quick decisions regarding what units to train, what orders to give to the trained units, how to gather and produce more resources, where to explore in the map, where to search for the objective and, of course, try to figure out where the enemy is and what their next move will be.

Classic examples

Released by Blizzard in 1998, *StarCraft* is one of the most popular RTS games to this day and has a long history of professional players competing against each other in worldwide championships (it was also one of the very first esports). The classic *StarCraft* gameplay in the remastered version was released in 2017 and is still one of the benchmarks of this strategy sub-genre, as well as *Warcraft III*, which was released in 2002.

Even though these games can be listed as the most popular of this genre, it is no doubt that other games were the pioneers in many forms of gameplay and mechanics that we define as RTS today. Developed by Westwood Studios and released by Virgin Games in 1993, *Dune II* was the very first RTS game to introduce resources gathering, base and unit buildings, construction dependencies, and different factions with unique weapons – all features that are now part of any RTS game. *Dune II* was not the first RTS game, but it was the game that helped define this genre.

Another RTS game that should be mentioned is the *Age of Empires* series, started in 1997, which was a combination of *Warcraft* and *Civilization*, the latter being a **turn-based strategy** (**TBS**) game.

Modern games

The RTS games that were released in the 1990s defined the base of the genre, but many other games expanded the features, gameplay, and mechanics, creating sub-genres such as **real-time tactics** (**RTT**) and **Explore, Expand, Exploit, Exterminate** (**4X**).

Dune: Spice Wars is a great example of a modern 4X RTS game that was released in 2022 as early access on Steam. Following in the footsteps of *Dune II*, this game contains all the great features that are expected from this genre with modern graphics. *Company of Heroes*, a World War II-based game, also pushed RTS games further with real-time physics and destructible environments.

RTS games are also the foundation of the very popular **multiplayer online battle arena** (**MOBA**) that was born as a *Warcraft III* mod in 2003 called *Defense of the Ancients (DotA)*. Years later, *DotA 2* (2013) and *League of Legends* (2009) were released and defined a new strategy sub-genre.

There are so many great classic and modern RTS games that we could cover an entire book on them, but I had the difficult task of selecting a few of them to illustrate their beginnings and evolution.

Now, we are going to look at the features that most of these games share.

Understanding the game and level design of an RTS game

Game design is the art of creating the idea and the rules that describe what the game is and, more importantly, making it a fun and remarkable experience for the player. **Level design** is a specialization of game design that is responsible for level creation.

We can define most of the gameplay and mechanics of a video game by looking at the three Cs: character(s), camera(s), and control(s). In the following sections, we will see how these three aspects of a game help tighten the gameplay because they all work together and should move alongside each other.

Character

Characters in an RTS game are represented by the units that the player can use in the game, as well as the enemies that are spread across the map. Usually, the main game plot is not attributed to one character in an RTS game but rather to the collective units that the player, as the commander, can control to perform actions. The player will think about the strategy first, and then use the units as a tool to achieve what was planned.

Units can be controlled to explore the map, attack enemies, defend the settlement, collect resources, and create buildings. The characters here are not important, nor do they have a great impact on the game, but it is part of the strategy that the player built.

In a few RTS games, there are characters such as heroes or infamous bosses that have distinct personalities that can drive the story of the game. In this case, the player is still controlling nameless units, but there is a hero to fight alongside or a greater evil to be defeated.

Camera

The **camera** is one of the most characteristic aspects of an RTS game. Besides the ability to move around the map at will, as we are going to see in the next section, the camera usually shows the top view of the territory, and all parts that are not explored yet are covered with fog; this means that the player can see the unexplored regions of the map and send units to explore them, clearing away the fog.

It is also possible to zoom in and out on the map, which helps give a macro view of the battlefield before you take micro-decisions to attack, defend, build, produce, and gather resources. The ability to move the camera quickly using the mini-map is very important and useful in the late stages of the match where a lot of things are happening at the same time across the territory.

An RTS map usually has a starting point (the blue X) and shows the enemy base that must be conquered or destroyed (the red X). Only the player's initial position is shown uncovered in the camera, and many other hidden objectives or resources are hidden in different locations on the map, so the player will find something interesting in any direction that they move.

Control

The player **control** is rather basic in an RTS game but it's this simplicity that helps players make the best and quickest decisions. Controlling the camera is vital to decide where to go or to monitor what the enemies are doing, and with easy access using the mini-map, players can control the camera quickly.

Besides the camera, the player can also control the troops by selecting one or more units and giving them a command – this command could be to gather nearby resources, attack an enemy, or just move to a position and wait idly for any enemy threat.

Selecting units or a building will display different options in the UI so that the player has all the information required to decide on the next steps. All control is usually done by using the mouse's left and right buttons and the cursor movement on the screen. Some games were created or adapted to consoles and these controls were translated to the gamepad. The same happens for mobile RTS games and touchscreen controls, which work great since they're very similar to mouse control.

Now that we know what an RTS game is and what kind of game and level design is involved, we need to define the game that will be developed by you throughout this book. The best way to define the scope of the project is by creating a simple game design document.

Creating a game design document

A **game design document** (**GDD**) is a document written by the game designer that defines the scope of the game, from the characters and story to the gameplay and mechanics used by the player. There are different approaches to elaborate a GDD, from straightforward text to dense documentation full of details and explanations. We are going to use a simple but effective method that consists of defining the bare bones of our game with enough details that will help us develop our game with all the features in mind.

In the following sections of this chapter, we are going to define the game that will be developed throughout this book: *Dragoncraft*.

Game outline

Dragoncraft is an RTS game where the player will defend a small village against the dragons that are scattered through the land. Before facing the mighty dragons and destroying their nests, the player needs to expand the small village by creating new buildings and training an army to both attack and defend the village. To grow the village, the player will need to gather and produce resources while defending the territory against orcs and exploring the lands beyond the village to find even more resources. It is important to create the right strategy to explore and defend before the dragons start to hunt for more food.

World

The level has a map showing land with a few forests, villages, and the dragon's nest. Villages can contain both enemies and extra resources for the player, so it's well worth the adventure to explore and raid them. Each level will have a map that increases the difficulty, giving the player more challenges.

When the game starts, the entire map is covered by clouds, so the player will have to send units to explore and review the map. The player must defend the village; otherwise, the enemies will kill all the units and take the resources away, making the player lose the game.

Characters

The player's army can be trained to be bigger and stronger based on what buildings and levels the player crafted in the village. There will be two basic types of soldiers: the Footman, a melee combatant, and the Wizards, who can cast powerful spells from a safe distance:

Figure 1.1: Mini Legion Footman PBR HP Polyart by Dungeon Manson, ©2022 Unity Technologies

The units can be generated and upgraded in the training camp using resources that the player collects. When both units are combined and used wisely, they can form a very versatile army.

Enemies and bosses

Hidden in other villages across the map, the enemies will wait for the player's units to approach their territory. The orcs will attack the units with no mercy and, if they defeat the player's units, they will follow the trail back to the player's village to raid it. There are a few errant orcs that will spawn in different locations and search for the player's village, attacking at first sight:

Figure 1.2: Mini Legion Grunt PBR HP Polyart by Dungeon Manson, ©2022 Unity Technologies

As the player starts to upgrade the units, the orcs will also become stronger by spawning new and improved versions. A few variables of the orcs with different colors and stats will keep the game challenging for the player.

The player's objective is to find and kill the dragon and destroy the dragon's nest. A few orcs will help the dragon, which is like the level boss, to defend the nest and attack the player. Each level will have a different dragon with different stats:

Figure 1.3: Dragon the Soul Eater and Dragon Boar by Dungeon Manson. ©2022 Unity Technologies

If the player is taking too much time to find the dragon's nest, a new dragon will be spawned to protect the nest while the other dragon goes to look for the player on the map. The dragons are quite strong and it will take a few unit upgrades and a large army to defeat them. Unlike the orcs, the dragon will not abandon its nest to look for the player's village if it kills all player units sent to explore the map.

Gameplay overview

Dragoncraft is an RTS game where the player starts in a small village with a couple of units and very limited resources. The player will need to expand their village by gathering resources and training more units to be prepared to find and kill the dragon hidden on the map. The player gives commands to the units to gather resources, build or upgrade a construction, and train more units.

When the game starts, only the village is visible in the top-down camera, and the player must select and command the units to go explore the map and clear the way. The mini-map on the UI is a great tool to quickly move the camera to a desired location on the territory and will be very helpful in the late game.

Enemies, controlled by the game's AI, are hidden across the map and will attack as soon as they see the player's unit approaching. Also, from time to time, a few enemies will start to explore the map and find the player's village to raid and loot the resources. The player must explore and find the dragon hidden on the map, defeat it, and destroy the nest. If they take too much time to find the dragon, the dragon will start to look around and explore the map, but the nest will be very well guarded by another dragon.

The player oversees the village and gives out orders to make the units perform activities. The player mainly completes input via the mouse, in which they can do the following actions:

- Left-click to select a unit or building

- Left-click and drag to select units

- When the units are selected, right-click to set a movement target (when the target is in a valid position) or attack an enemy

- When units or a building is selected, the UI at the bottom of the screen will change to display information and extra actions available

- Moving the mouse cursor in any direction will also move the top camera

Besides the mouse, the player can use the *Escape* key to pause the game, and the *Space* key to move the camera back to the village.

Game experience

The player must have the feeling that they need to think and act quickly as the enemies are doing the same to raid the village. All tools and resources available should be useful for the player to make decisions and see the outcome so they can plan the next steps.

The game should be balanced in such a way that a series of wrong decisions may lead the player to lose the game, but at the same time, the right decisions lead them to victory. The victory path should be challenging and not a shortcut that will make the player so strong that nothing can defeat them. In the end, the player must feel rewarded by the experience and want to play again but on a slightly more difficult map.

Mechanics and modes

The gameplay mechanics are the classic RTS actions that the player must choose wisely because each one takes some time to complete, and the units can't be interrupted once they start an action. The selected units can perform different actions, such as the following:

- Gather resources from a location and bring them to the main village building
- Build or upgrade constructions
- Attack an enemy
- Move and stay idle in one specific spot until another command is given

There are different types of buildings and they can all be upgraded. These buildings are required to store resources, train new units, and unlock new upgrades. Besides the main hut, all other buildings must be constructed using at least one unit, and the cost of resources is displayed in the UI. The buildings are as follows:

- **Town Hall**: This building is responsible for storing resources. Upgrades will increase the number of resources received and unlock other buildings.
- **Barracks**: This building is responsible for the units' training. Upgrades will unlock different units such as the Wizard for ranged combat and stronger units to be trained.
- **Defense Tower**: This building is used for defense. If the enemies enter the tower area, a ranged attack will hit them.
- **Blacksmith**: This building is used to craft strong weapons and armor for the units, giving them more means of attack and defense.

Once the player finds and defeats the dragon, and destroys the dragon's nest, a new game will be offered to the player, where they can choose to play the same map again or the one that was unlocked by completing the current one. Each map is predefined, and no random levels are generated.

The GDD is a great resource for describing what the game is about and the gameplay and mechanics that make it a great game. It is also important to let the game developers know what to create and how everything connects to make a memorable experience for the player.

Summary

In this chapter, we learned what an RTS game is and what the main gameplay mechanics that define this genre are. We also looked at a few examples of such features that led to our own GDD, which contains the outline of the game that will be developed through this book: *Dragoncraft*.

In *Chapter 2, Setting Up Unity and the Dragoncraft Project*, we are going to download and install the Unity editor, learn how to set up Visual Studio Code to work with the engine, and how to create a new project and organize all assets that will be downloaded and used in later chapters.

Further reading

For more examples of RTS games and their history, check out these links:

- *Real-time strategy*: `https://en.wikipedia.org/wiki/Real-time_strategy`
- *List of real-time strategy video games*: `https://en.wikipedia.org/wiki/List_of_real-time_strategy_video_games`

You can learn more about game design by reading these excellent books:

- Schell, Jesse. *The Art of Game Design: A Book of Lenses.* 3rd ed. (2019) CRC Press
- Rogers, Scott. *Level Up! The Guide to Great Video Game Design.* 2nd ed. (2014) Wiley
- Fullerton, Tracy. *Game Design Workshop: Playcentric Approach to Creating Innovative Games.* 4th ed. (2018) CRC Press

2

Setting Up Unity and the Dragoncraft Project

Today, many different game engines can be used to develop games, and most of them are free to use or have a free license until you start making thousands of dollars in profit. Even with many options available, the **Unity engine** is probably the most popular choice for both new and experienced game developers, capable of making projects from small games, all the way up to AAA games with advanced features that push the hardware to its limits.

In this chapter, you will learn how to download and install the Unity game engine, including the recommended version and modules to install, as well as how to configure **Visual Studio Code** as the default **integrated development environment** (**IDE**) and install the required extensions. This chapter will also show you how to download and import all content required for building our RTS game, *Dragoncraft*, from the **Unity Asset Store**.

By the end of this chapter, you will have learned how to set up and organize a Unity project to make it easier to work on, as well as how to keep it organized as we advance through the development of the game by adding code and assets.

In this chapter, we will cover the following topics:

- Downloading and configuring the Unity Editor
- Using Visual Studio Code as the default IDE
- Importing and organizing assets from the Unity Asset Store

Technical requirements

To complete this chapter and create *Dragoncraft*, here is the software you need to install:

- Visual Studio Code for the C# programming
- Unity Engine 2023.1 (or later)

The project setup for this chapter, along with the imported assets, can be found on GitHub at `https://github.com/PacktPublishing/Creating-an-RTS-game-in-Unity-2023`. It is important to note that sometimes code blocks are truncated to focus on the specific code/topic being discussed, so it is advised that you use the code on the GitHub repository to follow along with the project.

The assets from the Unity Asset Store that have been used in this project can be found at `https://assetstore.unity.com/lists/creating-a-rts-game-5773122416647`.

Downloading and configuring the Unity Editor

To get started, we will need to install **Unity Hub**, which is the entry point for installing the **Unity Editor** so that we can create and load projects. Head to `https://unity.com/download` and click on the **Download for Windows** button if you have a Windows setup or the **Download other versions** button for the macOS installer and Linux instructions.

Once downloaded, follow the installer's instructions to add Unity Hub to your system. Next, click on the **Installs** tab and then click on the **Install Editor** button. From the following screen, select the latest version available for Unity, which is 2023.1 at the time of writing, or any later editions:

Figure 2.1 – Install Unity Editor

As soon as you click on the **Install** button, you will be presented with the following screen, where you can select the modules you would like to install:

Figure 2.2 – Unity modules

By default, Unity will install the desktop module, which is mandatory, and have pre-selected both Visual Studio Community and Documentation to be installed.

Clicking on the **Install** button will download and install all the selected modules for the desired Unity version. This can take some time, depending on your internet speed:

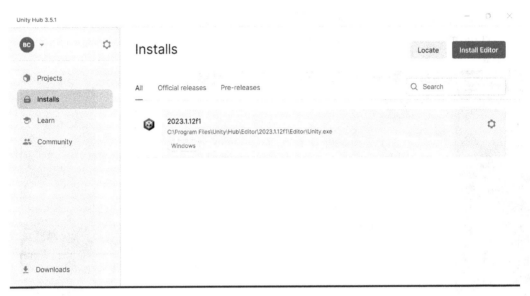

Figure 2.3 – Unity 2023.1 installed

> **Note**
>
> At this point, no other modules are required. However, once installed, any extra module can easily be installed using Unity Hub by clicking on the configuration (cog) icon to the right of the Unity Editor.

While this process is happening, you can proceed to the next section and install our IDE.

Using Visual Studio Code as the default IDE

Visual Studio Code is a lightweight but powerful IDE that provides extensions to expand the supported programming languages and technologies you want to use. It is available for Windows, macOS, and Linux. Go to https://code.visualstudio.com/ and click on the **Download** button. Select the option that matches your operating system:

Figure 2.4 – Visual Studio Code download

Once downloaded, unzip the file and double-click it to launch the IDE. On the left-hand side of the screen, select **Extensions** and search for C#, as shown in the following screenshot. Click **Install** to add C# support to Visual Studio Code:

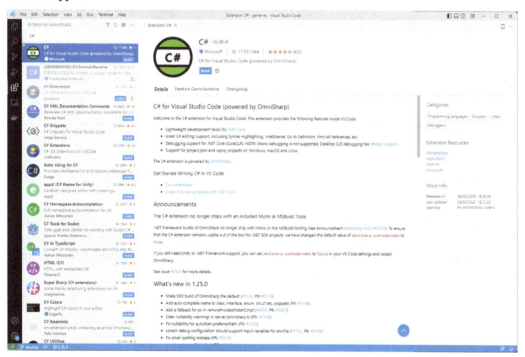

Figure 2.5 – Visual Studio Code C# extension

Now that Visual Studio Code has been installed and configured, we can create a new Unity project and set it up as the default IDE.

Launch Unity Hub and click on the **New Project** button in the top-right corner. Then, select the **3D Core** template, which is just an empty project with pre-configured settings for 3D projects. Next, type your project's name, which in this case is `Dragoncraft`, and select the desired location for the project to be created:

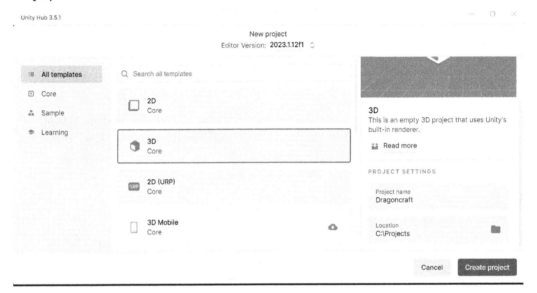

Figure 2.6 – New project settings

After clicking on the **Create project** button, Unity will start to create and set up the project for the first time, which can take a few minutes depending on your system.

The last step is to select Visual Studio Code as the default IDE via Unity Editor. To set it up, go to **Edit | Preferences** (on macOS, the **Preferences** option is located in Unity's main menu). On the opened screen, select **External Tools** from the left-hand side; then, from the **External Script Editor** option on the left side, select **Visual Studio Code** from the list of options:

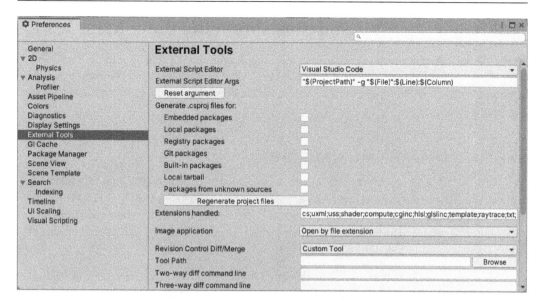

Figure 2.7 – External Tools

Visual Studio Code is a professional tool that can be used to develop any kind of project. However, you can also use Visual Studio 2022, which is a good, free IDE for personal use. However, keep in mind that it is heavier in both size and performance and, in this book, we will not be using any specific feature of Visual Studio, so Visual Studio Code is the best option for our *RTS Game* project.

If you wish to install Visual Studio 2022 and do not have it already, it is available on both Windows and macOS: `https://visualstudio.microsoft.com`. Follow the instructions at the aforementioned link to install the IDE.

Once installed, you can select it as the default IDE via the Unity Editor by going to the same settings screen as shown in *Figure 2.7*. You might need to close and re-open the project to see Visual Studio in the list of options or use the **Browse** option to look for the application.

Now that we have downloaded the Unity Editor, the new project has been created, and we have set up Visual Studio Code as the default IDE, we can start to import and organize the assets required to build our game.

Importing and organizing assets from the Unity Asset Store

Now that all the configuration is done for both the Unity Editor and the IDE, we can use our empty project to get started with our RTS game development. This book will cover all the coding aspects of creating an RTS game, but since we are not going to create any art, the free assets from the Unity Asset Store will be used.

Go to the Unity Asset Store (`https://assetstore.unity.com`) and create a free account that will allow you to download and import the assets in the Unity Editor.

Next, access this public list, which contains all the assets needed to develop our RTS game in this book (`https://assetstore.unity.com/lists/creating-a-rts-game-5773122416647`) and click on the **Add to My Assets** button for each of the six assets. Every time you click that button, a pop-up message stating **Unity Terms of Service** will be displayed for you to accept the terms of service before the asset is added to your account.

Once added, you have the option to click on the **Open in Unity** button. If you have the Unity Editor open, you will be shown the package manager and the asset that's been selected, ready to be downloaded and then imported:

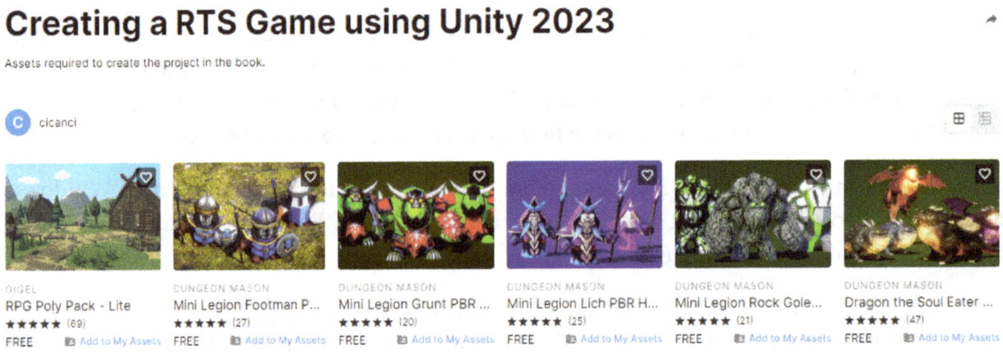

Figure 2.8 – Asset list in the Unity Asset Store

You should be able to see all assets in the Unity Editor, on the **Package Manager** screen, as shown here:

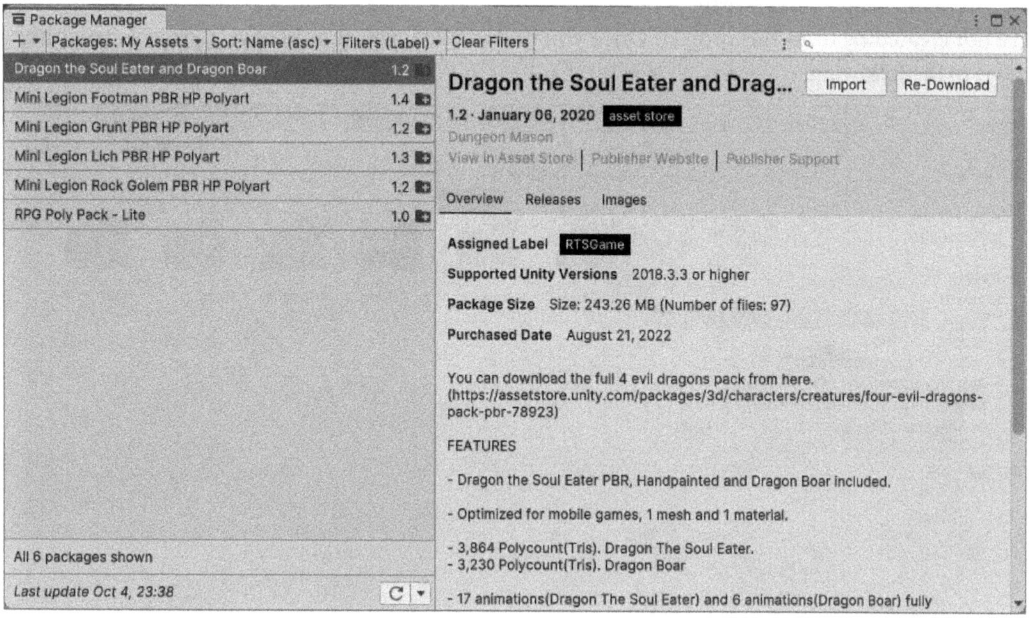

Figure 2.9 – Downloading and importing assets from the Package Manager screen

You will have a **Download** button next to each asset that you need to click so that Unity will download that package into your system but not into your project. As soon as the download is finished, you will be able to see an **Import** button, which will import the package into your opened project. This needs to be done for each package so that all the assets are imported into the project.

If everything worked correctly, you should have the following folders inside your Assets folder:

- FreeDragons
- Mini Legion Footman PBR HP Polyart
- Mini Legion Grunt PBR HP Polyart
- Mini Legion Lich PBR HP Polyart
- Mini Legion Rock Golem PBR HP Polyart
- RPGPP_LT
- Scenes (this one is created by default when you create a new project)

Now, to keep it organized, let's create a new folder by right-clicking **Assets** in our **Project** window. Then, from the menu, select **Create | Folder** and name it ThirdParty so that we know those assets were not created by us in this project. Once you've done this, you should have a project structure similar to the following:

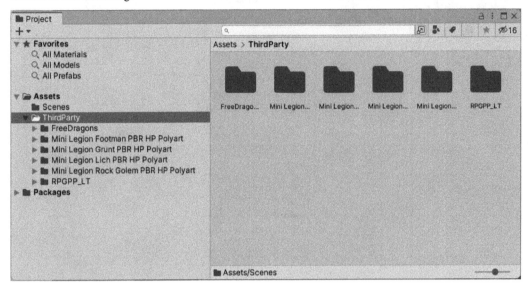

Figure 2.10 – Project structure after importing the necessary assets

With all of this done, we are ready to get started with the first coding part of *Dragoncraft* in the next chapter. You will probably do most of the steps shown in this chapter only once since that is all you need to develop any game using the Unity Editor and Visual Studio Code as the default IDE. You can always go back to Unity Hub and download new versions or edit an installed version to add more modules, such as Android and iOS, for mobile game development.

Summary

In this chapter, we learned how to download, install, and set up the Unity Editor using Unity Hub. We also downloaded and installed Visual Studio Code, imported the required extension, and configured it as the default IDE for the Unity Editor. In the end, we imported all the free assets required from the Unity Asset Store into our newly created project and organized it.

In *Chapter 3, Getting Started with Our Level Design*, we are going to start coding our map editor and create a few levels using this tool, ScriptableObjects, and the assets we imported from the Unity Asset Store. We will also configure the main scene, lights, camera movement, and mini-map navigation – one of the key gameplay features of our RTS game.

3

Getting Started with Our Level Design

RTS games have a few characteristics and requirements that are very specific to them – the camera setup and movement have an industry standard to follow with some variations; the levels have many assets that are used to set the mood and predefine the available paths for the player to explore; and also, due to the high quantity of maps and variations, a map editor tool is required to speed up the level design, matching the rules defined by the game design and the gameplay mechanics.

This chapter will introduce you to the first scripts that will be required to write C# code to achieve the features and requirements of our game as defined in the previous chapter. You will learn how to set up a game scene with the proper light and camera settings for an RTS game, as well as how to create and use **Prefabs**, which are reusable assets that act as a template, to develop maps for the game. We are also going to explore one of the most flexible and useful resources available on Unity, **ScriptableObjects**, a data container that will be crucial for developing our map editor tool and will make it easier to build different maps.

By the end of this chapter, you will know how to use many useful resources and APIs from Unity that are applied to not only RTS games but all other kinds of games too. Everything developed in this chapter forms the building blocks that we will use to create the foundation of our *Dragoncraft* game, and introduces the tools that you can use later to expand the project even further with your own levels.

This chapter will cover the following topics:

- Setting up the base scene on Unity for an RTS game
- Creating the first map layout using Prefabs
- Using ScriptableObjects to configure the map
- Creating a map editor to speed up map creation
- Controlling the camera in RTS games

Technical requirements

The project setup in this chapter with the imported assets can be found on GitHub at `https://github.com/PacktPublishing/Creating-an-RTS-game-in-Unity-2023/tree/main/Dragoncraft/Assets/Chapter03`.

All scripts and scenes created in this chapter are available there, but we are also going to use the assets imported in the previous chapter, which are in the `Chapter02` folder.

Setting up our base scene

Before we create our first level, we need to set up the base scene of the project, which will be used by our map editor to populate the map with the assets we are going to define in the configuration for each level. In addition to the base scene for every new level, we will also create a Playground scene, which, as the name suggests, will be used a lot during the game production so we can develop and test features freely without breaking our levels or being limited to the configuration we have.

Scenes

When we created the new Unity project following the instructions from the previous chapter, Unity automatically created a first scene for us called `Sample Scene` inside the **Scenes** folder, which is basically an empty scene with a standard main camera and **directional light**. Right-click on the scene and select **Rename** – we'll change it to `Playground`, as you can see here:

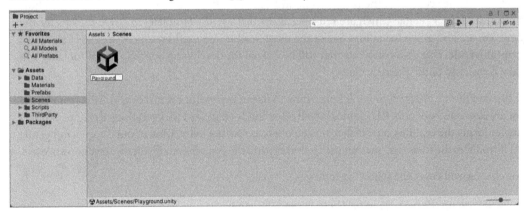

Figure 3.1 – Renaming the scene to Playground

If you do not have a scene or the **Scenes** folder already, you can create both by right-clicking on the **Assets** folder and selecting **Create | Folder**, and giving it the name `Scenes`. Then, right-click on the newly created folder, select **Create | Scene**, and name it `Playground`.

Now that we have our scene for the Playground, let's set it up properly so we can create the scene for our first level from this one without having to do this configuration more than once.

Lights

Usually, the **Light** settings generated by default when creating a new scene are all properly set up for most types of games and we do not need to change them at the beginning of the development. Later, we might have multiple light sources with different types to have a more robust illumination in our scene, but for now, we can leave the light as a single directional light in the scene.

You can inspect the **Light** settings in the scene by left-clicking on the **Directional Light** object in the **Hierarchy** window – the settings will be displayed in the **Inspector** window. For our RTS game, the default settings work very well out of the box; just make sure the settings are the same as in the following figure:

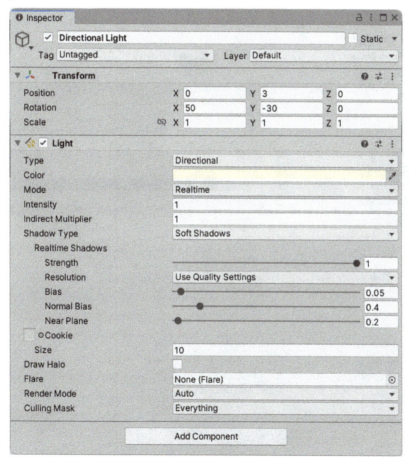

Figure 3.2 – Directional Light settings

For the **Directional Light** type, the only **Transform** property that changes the way the light reflects and casts shadows is **Rotation**. You can play with the values to find what looks better for your scene; however, the values shown in *Figure 3.2* are the ones we are going to use for all levels in our RTS game.

Camera

The camera is one of the most important items in an RTS game because it is also used by the player as a tool to help make strategic decisions since they can move it around the map during gameplay. We'll talk about camera movement later in this chapter, but for now, we are just going to set up the configurations that will not change while moving or zooming around the map, which are the **Projection** and **Transform** properties.

In the **Hierarchy** window, left-click on the **Main Camera** object to show the details in the **Inspector** window. Now, set the **Transform** values for both **Position** and **Rotation** to match the values shown here:

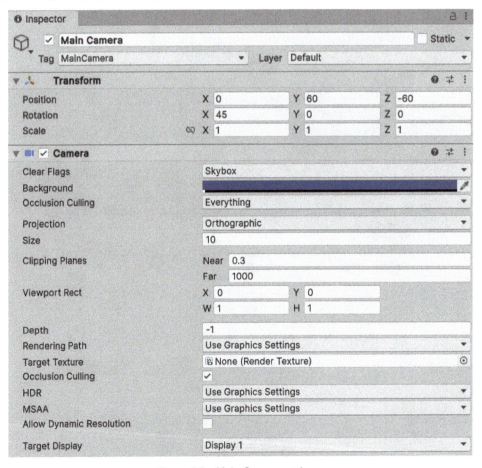

Figure 3.3 – Main Camera settings

The next couple of changes must be made to the **Camera** component, which is already attached to the **MainCamera** object. Change **Projection** to **Orthographic** – using this projection, the perspective is removed and objects do not get smaller with distance (which works best for RTS games). Then, set the **Size** property as 10 – this will be used to adjust the player's camera zoom during gameplay. All the other settings can be kept as the default values.

Ground foundation

The Playground scene is almost ready to be used, but we are still missing one last thing to make it usable for developing and testing our levels: adding a **plane** to be the ground foundation of the scene. A plane is a built-in 3D object that represents a flat surface with a Mesh Renderer component, which can be used to set textures or materials, and a Mesh Collider to detect collision.

> **Note**
>
> The Mesh Renderer is a component that renders a mesh, which is the main graphic primitive of Unity. The Mesh Collider generates a physics collider based on the mesh attached to the component.

To add a plane to the current scene, go to **GameObject | 3D Object | Plane**. Alternatively, you can right-click on the **Hierarchy** window and select the **3D Object | Plane** option.

Once added, make sure to set the **Transform** values for both **Position** and **Rotation** to (0, 0, 0) and **Scale** to (10, 10, 10) so that the plane is big enough to place assets and build the level. **Transform** is a Unity component that determines the position, rotation, and scale of a GameObject in the scene, and we are going to use it many times to position the GameObjects in the 3D world.

The following figure shows the expected outcome (do not forget to save the scene after adding the plane and making the changes):

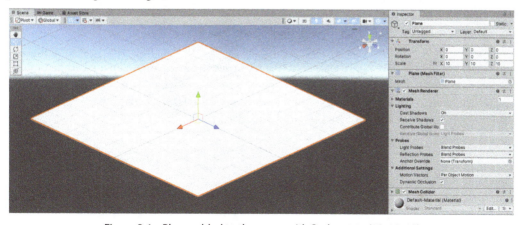

Figure 3.4 – Plane added to the scene with Scale set to (10, 10, 10)

For the `Playground` scene, we can leave the plane white to make it easier to identify that it is a test or development scene, but for the real levels, we will be creating materials to change the color and we will create different map sizes by tweaking their scale (the scale of the plane will be referred to as our map limit, which is what the player will be able to see and interact with without moving out of the boundaries).

Level scene

Now that we have our `Playground` scene, we can make a copy of it to set as our first game level. To do that, select the scene in the **Project** view and press *Ctrl + D* on Windows (or *Command + D* on macOS) to duplicate it. Then, rename the new copy `Level01`. At this point, you should have two scenes, as shown here:

Figure 3.5 – The Playground and Level01 scenes

Double-click on the **Level01** scene to open it – you can see the name of the opened scene in the **Hierarchy** window to make sure it is the correct one. To make it different from the `Playground` scene, we are going to add a green color to the plane using a **Material** asset, which has a reference to a **Shader** object and is used to describe the appearance of surfaces.

In the same way as we created the **Scenes** folder, create a new folder called `Materials`. Once you have that new folder, left-click on it and select **Create | Material**, then name it `GreenGrass`.

Select the newly created material, go to the **Inspector** window, and right-click on the white field on the right of the **Albedo** property, which will open a new window with a color wheel. In this window, type `669966` in the **Hexadecimal** field, and you should see a green color appear:

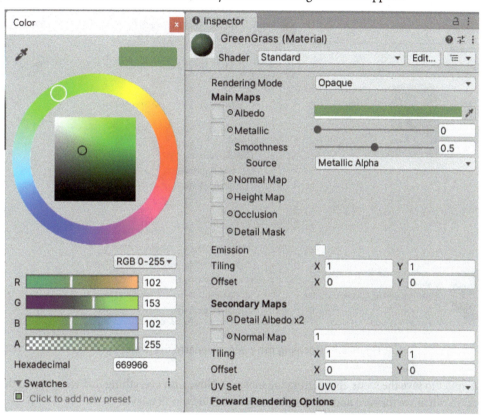

Figure 3.6 – Setting the material color

Finally, go back to the **Hierarchy** window and select **Plane** to see the details in the **Inspector** window. In the **Mesh Renderer** component, **Plane** has `Default-Material` as the material used. Click on the white circle on the right of the material name to open the **Select Material** window and search for the `GreenGrass` material that we just created. After selecting it, the plane will become green. This is the most common way to change the color of a 3D object, and you can have multiple materials to use different colors as well.

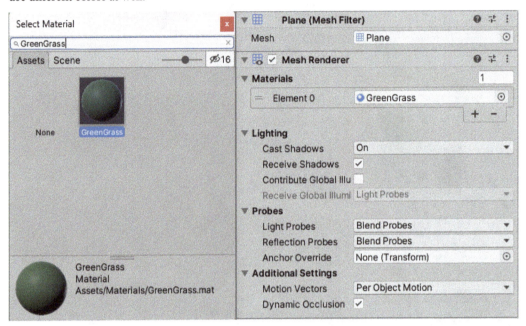

Figure 3.7 – Applying the green material to the plane

Do not forget to save the scene after setting the material! Now, with everything that we set up in this section, we can move forward and start adding assets to build our first level.

Creating the first map layout using Prefabs

Using `Level01`, we are going to add a few Prefabs from the `RPGPP_LT` package imported into the project in the previous chapter. For now, we are not going to use the `Dragon` package nor any of the Mini `Legion` packages because they are not part of the level design; they are part of the design of the gameplay that will come in the next chapters of the book.

In the **Project** view, expand the **ThirdParty** folder, and then the **RPGPP_LT** folder:

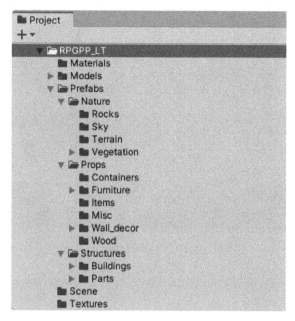

Figure 3.8 – The Prefabs structure inside the RPGPP_LT assets package

Here, you will find a well-organized structure with many assets that we are going to use to create our levels. You can ignore all internal folders and just focus on the **Prefabs** folder – it is divided into categories and subcategories, which contain everything we need. Moving forward, Prefabs will be mentioned by their location within these folders or the filename.

Some of the Prefabs can be easily used by dragging and dropping them into the scene or the **Hierarchy** window. Before adding Prefabs to the scene, make sure that you are looking in the **Scene** view and at the correct orientation; otherwise, the assets will not be in the desired position when playing the game in the **Game** view. To do so, look at the top-right corner of the **Scene** view and adjust the orientation to be X on the right and Z at the top, as shown in the following figure:

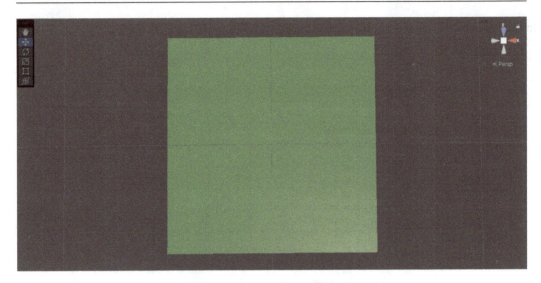

Figure 3.9 – The Plane GameObject centered in the Scene view and the correct orientation

You can adjust the orientation by clicking and holding the right mouse button and moving it in the **Scene** view until the orientation is correct. To see the plane as in *Figure 3.9*, double-click on the **Plane** GameObject in the **Hierarchy** window and Unity will center the GameObject in the middle of the **Scene** view.

Now, we can start to add the Prefabs into the scene. For now, we are going to ignore the **Props** folder because it only has assets such as Furniture and small decorations, which you can add later to have more details in the scene. Instead, we will focus now on the **Nature** and **Structures** folders.

First, expand the **Nature** folder and click on **Terrain.** There are a few nice terrains to use, but we will select the one called rpgpp_lt_terrain_grass_01; drag and drop it into the **Hierarchy** window or the **Scene** view. Once there, you can use the right mouse button and the blue and red arrows to move it. If you do not see both arrows as in the following figure, select the second option in the Toolbar in the upper-left corner of the **Scene** view or press *W* on your keyboard, which is a shortcut for the **Move** tool.

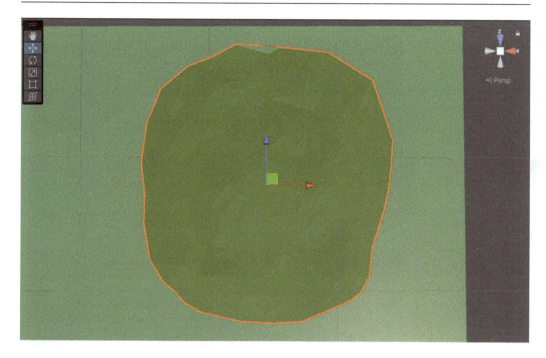

Figure 3.10 – Moving Prefabs in the scene

Feel free to explore the available Prefabs and experiment with how they look in the scene. Good level design takes time and patience to be achieved and requires practice and testing a lot of possibilities.

Keep in mind that, in the next chapter, we are going to work on a feature for the mini-map and the hidden map that will only be revealed if the player decides to send the units to explore a covered region of the map. At the end of this chapter, we will create a script to move the camera, and you will be able to navigate both vertically and horizontally to explore the uncovered map.

The following figure presents a suggestion for the first level, which is a small, simple map with two different villages and a small hill between them so the player can find a path:

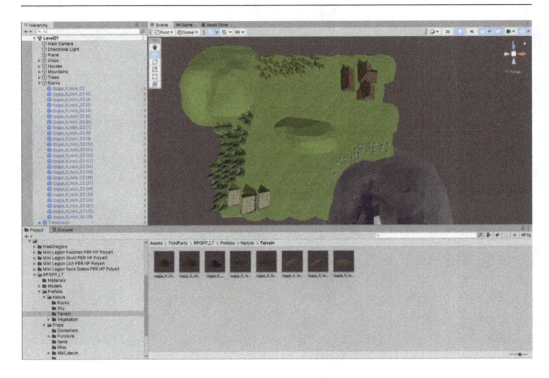

Figure 3.11 – Suggestion for one small, simple level

Also notice that in the **Hierarchy** window, all the scene elements are organized by type, which makes them easier to find. When placing a Prefab on the scene, you can always use the shortcut *Ctrl + D* on Windows (or *Command + D* on macOS) to duplicate it, which makes it faster and easier to add more of the same Prefab.

You can also play with the **Y** coordinate for bigger items such as the mountains and hills to reduce the visible size. If it is still too big for the scene, you can change the scale from the default value of **1** to 0.5 for **X**, **Y**, and **Z** in the **Scale** property in the **Transform** component, which becomes visible when selecting the object in the **Inspector** window of the Unity Editor.

The scene shown in *Figure 3.11* can be found in the project in the GitHub repository of the book, in the Chapter 03 folder (https://github.com/PacktPublishing/Making-a-RTS-game-with-Unity-2023/tree/main/Dragoncraft/Assets/Chapter03). While progressing through the book, we are going to create or use pre-created maps from the repository so we can cover other parts of the RTS game development. Even though all scenes used in this book are presented in the project, feel free to create your own or modify them to make them yours. If something does not look right, you can always come back to the project scenes as a reference.

To make the level design process faster, we are going to next see how to create a custom group of Prefabs that can be added and positioned in the scene instead of adding Prefabs one by one.

Creating a custom group of Prefabs

Another way to save some time while developing new maps is to create Prefabs by grouping other Prefabs together. First, create a new folder called `Prefabs` in the project, which will be the place where we are going to store the created Prefabs. Now, in the **Hierarchy** window, right-click anywhere below the **Rocks** GameObject and, in the menu, select **Create Empty** – you can name this new GameObject `TreeGroup`.

Next, select a few Prefabs together in the scene (for example, three trees), then drag and drop them into the new **TreeGroup** GameObject, as shown in the next figure. After that, select **TreeGroup** and drag and drop it into the **Prefabs** folder; this will create the new Prefab and change the color of the **TreeGroup** GameObject's icon to blue, indicating that it is now a Prefab.

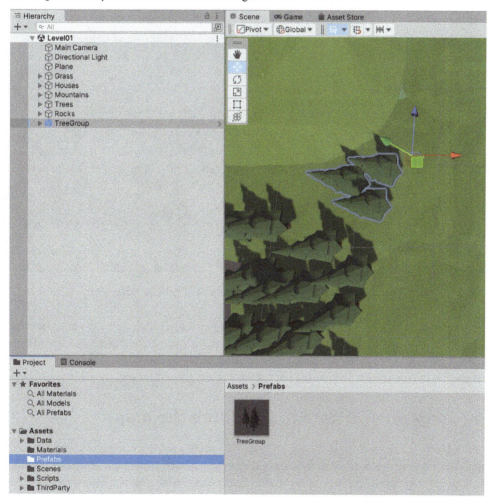

Figure 3.12 – Creating a Prefab by moving a GameObject into a Project folder

In this example, there is only one Prefab created by grouping others in the same Prefab, but this is a very useful resource from Unity that allows us developers to create complex groups of Prefabs and reuse them easily across the project. When creating the level configuration using ScriptableObjects, remember that you can also use these groups of Prefabs instead of setting up small objects one by one in the scene.

You can always get very creative to build a level using the Prefabs available in the RPGPP_LT package. Besides the assets used in the previous example, there are a lot of props to decorate the level and make it rich and unique. The package also includes a great sample scene, demonstrating all the Prefabs included. You can also use this scene as a starting point to build your own levels; just remember to duplicate the scene or copy the Prefabs into your scene by selecting and copying them from the **Hierarchy** window and then pasting the Prefabs into your own scene.

Figure 3.13 – The sample scene from the RPGPP_LT package

Now that we have learned how to build the level by dragging and dropping Prefabs into the scene, we are going to see next how we can create a configuration file using ScriptableObjects, which will speed up the map creation process. Remember that you can always mix the manual process of placing the Prefabs into the scene with configuring part of it using the map editor tool we are going to build.

Using ScriptableObjects to configure the map

ScriptableObjects is a data container created by Unity that can be used to store data and share it across the game, keeping only one instance of the data in memory. Although it has a few limitations – with a lack of dictionary support being one – it is still very flexible, easy to use, and quick to read data. Due to these benefits, it is the recommended way of storing configuration data inside a game, instead of using other data types such as JSON or plain text, for example.

Another great advantage of using ScriptableObjects is that you define the data format as a standard C# class, which makes it quite transparent to read and write data, without having to write code to deal with I/O operations or having to use an external solution.

Here, we are going to create two different ScriptableObjects: one responsible for the game configuration and the other responsible for the level configuration. We will also add a custom editor for the level configuration, to render the properties in the **Inspector** window in a way that is easier to manage. Let's get started.

Creating ScriptableObjects for game configuration

For game configuration, we are going to create three script files: the ScriptableObject with a list of Prefabs, the item class (which will have a Prefab reference and a type), and the item type. Before we start creating the classes, create a new folder called Scripts inside the **Assets** folder. Then, inside the **Scripts** folder, create another one called Configuration. All three scripts will be created in the Configuration folder.

Right-click on the **Configuration** folder, select **Create | C# Script**, name it LevelItemType, and open it in your IDE. This will be an enum class with options that will be used in the next class that we create (enums are very useful, and we will see them a lot through the book).

Now, replace the content of the script with the following code:

```
using System;
namespace Dragoncraft
{
    [Serializable]
    public enum LevelItemType
    {
        None,
        Tree,
        House,
        Rock
    }
}
```

The Serializable attribute must be present on all enums and classes that are going to be used in the ScriptableObject; otherwise, Unity will not be able to store the data for that specific type. Enums are serialized as numbers, starting at zero (unless specified), and deserialized back to the enum class type, making it very easy to use in the code without having to cast the data to the enum type by yourself.

A good practice to organize code (which we are going to adopt moving forward) is to always add a namespace to every script that we create; for more granularity, we could have the namespace as Dragoncraft.Configuration, for example, but to keep it simple, just Dragoncraft is enough for now.

The next class we are going to create is the item definition. Using the same steps and folder as the previous script, create a new one called `LevelItem`. Open the newly created script and replace the content with the following script:

```
using System;
using UnityEngine;
namespace Dragoncraft
{
    [Serializable]
    public class LevelItem
    {
        public LevelItemType Type;
        public GameObject Prefab;
    }
}
```

`LevelItemType` is the enum class we just created – here, it is used to define the level item type, so we easily find it by type. We also have a `GameObject` variable that has a reference to the Prefab that is linked to the type; for example, `LevelItemType.Rock` will have a reference to a Prefab of a rock.

The last script for the game configuration we are going to create is the actual ScriptableObject, in which the main objective is to hold a list of level items that can be used later to instantiate all the Prefabs in the map. Create a new script in the same location, name it `LevelConfiguration`, and open it in your IDE. Then, replace the content of the file with the following script:

```
using System.Collections.Generic;
using UnityEngine;
namespace Dragoncraft
{
    [CreateAssetMenu(menuName = "Dragoncraft/New Configuration")]
    public class LevelConfiguration : ScriptableObject
    {
        public List<LevelItem> LevelItems = new List<LevelItem>();

        public LevelItem FindByType(LevelItemType type)
        {
            return LevelItems.Find(item => item.Type == type);
        }
    }
}
```

This script is quite simple but introduces a few important things we are going to use in this book.

The `CreateAssetMenu` attribute is a handy shortcut that allows Unity to create a menu item that will create a ScriptableObject of this class type. `menuName` is the name and path of the menu item that will be visible when you click on any project folder to create a new asset. This attribute alone is not enough to create the asset; the class must inherit from the `ScriptableObject` class, which will allow Unity to create the proper asset type.

Since the ScriptableObject does not support the `Dictionary` type due to serialization limitations, we need to use the `List` data type, and in this case, we are creating and initializing a `List<LevelItem>` variable that will be a container for all level items we have configured to use. By using the Find API, it is quite easy to filter the list to get the item we are looking for when calling the `FindByType` method. The parameter for the `Find` method in the list object is a predicate that has a variable representing an item of the list, also named `item` in this case, and the condition, which is any item that has a type equal to the one we are looking for.

Now, it is time to create the new ScriptableObject based on that class. First, create a new folder inside **Assets** called `Data`, and another folder inside **Data** called `Configuration`. Right-click on the **Configuration** folder on the menu and select **Create | Dragoncraft | New Configuration**, which should be the very first option, and then name the file `DefaultConfiguration`. When selecting the new file, we can see that, in the **Inspector** window, there is a **Level Items** list with a size of **0**. Enter 3 instead, so that three items will be added to the list. Then, click on the triangle near each element to expand the content and fill the items using the same configuration as in the following figure:

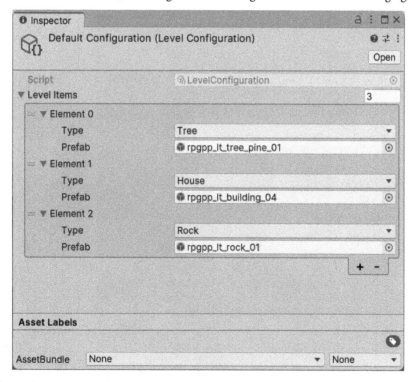

Figure 3.14 – The default configuration for the ScriptableObject

Remember that you can always click on the white circle on the right of the Prefab field to open a popup, click on the **Assets** tab, and search for the asset by name. You should have one Prefab with that exact name for each type. If you have more than one element in this list with the same type, only the first element will be considered when trying to find by type. Any other extra elements with the same type will be ignored, so it is important to create more types in the enum class if you want to have multiple trees, for example.

We can always add more types and Prefabs here; you only need to add more types in the LevelItemType enum class.

Creating ScriptableObjects for level configuration

The level configuration is a bit more complex than the game configuration because we are going to write a custom inspector for this ScriptableObject. This will also have three different files: the class for the slot information, the ScriptableObject itself, and the custom editor, which will change the default style of the level configuration file.

These files will be in a different location so, inside **Configuration,** create a new folder called Level. Then, create a new C# script with the name LevelSlot and enter the following content:

```
using System;
using UnityEngine;
namespace Dragoncraft
{
  [Serializable]
  public class LevelSlot
  {
    public LevelItemType ItemType;
    public Vector2Int Coordinates;

    public LevelSlot(LevelItemType type, Vector2Int coordinates)
    {
      ItemType = type;
      Coordinates = coordinates;
    }
  }

}
```

In this new class, which is serializable to be used by the ScriptableObject next, we again have LevelItemType, but this time, it will be used to find the Prefab to be instantiated in this slot. Previously, LevelItemType was used as a global game configuration to be reused by the whole project, but now, LevelItemType has the information to tell the game what is expected in one specific slot.

Plus, it has `Vector2Int` coordinates, which differs from `Vector3`, in that this object has only X and Y coordinates, and the values are integers. If you imagine the map as a 2D board, like a chess board, X and Y are the coordinates that we are going to use when instantiating the Prefabs.

The `LevelSlot` class also has something different in the definition, which is a constructor for the `LevelSlot` object that received two parameters (`itemType` and `coordinates`), the same two variables present in this class. This constructor indicates that an object of this class can only be created if you provide both parameters. The default class constructor without parameters is available in every class, but as soon as you create a constructor with parameters, that default constructor is not available anymore unless you also add it to the class without parameters. In our case, we do not need it, so the only constructor for this class will be the one using the `itemType` and `coordinates` parameters.

The next class to be created in the same folder is `LevelData`. This is the `ScriptableObject` class responsible for the level configuration; it is quite simple, but later, it will become powerful with a custom editor for the **Inspector** window. For now, replace the content of the script created with the following code:

```
using System.Collections.Generic;
using UnityEngine;
namespace Dragoncraft
{
    [CreateAssetMenu(menuName = "Dragoncraft/New Level")]
    public class LevelData : ScriptableObject
    {
        public List<LevelSlot> Slots = new List<LevelSlot>();
        public int Columns;
        public int Rows;
        public LevelConfiguration Configuration;
    }
}
```

Like the game configuration file, here, we also have a data container, `List<LevelSlot>`, for the slots in the map; this will be a list of slots and coordinates that later will be used to instantiate all assets. `Columns` and `Rows` are the sizes of the content in the map – if you configure this file as 3 columns and 3 rows, and all slots have trees, this will be used to instantiate 9 trees positioned in 3 rows and 3 columns on top of the map.

The last `LevelConfiguration` property in the `LevelData` class is a reference to the ScriptableObject that we created in the last section, *Creating ScriptableObjects for game configuration* – it has the default configuration for Prefabs we are going to add in the level map. This approach gives us the flexibility to have multiple level configurations with different sets of assets for each level map.

Let us create our first `LevelData` asset file now. Inside the **Data** folder, which we created in the first ScriptableObject, create a new folder called `Level`, and right-click on the new **Level** folder to create a new `LevelData` asset by selecting the **Dragoncraft | New Level** option. Name the new file `Level01` and click to display the content in the **Inspector** window; you should be able to see an empty configuration file as in the following figure:

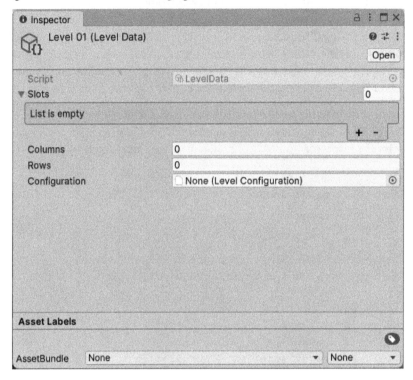

Figure 3.15 – The level configuration ScriptableObject without the custom editor

We are going to leave it without any configuration for now because we are going to add the custom editor script first to make it easier to initialize, visualize, and update the data in the following section.

Adding a custom editor for the level configuration

The next class we are going to create is very different from a regular class because it is only meant to be used in the Unity Editor. This means that this class will not be present in the game and will work as a nice tool to help us develop our game. For this reason, we need to create a folder called `Editor` inside the **Scripts** folder and add a new C# script with the name `LevelDataEditor`.

When we create a ScriptableObject, all properties defined as public class variables will be automatically visible in the **Inspector** window when selecting the object created. This is very helpful, as we saw in the last couple of examples, but it is also limited to what can be rendered by default. However, it is possible to create a custom appearance and even add buttons to the ScriptableObject using a custom editor, as we are going to do next.

Open the script that we just created to edit and replace the content with the following code:

```
using System.Collections.Generic;
using UnityEditor;
using UnityEngine;
namespace Dragoncraft
{
    [CustomEditor(typeof(LevelData))]
    public class LevelDataEditor : Editor
    {
        public override void OnInspectorGUI()
        {
            LevelData levelData = (LevelData)target;
            AddLevelDetails(levelData);
            AddLevelSlots(levelData);
            AddButtonInitialize(levelData);
            AddButtonUpdate(levelData);


        }
    }
}
```

There are a few key elements highlighted in this code that are required to make the custom editor work in the ScriptableObject.

First, we have the CustomEditor attribute, which has one parameter – the type of the class we want to create the custom editor for. In this case, we are going to create a custom editor for the LevelData ScriptableObject class. It is important to note that this LevelDataEditor class does nothing on its own; it only extends the visuals and functions of another class. And, because of the type that is in the attribute, it will work only for the specified class.

Since this is an editor tool, we need to inherit from the Editor class. We already created this class inside a folder called **Editor,** which is one of the requirements so Unity will understand that it will not be part of the game. Also, the UnityEditor namespace at the beginning of the script only works for editor scripts and will cause a compilation error if you try to include it in a regular script.

There is only one method from the Editor class that we need to override here, OnInspectorGUI. This method is called internally by Unity to update the interface of the Editor class. Because we are overriding it, we can add custom code there so the rendered **graphical user interface (GUI)** will be drawn as we define it.

As we told the custom editor script that this class is tied to `LevelData`, we can access the actual ScriptableObject data using the `target` class variable. The four methods below the `target` class variable will be created in the following subsections – they are going to use the `target` variable to modify the ScriptableObject data as we interact with the custom editor we created.

Adding level details

The first method is `AddLevelDetails`, which is responsible for holding a reference to the ScriptableObject for the level configuration, the number of columns, and the number of rows in our plane. Copy the following code inside the class, after the `OnInspectorGUI` method:

```
private void AddLevelDetails(LevelData levelData)
{
    levelData.Configuration = EditorGUILayout.ObjectField("Level: ",
        levelData.Configuration,
        typeof(LevelConfiguration), false) as LevelConfiguration;

    levelData.Columns = EditorGUILayout.IntSlider("Columns: ",
        levelData.Columns, 1, 25);

    levelData.Rows = EditorGUILayout.IntSlider("Rows: ",
        levelData.Rows, 1, 25);
}
```

In this method, we are going to ask for the input of three different pieces of information: the level configuration ScriptableObject used as the Prefab reference, the number of columns, and the number of rows.

The first property, `Configuration`, is going to be an `EditorGUILayout.ObjectField` method that will be rendered in the **Inspector** window as a button to select the file from the project folder. The first parameter is the label text displayed before the button, with the `Level:` value. The second parameter is the `Configuration` property itself, followed by the type of class to be selected on Unity. The last parameter is the `false` value to indicate that we are not allowing scenes to be selected here. This object field is created and converted to a `LevelConfiguration` class and then set back to the same `Configuration` property used in the object field creation.

We are going to use `EditorGUILayout` in the next variables to read the value returned from the GUI object created. In a few cases, such as `Columns` and `Rows`, both are integers so we can use a method that will return the right type without needing to cast as we did before for `LevelConfiguration`, which was an object. `EditorGuiLayout.IntSlider` will create an input field that accepts and returns integer variables, using the first parameter as the label text. The nice thing about `IntSlider` is that it will render a slider in the **Inspector** window with a limit defined by the values of 1, as the lowest, and 25, as the highest possible values for that field. This way, we can guarantee that the value will be in the range defined by us, without needing to add conditionals for validation after getting the input value.

Both Columns and Rows have the exact same lines to get the value from the **Inspector** window and store it back to the ScriptableObject. These variables will be used next to define the size of our grid, which will have a list of values to be selected for each slot positioned in the plane.

Adding level slots

The following AddLevelSlots method will be used to loop over all rows and columns to add an EnumPopup layout property for each item. The EnumPopup layout is going to render a dropdown in the **Inspector** window that will allow us to select any enum of the LevelItemType type in each X and Y position, which is a coordinate made using the respective row and column positions:

```
private void AddLevelSlots(LevelData levelData)
{
    EditorGUILayout.LabelField("Level Item per position:");

    for (int x = 0; x < levelData.Rows; x++)
    {
        GUILayout.BeginHorizontal();
        for (int y = 0; y < levelData.Columns; y++)
        {
            LevelSlot slot = FindLevelSlot(levelData.Slots, x, y);

            slot.ItemType =
                (LevelItemType)EditorGUILayout.EnumPopup(
                    slot.ItemType);
        }
        GUILayout.EndHorizontal();
    }
}
```

Unlike the previous method, here, we are going to define the label text before the fields using LabelField; this will add the label text in one line, and the content will be right in the next one.

Since this is a grid, there are two loops, one for Rows and one for Columns, and since we start the loop per row, the content of each column will be placed between BeginHorizontal and EndHorizontal. These two methods are used to make sure the content created between them is placed horizontally, without creating a new row, unless we specify another pair of BeginHorizontal and EndHorizontal. If we were looping the columns first, we could use BeginVertical and EndVertical to perform the same thing but in a different order, meaning that we would start adding elements vertically to complete a column before moving to the next column.

Next, we have a couple of important lines inside the second loop: first, we will find a `LevelSlot` object for the given row and column coordinates, and then we render the `LevelSlot` enum as a drop-down property. The `FindLevelSlot` method was created here to simplify the code and will be created in the following code block, but its objective is to find a slot, considering the list of lists as the first parameter in the X and Y coordinates, which are the corresponding values of the current row and column, respectively, in the loop.

The `LevelSlot` object returned by the method is used in the next line to create an EnumPopup object that accepts an enum as the parameter and returns the updated value of the enum from the GUI. Here, we need to cast the return value to the `LevelItemType` enum class before assigning the value returned from the `FindLevelSlot` method to the `ItemType` property. The EnumPopup method will render in the **Inspector** window a list of all possible `LevelItemType` values with the corresponding value for each slot if there is one stored in the ScriptableObject.

Now, we are going to define the method responsible for finding the slot in the list provided considering the X and Y coordinates. If the slot is not found, a new one is created and returned. This method will always return a `LevelSlot` object, either an existing instance or a new object:

```
private LevelSlot FindLevelSlot(List<LevelSlot> slots, int x, int y)
{
    LevelSlot slot = slots.Find(i => i.Coordinates.x == x &&
                                     i.Coordinates.y == y);
    if(slot == null)
    {
        slot = new LevelSlot(LevelItemType.None,
                             new Vector2Int(x, y));
        slots.Add(slot);
    }
    return slot;
}
```

The `FindLevelSlot` method used in the preceding code is actually very simple. First, we try to find a valid `LevelSlot` variable in the slots list for the X and Y coordinates. If it is not found, the object will be `null`; then, we need to create a new `LevelSlot` object with the `LevelItemType`. None type and the X and Y coordinates as a `Vector2Int` object. After creating a new object, we need to add the new `LevelSlot` object to the list of slots, so that next time, we will be able to find it for that specific X and Y coordinate. Whether it was found or newly created, the slot is returned at the end of the method.

With that last code to find a slot, we finished the grid logic and rendering. Now, the last thing that we are going to create in our custom editor is a couple of buttons: one to initialize the list of objects and another to update and save all our changes in the ScriptableObject.

Adding buttons to the custom editor

So far, we only added data fields that can be manipulated in either the **Inspector** window or the custom editor class. Now, we are going to add a couple of buttons that will initialize our grid based on the number of rows and columns and then update (that is, save) the changes in the ScriptableObject file:

```
private void AddButtonInitialize(LevelData levelData)
{
    if (GUILayout.Button("Initialize"))
    {
        Initialize(levelData);
    }
}
private void AddButtonUpdate(LevelData levelData)
{
    if (GUILayout.Button("Update"))
    {
        EditorUtility.SetDirty(levelData);
        AssetDatabase.SaveAssets();
        AssetDatabase.Refresh();
    }
}
```

The AddButtonInitialize method adds a Button element in the **Inspector** window using GUILayout. The only parameter used here is the label that will be displayed in the button, which is Initialize. When the button is clicked in the **Inspector** window, the if validation will be true and the content of it will be executed. In the AddButtonInitialize method, we are executing a method called Initialize, which will be created in the next code block.

The AddButtonUpdate method has the same button logic but, this time, it is a button with the label Update. If clicked, a sequence of methods from the Unity API will be executed to update and save the ScriptableObject. The SetDirty method will tell the engine that the ScriptableObject (here with the levelData reference) changed and will be marked as not saved. After that, the SaveAssets method will write in the disk all assets that were not saved yet, which will be our ScriptableObject. Finally, Refresh will reimport into the engine all assets that changed. These three lines ensure that the ScriptableObject we are editing will be saved and updated in the engine.

Next, the following method, `Initialize`, is the one that will be called when the **Initialize** button is clicked on the custom editor – it clears the current slots and adds one for each combination of row and column based on the respective numbers set in the custom editor. Besides the initialization, this method also performs a resize operation when we need to increase or decrease the number of rows and columns in the custom editor:

```
private void Initialize(LevelData levelData)
{
    levelData.Slots.Clear();

    for (int x = 0; x < levelData.Rows; x++)
    {
        for (int y = 0; y < levelData.Columns; y++)
        {
            LevelSlot levelSlot =
                new LevelSlot(LevelItemType.None,
                new Vector2Int(x, y));

            levelData.Slots.Add(levelSlot);
        }
    }
}
```

The `Initialize` method will be collected only if the button with the same name is clicked in the **Inspector** window. The script is a simple initialization that will first clear the list of slots before looping through all rows and columns, creating and adding a new `LevelSlot` object for each position in the grid with the default value of `LevelItemType.None`.

Note that this method will also reset an existing grid and remove all previous slots and add new ones. This is useful when the size of the rows and/or columns is reduced, and we need to remove the extra slots not used anymore.

The file is not saved after this method initializes or resets the slots, so we need to click the **Update** button, in the **Inspector** window as shown in the image below, to make sure all changes are written to the ScriptableObject and saved. Once the script is saved, we can select our **Level01** asset and see how it looks now in the **Inspector** window:

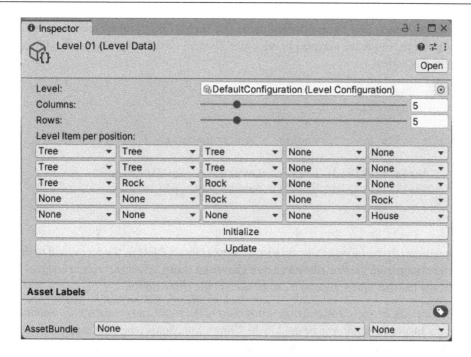

Figure 3.16 – The LevelData ScriptableObject with the custom editor applied

With that last method, we conclude our custom editor script. It might look complex, but it is quite simple after breaking down every method and the important lines. This is a powerful resource to create great-looking tools in Unity that can increase usability and reduce the workload of tasks, such as creating a level in our case.

Feel free to create as many `LevelData` ScriptableObjects as you want and explore the configuration possibilities using the custom editor script we just created. Now that we can create level configurations and level data, the next step is to use these ScriptableObjects to add Prefabs to our map and generate a level based on it.

Creating a map editor to speed up map creation

Now, it is time to gather both configuration files – **DefaultConfiguration** (`LevelConfiguration`) and **Level01** (`LevelData`) – and put them into use in a script that will initialize and place Prefabs based on the setup we made on each ScriptableObject. The script will be a component that can be attached to any GameObject in the scene and will need a couple of references to work: the `LevelData` asset we want to use, and the plane used to place the Prefabs in the scene.

So, first, create a new folder named **Level** inside the **Script** folder, then create a script called LevelComponent inside the **Scripts | Level** folder. Then, open the script and replace the content with the following code:

```
using UnityEngine;
namespace Dragoncraft
{
    public class LevelComponent : MonoBehaviour
    {
        [SerializeField] private LevelData _levelData;
        [SerializeField] private GameObject _plane;
    }
}
```

We are going to attach the LevelComponent class to a GameObject in the scene, and to do that, the class is inheriting from the MonoBehaviour class from Unity. When we create a script in Unity, the class already has that inheritance so it can be used in the Unity Editor.

Another characteristic of MonoBehaviour is that most of the public class variables are automatically available in the **Inspector** window to be used in the Unity Editor. Some types that cannot be serialized by Unity are not supported, and that usually applies to classes that do not support this attribute or complex data structures such as dictionaries.

However, we are not going to add the class variables as public but, instead, declare them as private so they are only visible within the class. In this case, to make them visible in the **Inspector** window, we need to tell Unity to do so by adding the SerializeField attribute before each class variable. The advantage here is that we have control over the visibility of the class variables, but we still have the possibility to make them visible in the **Inspector** window if we decide that the value of one specific class variable could be changed in the Unity Editor.

LevelData is a reference to the level configuration we want to use in this class as the map setup. Once added to the script and attached to the GameObject in the scene, we will be able to select a file by clicking on the LevelData field in the **Inspector** window or dragging the level configuration file from the project and dropping it into the same field. GameObject here is a reference to the plane in the scene that we added in *Chapter 2, Setting Up Unity and the Dragoncraft Project*. You can also drag and drop the reference here, but this time, it is from the scene instead of a file in the project. Both class variables are required to make this component work properly.

Now, we are going to add the following method in the class, which will override the internal Start method in the MonoBehaviour class. Here, we will use the GameObject of the plane to calculate the size of our map, start position, and offset between the Prefabs that we are going to instantiate in the next method:

```
private void Start()
{
```

```
    if (_levelData == null || _plane == null)
    {
        Debug.LogError("Missing LevelData or Plane");
        return;
    }

    Collider planeCollider = _plane.GetComponent<Collider>();
    Vector3 planeSize = planeCollider.bounds.size;

    Vector3 startPosition = new Vector3(-planeSize.x / 2, 0,
                                        planeSize.z / 2);

    float offsetX = planeSize.x / _levelData.Columns - 1;
    float offsetZ = planeSize.z / _levelData.Rows - 1;

    Initialize(startPosition, offsetX, offsetZ);
}
```

Since both _levelData and _plane can be null if not selected in the **Inspector** window, we need to check whether any of them are invalid so we can use LogError to print an error message in the Unity console, saying that something is missing in this component. To avoid any errors due to invalid or null values, the return keyword will take care to skip the rest of the method if the condition of the null check is true.

One way to determine the size of the plane is to get its Collider component and check the size of the boundaries. If the plane is not null, it will always have a Collider component unless it was removed manually from the GameObject. The size is Vector3, which means it has the X, Y, and Z coordinates. However, since this is a 2D-like view of the board, we are only considering X and Z, ignoring the value of Y, or setting it to 0.

The Initialize method, which is going to be defined in a moment, needs three different parameters: startPosition to instantiate Prefabs, and offsetX and offsetZ, which will be used to add some distance between Prefabs. The startPosition parameter might look confusing, but it is just half of the plane size. Since the origin position (0, 0, 0) is in the center of the place, we need to move half of the size to the left in the X coordinate (thus the negative sign) and half of the size up in the Z coordinate (note that we are not using the Y coordinate, which is why the value is 0). The result of this calculation is that the upper-left corner of the screen becomes the starting position to instantiate Prefabs.

Next, we have both offsetX and offsetZ, which are the distances between each Prefab that is going to be instantiated. To calculate that, we divide the size of the plane by the number of columns and rows, respectively, but subtract 1 from each value because we are starting the position from 0. With these three variables (startPosition, offsetX, and offsetZ) calculated, we can move forward with the declaration of the following method:

```
private void Initialize(Vector3 start, float offsetX, float offsetZ)
{
    foreach (LevelSlot slot in _levelData.Slots)
    {
        LevelItem levelItem =
            _levelData.Configuration.FindByType(slot.ItemType);

        if (levelItem == null)
        {
            continue;
        }

        float x = start.x + (slot.Coordinates.y * offsetX) +
            offsetX / 2;
        float z = start.z - (slot.Coordinates.x * offsetZ) -
            offsetZ / 2;

        Vector3 position = new Vector3(x, 0, z);
        Instantiate(levelItem.Prefab, position, Quaternion.identity,
            transform);
    }
}
```

The objective of this method is to go over all slots in the level configuration and instantiate the correct Prefab based on its type in a position that we are going to calculate using the parameters of this method.

In the first line inside the foreach loop, we are going to find the LevelItem object for the current Configuration Scriptable Object based on the LevelItemType enum; this can be done by calling the FindByType method, which we defined in the *Creating ScriptableObjects for game configuration* section, inside the game configuration class. If the LevelItem object is not found, the value of the variable will be null so we can continue our loop to the next iteration.

The calculation for both X and Z can be a bit tricky because we are using the start position of X and Z, but then in the X calculation, the Y slot coordinate is used instead of X. And, in the Z calculation, we are using the X slot coordinate. This is because, in the level configuration, the X slot coordinate corresponds to the horizontal position and the Y slot coordinate corresponds to the vertical position.

When calculating X – for example, from the slot coordinates (0, 0) to (0, 1) – we are moving horizontally but keeping the same value for the X slot coordinate; what changes is the vertical position. In this case, we use the variation of the Y slot coordinate to calculate X, multiplied by the corresponding offset, which is the distance between each Prefab. The last part of the calculation is the addition or subtraction of half of the distance between each Prefab so it will be instantiated centered in the corresponding slot.

Also, note that we add values in the X calculation while we subtract values in the Z calculation. This is due to the starting position in the upper-left corner of the plane, as we saw in the Start method. We need to add to move horizontally and subtract to move vertically. Both X and Z are used to calculate the position, and Y is 0 because we are not using it.

Finally, with everything calculated, we can add a copy of the Prefab using the Instantiate method from Unity. This method requires a few parameters: the Prefab that we want to create a copy of and instantiate, the position, the rotation, and the parent of the new object. The Prefab is a reference in levelItem, and the position was calculated in the previous lines. In this method, we are not changing the rotation, so the default value of it will be the Quaternion.Identity constant. The last parameter, the parent of the new object, is used to keep the scene organized and have all the new objects under the GameObject that has this script in the **Hierarchy** window instead of having them all over the list.

Now, you can left-click in the **Hierarchy** window and select the **Create Empty** option to add a new GameObject and rename it Level to keep it organized. Next, click on **Add Component**, type in Level Component, and left-click on the component name to add it to the GameObject. Now, click on the **Level Data** field to select the **Level01** level configuration created before, and drag and drop the **Plane** object from the **Hierarchy** window to the **Plane** field in the component. As a result, you should have something like the following figure:

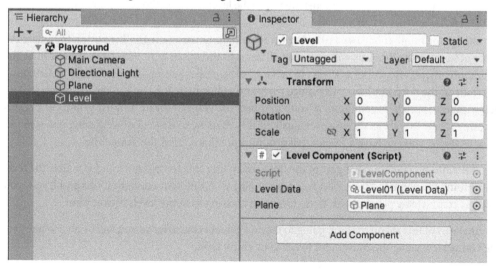

Figure 3.17 – The Level object with Level Component

Once everything is set up and saved, you can run the scene by clicking on the **Play** button in Unity. You should see something like *Figure 3.18* in your **Scene** view. This is based on the level configuration we created earlier, where the Prefabs were instantiated in the correct coordinates and with a proportional distance between them.

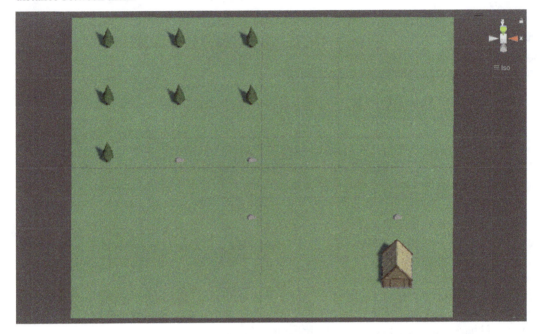

Figure 3.18 – The Prefabs instantiated in the level based on the configuration

If you change the number of rows and columns, remember to use the **Initialize** button to update the number of items in the list. This will also reset all items to the None type, which is required since the initialization process also works as a reset to make sure the proper settings are used when instantiating the Prefabs in the correct places. The more rows and columns you have, the more objects will be added with less space between them. Also remember that you can create and use groups of Prefabs by creating a new Prefab, as we saw previously in this chapter, so instead of rendering one tree per slot, for example, you can have a Prefab that groups more than one at the same time.

Congratulations on reaching this part of the book and being able to configure and visualize the level while running the game in Unity. Feel free now to play with different configurations and levels; you should get comfortable playing with these tools and creating as many levels as you want.

In the final part of this chapter, we are going to create a camera controller so you will be able to navigate in the **Game** view using the movement to see your entire level.

Controlling the camera

The main camera has two different movements that the player can control: zooming in and out and panning vertically and horizontally. Although the player can perform both actions, there are limits defined by us, so any movement and zoom will display only assets and areas that the player can really see and interact with, hiding the edges of the world.

We are going to write a quite simple, yet flexible script to add a component to the main camera in the scene. The parameters exposed by the component in the **Inspector** window will make it easier to tweak and find the best values for each attribute.

So, create a new folder inside **Scripts** called Camera, and create a new script there called CameraController.cs. Now, first, we are going to define all the properties in this class, as shown in the following script (notice that although the properties are private, the [SerializeField] attribute will expose them in the **Inspector** window so we can easily update the values):

```
using UnityEngine;
namespace Dragoncraft
{
    [RequireComponent(typeof(Camera))]
    public class CameraController : MonoBehaviour
    {
        [SerializeField] private float _borderSize = 50f;
        [SerializeField] private float _panSpeed = 10f;
        [SerializeField] private Vector2 _panLimit =
            new Vector2(30f, 35f);
        [SerializeField] private float _scrollSpeed = 1000f;
        [SerializeField] private Vector2 _scrollLimit =
            new Vector2(5f, 10f);
        private Vector3 _initialPosition = Vector3.zero;
        private Camera _camera = null;
    }
}
```

As we did earlier, all classes created for the game will have the Dragoncraft namespace to keep the code organized. The class is still MonoBehaviour but, this time, we also included a new attribute called RequiredComponent and the type of camera, which means that this script requires that the GameObject already has a camera script attached to it and will not let anyone add this script to a GameObject without this requirement. This is very useful for two reasons:

- It enforces the dependency on another GameObject – in this case, the Camera component.

- Since the GameObject is a requirement for the class, we do not need to validate whether the component exists in the class code. Here, the Camera component is always a valid GameObject that can be used without checking whether it is not null, for example.

Next, we have our first variable to control the camera, the _borderSize variable with a default value. This is the size of an imaginary border in pixels, which we will consider as the zone eligible to move the camera if the mouse cursor is in this area determined by the border size from each side of the screen.

As mentioned earlier, the player can perform two different actions with the camera: zooming and panning. The _panSpeed variable is, as the name suggests, the velocity that the camera will move horizontally and/or vertically. The _scrollSpeed variable has the same usage but uses the mouse scrolling wheel and applies to the camera zoom action. These two variables also have a pair of limits that will help us to limit the player movement and zoom within our definition. Both _panLimit and _scrollLimit are Vector2 variables, so we have the lower and upper boundaries of each action. The default value for all these variables is based on our Playground scene size and can be tweaked on each scene if the values are not working properly.

The last couple of variables are not exposed because they should not be changed outside the script since they are there only to help us. The _initialPosition variable is required so we know where the camera started when the game is running, and we can consider this information when moving the camera. While we do not need the Camera component for the panning action (since we already have access to the Transform property with the script attached to the GameObject), we do need the _camera variable for the zoom action because we are going to modify one property there when the player uses the mouse scrolling wheel. Now, add the following Start method to initialize both variables:

```
private void Start()
{
    _initialPosition = transform.position;
    _camera = GetComponent<Camera>();
}
```

As mentioned before, we do not need to check whether the Camera component exists in this case because of the RequiredComponent attribute in the class, so it is safe to get the component and store the reference in the _camera variable. The _initialPosition variable will store the initial position of the camera to be used later.

Now that we have the script with all the variables needed, we can add it to the camera. Select the **MainCamera** object from your **Hierarchy** window and, in the **Inspector** window, click on the **Add Component** button and search for the Camera component script we have just created. Once added, you should be able to see the component in your **MainCamera** GameObject, as shown here:

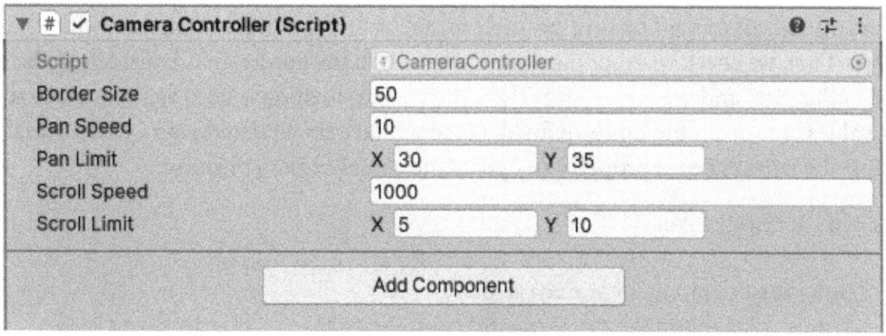

Figure 3.19 – The CameraController component attached to the MainCamera GameObject

Back to the code, we are going to create a method to update the information required for the zooming action to work. This method is UpdateZoom and will be called later, but for now, we are going to understand what is required to perform the zooming action:

```
private void UpdateZoom()
{
    float scroll = Input.GetAxis("Mouse ScrollWheel");
    scroll = scroll * _scrollSpeed * Time.deltaTime;
    _camera.orthographicSize += scroll;
    _camera.orthographicSize = Mathf.Clamp(_camera.orthographicSize,
        _scrollLimit.x, _scrollLimit.y);
}
```

First, we need to know how much the player scrolled the mouse wheel, and, to get that information, we can use the Unity Input.GetAxis API. This method will return a float value in the range -1...1, which we can use to find our own value by multiplying the scroll value by our scroll speed, and the Unity Time.deltaTime API, which is the time in seconds from the last frame to the current one. The delta time is very useful to calculate time or to smooth movements such as scrolling, for example. We are also going to use it to smooth the camera panning in the next method.

The last two lines in this method are there to apply the calculated scroll value to orthographicSize, making the zoom in and out action, and clamping that value between our defined scroll limits. The clamp method from the Mathf API has three parameters: a value, the lower limit, and the upper limit. If the value exceeds any of the limits, the value will be ignored and, instead, the limit value will be returned from the method. This is very useful to make sure the value is in the required range without having to write conditional code to validate it.

The following UpdatePan method is responsible for the panning movement update, which has three different parts. First, we check whether the mouse cursor is in the border area, considering the border size defined in this class and the screen size. Then, the clamp method is used again to make sure our movement values are within the limits defined. Once we have the updated position variable, we set its value to the transform property to update the actual camera position:

```
private void UpdatePan()
{
    Vector3 position = transform.position;
    if (Input.mousePosition.y >= Screen.height - _borderSize)
        position.z += _panSpeed * Time.deltaTime;
    if (Input.mousePosition.y <= _borderSize)
        position.z -= _panSpeed * Time.deltaTime;
    if (Input.mousePosition.x >= Screen.width - _borderSize)
        position.x += _panSpeed * Time.deltaTime;
    if (Input.mousePosition.x <= _borderSize)
        position.x -= _panSpeed * Time.deltaTime;

    position.x = Mathf.Clamp(position.x,
        -_panLimit.x + _initialPosition.x,
        _panLimit.x + _initialPosition.x;

    position.z = Mathf.Clamp(position.z,
        -_panLimit.y + _initialPosition.z,
        _panLimit.y + _initialPosition.z);


    transform.position = position;
}
```

To find out where the mouse cursor is, we will use the Unity Input.mousePosition API, which is Vector3 with X, Y, and Z values. This could look a bit confusing since we are using the X and Z coordinates, and ignoring Y. That is because, when we are looking at the screen, Y is the depth, which we are not changing, while X is our horizontal axis, and Z is the vertical axis.

The following figure shows the camera axis, the Unity orientation, and our green plane. The green arrow is the Y depth axis, which is not updated since we only need to change the camera position in the X axis (red arrow) and the Z axis (blue arrow) to have the panning movement over the plane.

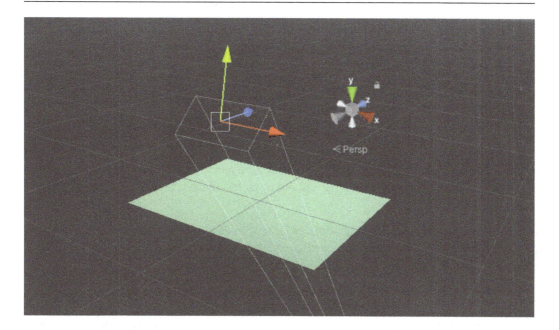

Figure 3.20 – Camera orientation and the visible plane

Considering that the bottom-left corner of the screen is the position (0, 0) for the mouse position, we can calculate whether the mouse cursor is in our border size by checking whether its position is lower than the border size (for the left and bottom sides). If `Input.MousePosition.X` or `Input.MousePosition.Y` is higher than `Screen.Width` or `Screen.Height`, respectively, minus the `_borderSize` value, then we can increase the position on X and Z. However, if `Input.MousePosition.X` or `InputMousePosition.Y` is equal or less than `_borderSize`, we need to decrease the position on X and Z. By using those four conditionals in the first lines of the `UpdatePan` method, it is possible to validate whether we can move the camera or not.

Also, note that we are relying on `Time.deltaTime` once more to smooth the camera panning movement and we are controlling the velocity using `_panSpeed`. If the camera movement feels too slow or too fast, this is the variable you can play with to find values that would suit you better.

Using the `Mathf.Clamp` method again, we can also make sure that the camera will not move outside our determined limit. However, this time, we also consider the `_initialPosition` variable that was set in the `Start` method and used here so no matter where the camera is set, the clamping will consider the correct value. Without the initial position, the clamp would consider that the camera is always at position 0 (0, 0, 0) and it would stop (or not stop) moving at the wrong coordinates.

After all the math is done, we can safely set the new position to the `transform.position` property, which will move the camera to that new location. Unlike in the `UpdateZoom` method, here, we do not need to update the `Camera` component since the GameObject that has this script attached is the camera itself, so the `Transform` property is enough for the movement.

Finally, both camera actions have been created, and we can call both in the `Update` method, as shown here:

```
private void Update()
{
    UpdateZoom();
    UpdatePan();
}
```

Both the `Start` and `Update` methods are from the `MonoBehaviour` class and automatically call Unity when the game is running if the script containing both is attached to an active GameObject in the current scene.

Feel free to change and update all variable values for the camera until you find what works best for you, as the camera movement and zoom speeds could be very personal to each developer. The values here are just suggestions, and you can always revert to them if you prefer.

Keep in mind that if you update the values in the **Inspector** window while the game is running in the Unity Editor, you will lose all changes as soon as you stop it. You can take note of the values you change to update in the **Inspector** window when the game is not running, or right-click on the component and select **Copy Component**:

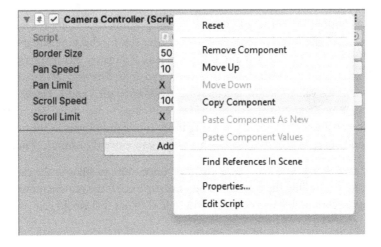

Figure 3.21 – Copying the component values while running the game in the Unity Editor

When the game is not running in the Unity Editor, you can right-click again in the component and select the **Paste Component Values** option. Remember to save the scene so the updated values will stay there. Also, it is worth mentioning that changing the values in the script has no effect after adding it to a GameObject. In this case, always update the values in the **Inspector** window and leave the values in the script. The default values are used only when adding the script to a GameObject.

Summary

Congratulations on reaching the end of this chapter; we did a lot of work here, which forms the foundation for our game. In this chapter, we learned how to set up the scene and create a level using Prefabs and how to group Prefabs into a new one.

Our first contact with C# scripts gave us an understanding of how to use ScriptableObjects, and we developed a simple tool to speed up the level creation. We also wrote code to control the camera movement based on the player's mouse position, considering the size of the plane as the screen limitation.

All of that forms the base of our game, the blocks that we will use and reuse throughout this book until we have our *Dragoncraft* game done.

In *Chapter 4, Creating the User Interface and HUD*, we are going to build the game UI and Prefabs that will be used to inform the player of all the actions and outcomes of the strategic decisions made during the game. You will be introduced to the best practices of creating a UI in Unity and how to make it flexible to fit nicely on different screen sizes and aspect ratios. Finally, we are going to see how to create and use visual effects and sound effects as feedback for the player's actions.

Further reading

We covered a lot of Unity features in this chapter that might be new to you, so remember to always check the official documentation to read more about them:

- *Camera Component*: `https://docs.unity3d.com/2023.1/Documentation/Manual/class-Camera.html`

- *ScriptableObject*: `https://docs.unity3d.com/2023.1/Documentation/Manual/class-ScriptableObject.html`

- *Asset Database*: `https://docs.unity3d.com/2023.1/Documentation/Manual/AssetDatabase.html`

- *Prefabs*: `https://docs.unity3d.com/2023.1/Documentation/Manual/Prefabs.html`

- *Custom Editors*: `https://docs.unity3d.com/2023.1/Documentation/Manual/editor-CustomEditors.html`

- *Transforms*: `https://docs.unity3d.com/2023.1/Documentation/Manual/class-Transform.html`

- *Materials*: `https://docs.unity3d.com/2023.1/Documentation/Manual/materials-introduction.html`

There is some great documentation provided by Microsoft to learn more about some of the C# features that we used in this chapter, with some more examples of usage to help you understand other use cases:

- *C# Naming Conventions*: `https://learn.microsoft.com/en-us/dotnet/csharp/fundamentals/coding-style/coding-conventions`

- *Serializable Attribute Class*: `https://learn.microsoft.com/en-us/dotnet/api/system.serializableattribute`

- *Declare namespaces to organize types*: `https://learn.microsoft.com/en-us/dotnet/csharp/fundamentals/types/namespaces`

- *List<T> Class*: `https://learn.microsoft.com/en-us/dotnet/api/system.collections.generic.list-1`

4
Creating the User Interface and HUD

RTS games are mostly played using actions and commands from the game interface, so it is very important to have a clean and scalable UI to help the player master the game. In this chapter, we are going to draw inspiration from common UI elements in classic RTS games and create our own for the *Dragoncraft* game.

We are going to build the foundation for the menu options, resource counters, player actions, unity and enemy details, and the minimap, one of the core mechanics of RTS games. A flexible and adaptable UI component will be created and configured to have a great performance with low draw calls and will be responsive to different screen ratios. This chapter will also introduce you to Render Textures, multiple camera usages, and our first level manager.

By the end of this chapter, you will have learned how to create UI elements and position them considering different screen sizes, as well as having learned how to interact with the UI elements and prepare the interface for dynamic objects and text information. You will also learn how to set up multiple cameras to render different parts of the world on textures, and how to load multiple scenes at the same time.

In this chapter, we will cover the following topics:

- Using Canvas for a responsive UI
- Setting up the UI and HUD using Prefabs
- Rendering the minimap
- Loading the UI scene additively

Technical requirements

The project setup in this chapter with the imported assets can be found on GitHub at `https://github.com/PacktPublishing/Creating-an-RTS-game-in-Unity-2023` in the `Chapter04` folder inside the project.

The assets from the Unity Asset Store used in this project can be found at this link: `https://assetstore.unity.com/lists/creating-a-rts-game-5773122416647`.

Using Canvas for a responsive UI

Creating a UI on Unity involves more than just adding components to the screen and positioning them where you like; with so many screen aspect ratios and sizes available on TVs, monitors, tablets, and smartphones, we need to make sure our UI will look great on all screen sizes. To this end, we employ a concept called responsiveness, which is very common in web design and enables the design or UI to adapt itself to the screen size, while always ensuring a base position and dimension that is in proper proportion relative to the screen aspect ratio.

This may sound confusing and feel like a lot of work, but Unity has great tools to help us design and configure the best-looking UI for our game. Before diving into these tools, let us first review the UI of one of the games that we are using as inspiration for our own project, *Dragoncraft*, which is *Warcraft III* by Blizzard. Here is a basic recreation of what its interface looks like:

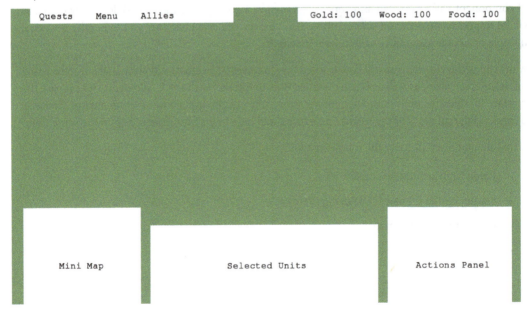

Figure 4.1 – Wireframe with the UI elements based on Warcraft III

Looking at the bottom of the screenshot, we can see three different zones of the UI, from left to right: the minimap, the selected units, and the actions panel. At the top of the screen, we can see the HUD with the **Quests**, **Menu**, and **Allies** buttons to the left, and the **Gold**, **Wood**, and **Food** resources to the right. These are the main elements of the UI and HUD in *Warcraft III*, and these will serve as our references as we develop the UI for our game in this chapter.

First, we need to create a new scene for the UI that will be loaded on top of each level scene. The UI will be a separate scene because we are going to use it on every level with a consistent layout, while the level scenes will be different from each other due to the various maps the game can have. Since this scene will be used only for the UI, we do not need light, therefore we can create a new scene and remove the GameObject with the light component:

1. Right-click inside the **Scenes** folder and select **Create | Scene**.
2. Name the new scene GameUI and double-click to open it.
3. In the **Hierarchy** view, right-click on the **Directional Light** GameObject and select **Delete**.
4. Right-click on the **Main Camera** GameObject and select **Rename**. Rename it UICamera.

Now we can set up the camera that will be used to render only the UI objects of the scene. One of the main settings to check here is the **Occlusion Culling** setting, which is responsible for rendering, or not rendering, specific layers set up on each GameObject. We are also going to remove the **Tag** that, by default, sets the camera as the main camera – which is something we do not want because this will only be the UI camera and cannot interfere with the main camera of our game from the level scenes.

So, to set up the camera UI, make the following changes to the default values relating to the main camera in the scene:

1. Left-click on **Tag** and select the **Untagged** option.
2. Left-click on **Layer** and select the **UI** option.
3. Make sure to set the **Transform Position** and **Rotation** fields to **(0, 0, 0)**.
4. Left-click on **Clear Flags** and select **Depth only**.
5. Left-click on **Occlusion Culling** and click on all options that have a check mark to deselect them and make sure the only one selected is the **UI**.
6. Left-click on **Projection** and change it to **Orthographic**.
7. In the **Depth** text field, replace the value **-1** with 0.
8. Left-click on the component under **Camera** called **Audio Listener**, and select the **Remove Component** option. The main camera on the level scene already has an Audio Listener and there can be only one on all active scenes, otherwise, Unity will spam messages telling you to remove any extra Audio Listener components.

The following screenshot shows the result:

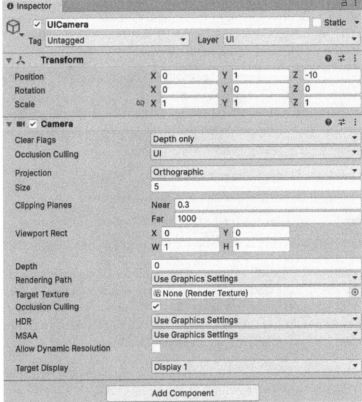

Figure 4.2 – The UI Camera settings

The **Layer** value of the GameObject is used together with the Occlusion Culling as we are going to see when we start to set up the Prefabs. In our current camera for the UI, only the GameObjects in the Layer UI will be rendered by the camera. Later, when we load this scene on top of level scene, we are going to remove the Layer UI from the main camera Occlusion Culling so each camera will be responsible for specific layers without rendering the content on each camera twice.

Before adding UI components to the scene, let us first set up our **Canvas**, which is the Unity component responsible for holding all the UI components and it must be configured to the responsiveness of our entire game UI. The Canvas works together with a couple of components: the **Canvas Scaler**, a component to control the scale and pixel density of the UI elements, and the **Graphic Raycaster**, used to Raycast against a canvas and find the UI elements. For our game UI configuration, we are going to set up only the Canvas and the Canvas Scaler components, like so:

1.   Right-click on the **Hierarchy** view and select the **UI | Canvas** option to add a new one to the scene.

2.   Left-click on the newly created Canvas to see the GameObject components in the **Inspector** view.

3. Left-click on **Render Mode** and select the **Screen Space - Camera** option.

4. Drag and drop the **UICamera** GameObject into the **Render Camera** field that has shown up after changing **Render Mode** to **Screen Space - Camera**.

5. Left-click on **Additional Shader Channels** and select the **Everything** option.

6. In the next component, **Canvas Scaler**, left-click on **UI Scale Mode** and select the **Scale With Screen Size** option.

7. After changing **UI Scale Mode**, a few fields will appear below it. In the **Reference Resolution** fields, type in 1280 in the **X** field and 720 in the **Y** field.

8. For the next option, **Screen Match Mode**, left-click to select the option **Match Width or Height**.

9. In the **Match** text field, change the value **0** to 1 by typing the new value or move the slider all the way to the right side.

The following screenshot shows the result:

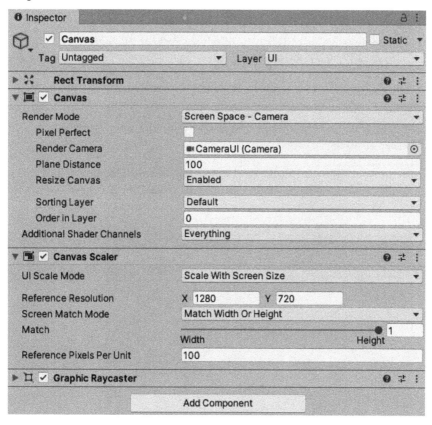

Figure 4.3 – The Canvas settings

UI Scale Mode will ensure that our UI will be scaled based on the screen size we defined in **Reference Resolution**. For our RTS game, we can consider the reference resolution as **high-definition (HD)**, which is also known as 720p – that is why **Y** or the height has a value of 720 pixels, and the corresponding value for **X** or the width is 1,280 pixels. We are also using **Screen Match Mode** to lock the resize of the Canvas based on matching the height or the width of the screen.

With both the camera and the Canvas all set up, we can finally start to add visual elements to our interface. However, since we will need art for the UI, there is another package that we need to import into our project from the Unity Asset Store, and it is called **GUI Parts**.

Go through the following instructions to add the GUI Parts package to your assets and see it in the Package Manager so you can import it into the project:

1. Open the `https://assetstore.unity.com/lists/creating-a-rts-game-5773122416647` link (you might need to log into your Unity Asset Store account created in *Chapter 2*).

2. In the list of assets, find **GUI Parts** and click on **Add to My Assets**.

3. Go back to the Unity Editor and click on **Window | Package Manager**.

4. Left-click on **Packages** and select the **My Assets** option to see all assets in your account.

5. Click on the **Download** button and then, once it has finished, click on the **Import** button. After you have downloaded it once, the button will change to **Re-Download**.

6. Once the package is imported into your project, move the **GUI_Parts** folder into the **ThirdParty** folder to keep all assets organized.

The following screenshot shows the **Package Manager** window with the new **GUI Parts** package selected (note that the button in the top-right corner reads as **Re-Download** because it was already downloaded once):

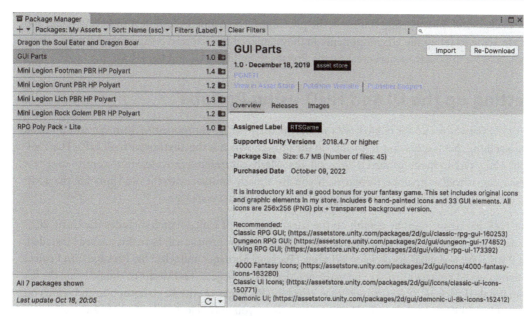

Figure 4.4 – Package Manager window with the GUI Parts package selected

As in the packages imported to build our levels, this package also has a great number of assets, more specifically **Sprites**, which are representations of 2D objects that we are going to use in our UI. Most of the assets will be resized to fit our needs, but it will all be done within the Unity Editor when adding the UI components, such as buttons or images. The following screenshot shows all the available assets in this package (these will be referenced by their filenames when creating new UI components):

Figure 4.5 – UI and HUD using the assets from the GUI Parts package

The background, buttons, and frames in the preceding screenshot are a perfect fit for our game and we are going to set up the UI components so they can keep the size and position on different screen sizes. Now we are ready to start creating the Prefabs to build the UI and the HUD.

Setting up the UI and HUD using Prefabs

With the Canvas set up for our UI, we can start adding the UI components. However, we still need one last thing that is quite easy to include in the project, which is the **TextMeshPro (TMP)**. The TMP is the Unity replacement for the UI Text and Text Mesh features and provides more control over text formatting as well as great performance. If you are already familiar with the old Unity UI, the TMP has almost the same names and features, but in a more robust package.

The TMP package comes by default on the latest versions of Unity, which includes the Unity 2023 version that we are using to develop our RTS game. Any new project will have this package installed, but if you want to double-check, go to **Window | Package Manager** and select the **Packages : In Project** option to see all packages that are currently included in our project. **TextMeshPro** should be on the list as an installed package. If you cannot see it, select the **Packages : Unity Registry** option and search on that list to add it by clicking on the **Install** button once you have selected the **TextMeshPro** package.

Having the TMP package installed is one part of the setup. To add our first UI component and finish the TMP setup we are going to create a new GameObject inside the **Canvas** GameObject and then add a **Button – TextMeshPro** UI component as a child of the new GameObject. This is what you need to do:

1. In the **Hierarchy** view, left-click on the **Canvas** GameObject and select **Create Empty**.

2. Name the new GameObject Menu.

3. Left-click on the **Menu** GameObject and select **UI | Button – TextMeshPro**.

4. A new **TMP Importer** window will be displayed to you. Click on the **Import TMP Essentials** button.

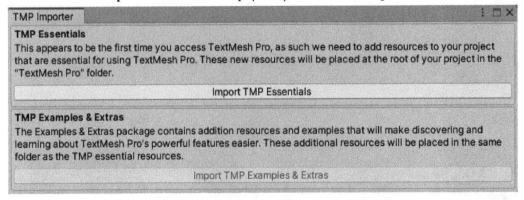

Figure 4.6 – The message asking to import TMP essentials

5. Once Unity is done importing the resources, the **Import TMP Essentials** button will be disabled, and you can close the screen by clicking on the **X** button. We do not need the TMP Examples & Extras package for our project.

6. Rename the button to MenuButton to keep all UI components with proper names since we are going to have a lot of them on the Canvas.

That is all we need to do before starting to use the TMP UI components. Also remember to keep the **TextMesh Pro** folder that was added to the project because the TMP will need those resources internally to make our UI components look good when running the game.

One very important thing that applies to all UI components is changing the **Rect Transform** properties in the parent GameObject, which will affect the child GameObjects. For instance, we just added a new GameObject called **Menu** as a child of the **Canvas** GameObject, and we added the **MenuButton** button as a child of the **Menu** GameObject. In this case, the **Rect Transform** property of the **Menu** GameObject will affect the **MenuButton** GameObject. However, changing the child properties does not change the parent properties.

Let's change the configuration of both the **Menu** and **MenuButton** GameObjects. Left-click to select the **Menu** GameObject in the **Hierarchy** view, then open the **Inspector** view. From there, do the following:

1. Left-click on the Anchor Presets represented by the square on the left side of the **Pos X** field – it has a default label of **center** on top and **middle** on the left side. When clicking on that square, a few options will be presented. Select the square that is in the bottom-right corner of the Anchor Presets, represented by blue arrows that show the content will expand both horizontally and vertically, which is called **stretch** in either direction.

2. Change **Pivot X** to 0 and **Pivot Y** to 1.

3. Once **Pivot** is changed, the position will have new values. Change the **Pos X**, **Pos Y**, and **Pos Z** values to 0.

4. Left-click to select the **MenuButton** GameObject in the **Hierarchy** view and repeat step *1*. However, this time, select the option in the top-left corner labeled **left** on the top side and **top** on the left side.

5. Change **Pivot X** to 0 and **Pivot Y** to 1. After that, change the **Pos X**, **Pos Y**, and **Pos Z** values to 0.

The following figure shows both the **Menu** and **MenuButton** GameObjects after the configuration changes:

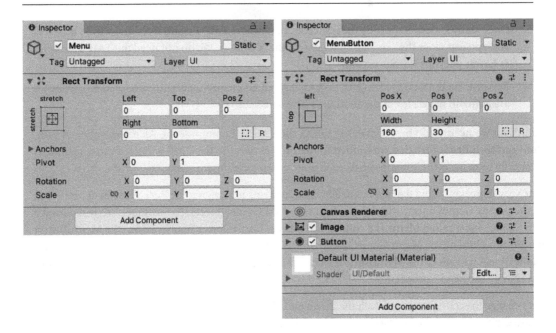

Figure 4.7 – The Menu properties (left) and the MenuButton properties (right)

Pivot is a property that tells the UI component where the coordinate (0, 0, 0) is set. Here, setting **Pivot** to (0, 1) will consider the coordinate (0, 0, 0) as the top-left corner. Then we use the Anchor Presets to set the **MenuButton** anchor at the top-left corner of the screen, and the **Menu** anchor as a stretch for both vertical and horizontal positions, which will make the UI component use the whole screen rather than a portion of it.

Now we can start to add a sprite to the **MenuButton** GameObject and an action to display another UI component that will have the menu options for the player to choose from. Our focus will be on the style and the correct position of the UI elements rather than the actual functionalities, which will come later in the book, to hook up the UI and the player actions.

Creating the MenuButton GameObject and popup

The sprite that we are going to use on the **MenuButton** GameObject is the same that will be used on all the other buttons. Therefore, it makes sense to create a Prefab out of the configured button so we can easily add more of them without having to do the same thing for each new button. Before creating the Prefab, we need to configure the **MenuButton** GameObject as shown in the following screenshot.

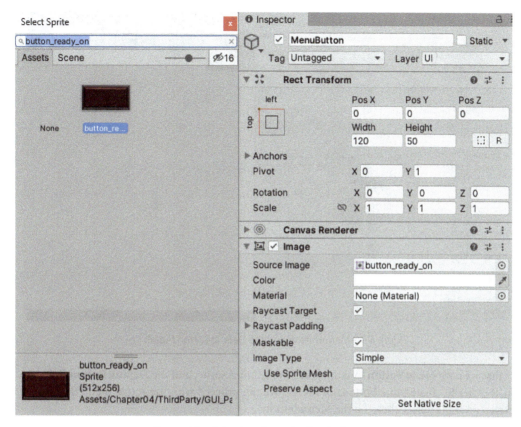

Figure 4.8 – Selecting the sprite for the button

In the **Image** component that is already attached to the **MenuButton** GameObject, left-click on the **Source Image** field and, in the new window that will open, search for **button_ready_on**. Double-click on the selected sprite to add it to the button. We are also going to change the button size. In the **Rect Transform** component, change **Width** to 120 and **Height** to 50.

Now that the button sprite and size are set, the next step is to adjust the text in the child GameObject of the **MenuButton** GameObject, which is a TMP text field with the default name **Text (TMP)**. Left-click to select the **Text (TMP)** GameObject and, in the **Inspector** view, we are going to change the text from **Button** to Menu. Leave all the other settings with their default values.

Now we can create the Prefab for the next buttons. First, in the **Project** view, create a new folder called UI inside the existing **Prefab** folder. Next, drag and drop **MenuButton** from the **Hierarchy** view into the new **UI** folder to create the Prefab. Finally, we can give the Prefab a better name by changing it to DefaultButton:

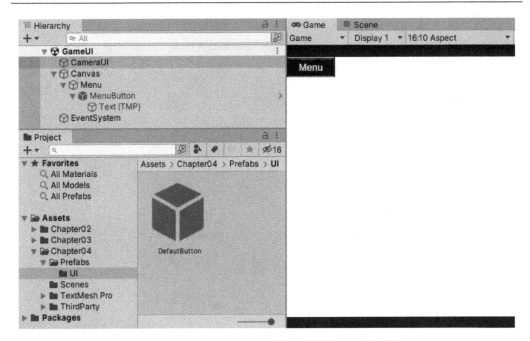

Figure 4.9 – MenuButton and the DefaultButton Prefab

Now that we have **MenuButton** looking good using a nice sprite, and the **DefaultButton** Prefab to make it easier to create more buttons in the UI, we can proceed with the actual menu that will be displayed once the player clicks on the **Menu** button.

We are going to add a pause menu pop-up panel with the options to either resume or exit the game. The pop-up panel will be displayed as soon as we click on the **Menu** button, with a transparent but slightly black background that will be on top of the game scene showing that the game was paused (at this time, we will not add the pause game logic yet, but the open and close menu actions will be added without any lines of code required). So, follow these steps:

1. Right-click on the **Menu** GameObject and select **Create Empty**, then name it PausePopup.

2. Select the **PausePopup** GameObject and, in the **Inspector** view, change **Anchor Preset** to stretch on both sides (as we did for the **Menu** GameObject). After changing **Anchor Preset**, change the **Left** property to 0, **Top** to 0, **Pos Z** to 0, **Right** to 0, and **Bottom** to 0.

3. With **PausePopup** selected, click on the **Add Component** button, search for **Canvas Renderer**, then double-click or press the *Enter* key to add it. No changes are required in this component.

4. Still with **PausePopup** selected, click on the **Add Component** button one more time, search for **Image**, and add the component.

5. On the **Image** component, click on the **Color** field to open the **Color** window. Change the values for **R**, **G**, and **B** to 0, and **A** to 155. You can close the window once the values are changed.

The following screenshot shows the **PausePopup** GameObject and all the configurations that were done in the preceding steps. On the left side, in the **Scene** view, we can see the black transparent rectangle that is covering the entire game scene.

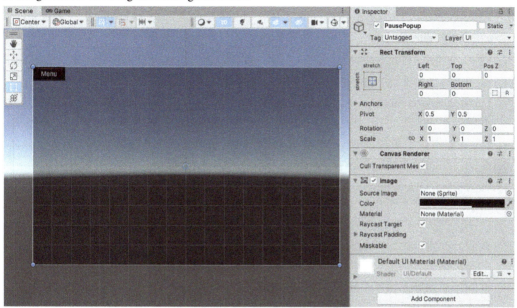

Figure 4.10 – The PausePopup GameObject set up

The **Image** component, when it has no **Source Image** Sprite, is a solid color that can be changed to any RGB value. The color also has an alpha channel that we used to make it slightly transparent so we can use this image on top of the game to display the **PausePopup** GameObject but the player will still be able to see some of the level behind it due to its transparency.

To help prevent mouse clicks on the map and UI while the **PausePopup** GameObject is visible, we will use the **Canvas Renderer** component to block any UI clicks that are behind it. In fact, if you just play the scene, you will notice that the **Menu** button cannot be clicked due to the **Canvas Renderer** component blocking the click event. This is exactly what we want and, together with the pause game functionality, we can make sure the player will have the game session stopped and the pause menu will be the only actionable item on the screen.

Now that we have the background to block UI interactions, we can create the actual pause menu pop-up panel and the two buttons, **Resume** and **Exit**:

1. Right-click on the **PausePopup** GameObject, select **Create Empty**, and name it `Background`.

2. Select the new **Background** GameObject and, in the **Inspector** view, change **Width** and **Height** to `256`.

3. Click on the **Add Component** button to add an **Image** component.

4. On the **Image** component, left-click on the small circle button located on the right side of the **Source Image** property. Then, in the new window, search for a sprite named **Mini_background**, and double-click to add it.

5. Right-click on the **Background** GameObject, select **Create Empty**, and name it `Frame`. This new GameObject must be a child of the **Background** GameObject.

6. Select the new **Frame** GameObject and, in the **Inspector** view, change **Width** and **Height** to 256.

7. Click on the **Add Component** button to add an **Image** component.

8. On the **Image** component, left-click on the small circle button located on the right side of the **Source Image** property. Then, in the new window, search for a sprite named **Mini_frame1**, and double-click to add it.

With the background and frame in place, we can add the two buttons, **Resume** and **Exit**. However, this time we are going to add them using the **DefaultButton** Prefab. We need to do the following steps twice, one for each button. We are going to consider the **Resume** button as the first button and the **Exit** button as the second button:

1. Drag the **DefaultButton** Prefab from the **Prefabs | UI** folder located in the **Project** view and drop it on top of the **PausePopup** GameObject in the **Hierarchy** view.

2. In the **Hierarchy** view, rename **DefaultButton** to `ResumeButton`. For the second **DefaultButton** GameObject, rename it to `ExitButton`.

3. Select the new button and, in the **Inspector** view, change the Anchor Presets to **middle** on the left and **center** at the top.

4. Change the **Pivot** values to 0.5 for **X** and 0.5 for **Y**.

5. Change **Pos X** to 0 and **Pos Y** to 32. Change **Pos Y** to -32 for the second button.

6. In the **Hierarchy** view, select the child GameObject called **Text (TMP)** and, in the **Inspector** view, change the text from **Menu** to `Resume`. Change it to `Exit` for the second button.

Once we have done these steps to add both the **Resume** button and the **Exit** button, our pause menu pop-up panel will look like the following screenshot. Notice that the buttons created from the **DefaultButton** Prefab have blue names, which indicates they are linked to a Prefab.

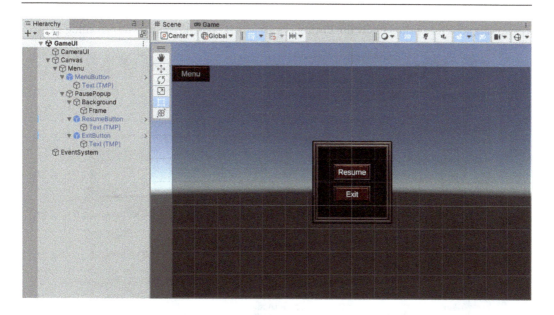

Figure 4.11 – The complete pause menu pop-up panel with the two buttons

The last change we are going to make to the pause menu is to add the action to open and close it by clicking on the **Menu** button, without writing a single line of code. We are also going to add the action to close the pause menu by clicking on the **Resume** button:

1. Left-click on the **MenuButton** GameObject and, in the **Inspector** view, find the **Button** component that is already attached.

2. The last part of the **Button** component is a box with the text **On Click ()** and two buttons to add, +, or remove, -, an item. Click on the + button to add a new item and notice that, after the new item is added, there is a field with the value **None (Object)**.

3. Drag the **PausePopup** GameObject from the **Hierarchy** view and drop it in the field that has the value **None (Object)**. The value of the field will change to **PausePopup**.

4. Left-click on the drop-down field with the value **No Function** and select **GameObject | SetActive (bool)**.

5. After selecting the preceding function, a new checkbox will appear on the right side of the field with the value **PausePopup**. Left-click on the checkbox to set the value to true.

We added a new event to the button, for when the player clicks on it, and assigned a function to be called when that happens. Note that the available functions will only appear on the list after dragging and dropping a GameObject to the event. Also, after selecting a function, a new field may or may not appear. In our case, since the function selected has a **bool** parameter, the checkbox was displayed there.

The following screenshot shows how the **Button** component is configured after the preceding steps:

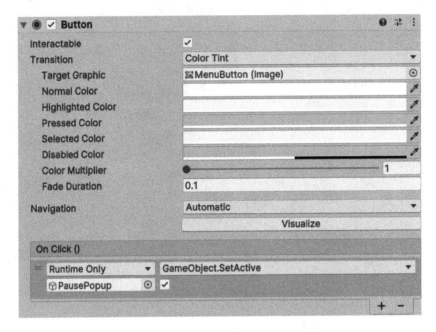

Figure 4.12 – The Button component with the On Click function set up

In this case, we did not need to write any code because there are some functions that are already presented in the **GameObject** class, such as **SetActive**, which we used. If we had any custom script attached to the **PausePopup** GameObject, and the script had any public functions, they would be listed there as well.

We are going to use a custom script in the next chapter, but for now, the available functions in the **GameObject** class are enough to show and hide the pause menu.

Now that the **MenuButton** GameObeject has the action to display the pause menu, we need to add the action to hide the pause menu to the **ResumeButton** GameObject. It is a bit similar to the previous steps but with a few differences:

1. Left-click on the **ResumeButton** GameObject and, in the **Inspector** view, find the **Button** component that is already attached.

2. Click on the + button to add a new item and notice that, after the new item is added, there is a field with the **None (Object)** value.

3. Drag the **PausePopup** GameObject from the **Hierarchy** view and drop it in the field that has the value **None (Object)**. The value of the field will change to **PausePopup**.

4. Left-click on the drop-down field with the **No Function** value and select **GameObject |
SetActive (bool)**.

5. Left-click on the **PausePopup** GameObject and, in the **Inspector** view, click on the checkbox
on the left side of the object name to disable it.

Here is the result:

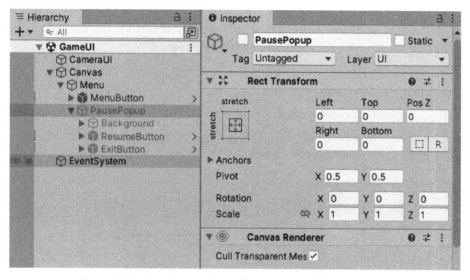

Figure 4.13 – PausePopup GameObject disabled in the hierarchy

The last step is required to make the **PausePopup** GameObject not visible since all GameObjects that
are disabled do not get rendered in the editor, nor in the game. Also note that when the GameObject
is not enabled, the GameObject itself and all the children GameObjects have a slightly faded name.

Now that we have the pause menu configured in the UI, we can run the scene in the editor and click
on the **Menu** button to display the popup with the two buttons. If we click on the **Resume** button,
the popup will be disabled but the **Game** view will still render – this will not be a problem once we
load this scene additively on top of the level scene later. Also, the **Exit** button will not have any action
at this moment because it will require some script code, which we are going to write later as well.

For now, let's move on to the next UI section, which is the resources counter.

Displaying the resources counter

Our game will have three different resources for the player to either gather or generate, and all resources
can be consumed or used to create a build or train a new unit. This next UI section is the counter for
each one of the resources: **Gold**, **Wood**, and **Food**. The logic to generate **Gold**, gather **Wood**, and
produce **Food** will be added later, in *Chapter 13, Producing and Gathering Resources*; for now, we are
going to focus on the interface that will display each resource's current amount.

Let's start by adding and customizing a new button in the interface, and then create a Prefab from it so the other two buttons can be easily created and modified. We are also going to create a new container for the resources counter, so they are all grouped in the same alignment using the **Horizontal Layout Group** component from Unity:

1. Right-click on the **Canvas** GameObject in the **Hierarchy** view, select **Create Empty**, and name it Resources.

2. Left-click on the new **Resources** GameObject to select it and, in the **Inspector** view, change the Anchor Presets to **top** on the left and **stretch** at the top.

3. Change the **Pivot** values to 1 for **X** and 1 for **Y**.

4. Change **Left** to 0, **Pos Y** to 0, and **Pos Z** to 0.

5. Click on the **Add Component** button and add the **Horizontal Layout Group** component.

6. After adding the component, change **Child Alignment** to **Upper Right** and uncheck the **Width** option of the **Child Force Expand** property.

The following screenshot shows the **Resource** GameObject after the changes we made in the previous steps:

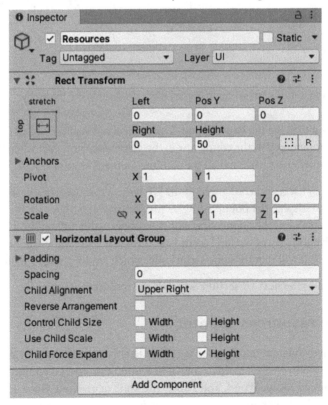

Figure 4.14 – The Resources GameObject settings

The **Horizontal Layout Group** component will enforce a horizontal alignment to all child GameObjects, and using the preceding settings, the alignment will start in the upper-right corner of the **Rect Transform** property and expand the height if the elements do not fit side by side.

Since we set our Anchor Preset to **stretch** at the top, the **Resources** GameObject will take all the available screen width and, because of the **Horizontal Layout Group** component, all buttons added are going to be automatically positioned starting on the right side of the screen. The preceding settings make it possible to always keep the buttons side by side, on the right side of the screen, at any screen width.

The **Child Force Expand** property will tell the component to expand to fill the space with the elements by **Width** or **Height**. Unchecking the **Width** checkbox will keep all elements aligned in the direction we want, otherwise, they would be centered across all available spaces, leaving empty gaps between the buttons. Once we add the buttons, you can return to this and play with the properties to see the effects on the elements.

Now we can add our first button, create a Prefab, and then add the other two buttons:

1. Right-click on the **Resources** GameObject, select **UI | Button – TextMeshPro**, and rename it to Gold.

2. Left-click on the new button and, on the **Inspector** view, click on the small circle button on the right side of the **Source Image** property in the **Image** component. In the new window, search for name_bar and double-click on the sprite to add it to the button.

3. On the **Rect Transform** component, change **Width** to 140 and **Height** to 40.

4. On the **Button** component, click on the three dots on the right side and select **Remove Component**. We do not need this component for the resources, as we do not need a click event (we are taking advantage of the **Button** component structure to create our resources counter).

5. In the **Hierarchy** view, expand the **Gold** button and left-click on the **Text (TMP)** child GameObject to select it – all the next changes will be performed in the components attached to it.

6. On the **Rect Transform** component, change the **Left** value to 10 and the **Right** value to 20.

7. On the **TextMeshPro – Text (UI)** component, change the text to Gold: 9999. This is a temporary value used to fill the text field.

8. On the same component, enable **Auto Size** by clicking on the checkbox and, in **Auto Size Options**, change **Min** to 10 and **Max** to 16. Change **Vertex Color** to white and **Alignment** to **Left**.

9. Drag and drop the **Gold** button from the **Hierarchy** view into the **Project** view, inside the **Prefabs | UI** folder. After creating the Prefab, rename it to ResourcesButton.

10. Finally, drag and drop the newly created **ResourcesButton** Prefab from the **Project** view into the **Hierarchy** view, on top of the **Resources** GameObject. Rename the GameObject to Wood and change the text to Wood: 9999.

11. Repeat the last step once more and this time rename the GameObject to Food and change the text to Food: 9999 (this value is temporary and will be dynamically changed in *Chapter 13, Producing and Gathering Resources*, based on the current inventory that the player has).

The following screenshot shows the current state of our UI:

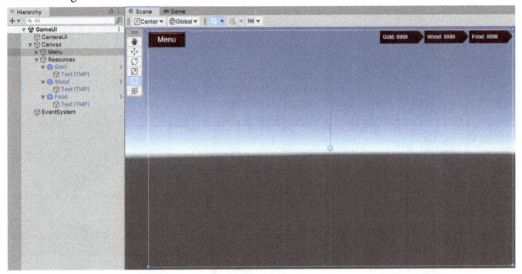

Figure 4.15 – The current UI state with the new resource counters on the right side

After adding the resource counters, we are pretty much done with the elements on the top portion of the screen and we can proceed with the bottom part, starting with the action buttons placed on the bottom-right side of the screen.

Adding the action buttons

The next UI section is a panel located at the bottom-right side of the screen and has buttons where the player can select one or more targets and perform the following actions: attack, defend, move, collect, build, and upgrade. All these actions can be executed by one or more units when they are selected by the player. For now, we need to prepare the UI and add the buttons. So, in the upcoming chapters, we can start to bind the buttons to their respective actions.

First, we are going to create the GameObject to hold all buttons and add the background and frame to it:

1. Right-click on the **Canvas** GameObject in the **Hierarchy** view, select **Create Empty**, and name it Actions.

2. Left-click on the new **Actions** GameObject to select it and, in the **Inspector** view, change the Anchor Presets to **bottom** on the left and **right** at the top.

3. Change the **Pivot** values to 1 for **X** and 0 for **Y**.

4. Change **Pos X** to 0, **Pos Y** to 0, and **Pos Z** to 0.

5. On the Rect Transform component, change **Width** to 280 and **Height** to 200.

6. Select the new **Background** GameObject and, in the **Inspector** view, change the Anchor Presets to **stretch** on the left and **stretch** at the top. Change **Left** to 0, **Top** to 0, and **Pos Z** to 0.

7. Click on the **Add Component** button and add an **Image** component to the **Background** GameObject.

8. After adding the **Image** component, click on the circle button on the right side of the **Source Image** property, and in the new window, search for the **Mini_background** sprite and double-click to add it.

9. Right-click on the **Background** GameObject, select **Create Empty**, and name it Frame.

10. Select the new **Frame** GameObject and change **Anchor Presets** to **stretch** on the left and **stretch** at the top. Change **Left** to 0, **Top** to 0, and **Pos Z** to 0.

11. Click on the **Add Component** button and add an **Image** component to the **Frame** GameObject. Click on the circle button on the right side of the **Source Image** property, search for the **Mini_frame0** sprite, and double-click to add it.

After following the preceding steps, we should see a rectangle in the bottom-right corner of the screen with a dark background and a gold frame. Now we are going to add three rows with two buttons on each one, with a total of six buttons in two columns.

We can create a container using the **Vertical Layout Group** component for the three rows of buttons, and another container on each of the button rows using the **Horizontal Layout Group** component so they fit side by side. Each button will reuse the **DefaultButton** Prefab.

The **Vertical Layout Group** and the **Horizontal Layout Group** components are very similar. The main difference is that while the first component places the child elements on top of each other, the latter places the child elements side by side. We are going to start creating the first row, which will make it easier to duplicate to create the second and third rows once it is done:

1. Right-click on the **Actions** GameObject in the **Hierarchy** view, select **Create Empty**, and name it Container.

2. Select the new **Container** GameObject and change the Anchor Presets to **stretch** on the left and **stretch** at the top.

3. Change the **Pivot** values to 0 for **X** and 1 for **Y**. After that, change **Left** to 15, **Top** to 15, and **Pos Z** to 0.

4. Click on the **Add Component** button and add a **Vertical Layout Group** component to the **Container** GameObject.

5. Right-click on the **Container** GameObject, select **Create Empty**, and name it Row1.

6. Select the new **Row1** GameObject and change **Width** to 256 and **Height** to 32.

7. Click on the **Add Component** button and add a **Horizontal Layout Group** component to the **Row1** GameObject.

8. Drag and drop the **DefaultButton** Prefab from the **Project** view into the **Row1** GameObject in the **Hierarchy** view. Do this twice to have two buttons as children of **Row1**.

9. Right-click on the **Row1** GameObject in the **Hierarchy** view and click on the **Duplicate** option. Do this twice so we have a total of three rows created. Rename **Row1(1)** to Row2 and **Row1(2)** to Row3.

10. Finally, we need to rename the duplicated buttons to have a better name and the proper text. In the first row, rename the left button to AttackButton and change the text to Attack, then rename the right button to DefenseButton and change the text to Defense. In the second row, rename the left button to MoveButton and the right button to CollectButton, with the respective Move and Collect texts. In the last row, the left button must be named BuildButton, and the right button should be named UpgradeButton, with the corresponding text as Build and Upgrade.

This time, we did not need to tweak any of the parameters for either **Vertical Layout Group** or **Horizontal Layout Group** and, as we saw in steps 8 and 9, it was very quick to add buttons and more rows of buttons without needing to adjust their position once we set the container size. Here, we can see the power of both Prefab usage and a flexible UI to speed up the development process.

Remember to double-check all the GameObjects' names and the buttons' text to make sure they match the following screenshot because later they will be referenced by their names when needed:

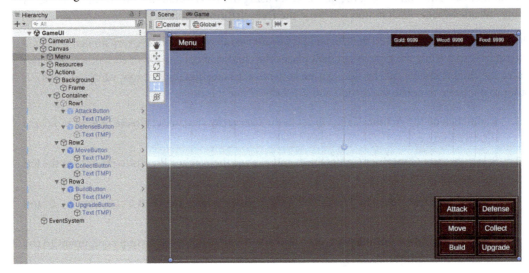

Figure 4.16 – The UI after adding the actions panel in the bottom-right corner of the screen

We are now approaching the two most complex sections of the UI. First, we are going to create a details panel where all the information from the selected objects such as units, buildings, enemies, and resources will be displayed.

Creating the details panel using a custom camera

The details panel is one of the most dynamic parts of the screen because we need to add the number of units selected and their **Health Points** (**HP**). We are going to create the UI elements now and prepare them so, later, in *Chapter 7, Attacking and Defending Units*, we can update the details panel using the selected units.

Besides these details, we are also going to have a small portrait of the object selected, using a custom camera to display the selected 3D model and an idle animation when applicable. Let's begin by creating the details panel and a Prefab that will be used to represent the selected units with their respective HP status:

1. Right-click on the **Canvas** GameObject in the **Hierarchy** view, select **Create Empty**, and name it Details.

2. Select the new **Details** GameObject and change the Anchor Presets to **bottom** on the left and **center** at the top. Change the **Pivot** values to 0.5 for **X** and 0 for **Y**, and for the positions, change **Pos X** to -40, **Pos Y** to 0, **Pos Z** to 0, **Width** to 480, and **Height** to 150.

3. Right-click on the newly created **Details** GameObject, select **Create Empty**, and name it Panel.

4. Select the new **Panel** GameObject and change the Anchor Presets to **bottom** on the left and **stretch** at the top. Change the **Pivot** values to 1 for **X** and 0 for **Y**, and for the positions, change **Left** to 150, **Pos Y** to 0, **Pos Z** to 0, **Right** to 0, and **Height** to 150.

5. Click on the **Add Component** button and add an **Image** component to the **Panel** GameObject. Click on the circle button on the right side of the **Source Image** property and, in the new window, search for the **Mini_background** sprite and double-click to add it.

6. Right-click on the **Panel** GameObject, select **Create Empty**, and name it Frame.

7. Select the new **Frame** GameObject and change the Anchor Presets to **bottom** on the left and **stretch** at the top. Change the **Pivot** values to 1 for **X** and 0 for **Y**, and for the positions, change **Left** to 0, **Top** to 0, **Pos Z** to 0, **Right** to 0, and **Height** to 150.

8. Click on the **Add Component** button and add an **Image** component to the **Frame** GameObject. Click on the circle button on the right side of the **Source Image** property, search for the **button_frame** sprite, and double-click to add it.

9. Right-click on the **Panel** GameObject, select **Create Empty**, and name it Container.

10. Select the new **Container** GameObject and change the Anchor Presets to **stretch** on the left and **stretch** at the top. For the positions, change **Left** to 20, **Top** to 15, **Pos Z** to 0, **Right** to 20, and **Bottom** to 15.

11. Click on the **Add Component** button and add a **Vertical Layout Group** component to the **Container** GameObject.

12. Right-click on the **Container** GameObject, select **Create Empty**, and name it Row1. Select the new **Row1** GameObject and change **Width** to 280 and **Height** to 50.

13. Click on the **Add Component** button and add a **Horizontal Layout Group** component to the **Row1** GameObject. Change the **Spacing** property to 7 and uncheck **Width** in the **Child Force Expand** property.

14. Right-click on the **Row1** GameObject, select **Duplicate** to make a copy, and name it Row2.

As we can see from the previous steps, the **Details** panel has a few combinations of property values that are required to have the correct position and size for all elements inside of the **Details** GameObject. We are using **Spacing** in the **Horizontal Layout Group** this time so the Prefabs that we are going to create next will have some extra space between them.

Now that the **Details** panel is set up, we can create a Prefab that will represent the selected unit and will be added to the **Row1** GameObject as temporary content while the units are selected by the player. Let us create the Prefab for the selected unit representation and duplicate it a few times to fill part of our **Details** panel:

1. Right-click on the **Row1** GameObject, select **UI | Image**, name it SelectedUnit, and change **Width** to 50 and **Height** to 50.

2. In the **Image** component, click on the circle button on the right side of the **Source Image** field and search for the **armor_icon** sprite.

3. Right-click on the **SelectedUnit** GameObject, select **UI | Image**, name it HPBar, and change the Anchor Presets to **bottom** on the left and **stretch** at the top. Change the **Pivot** values to 0.5 for **X** and 0 for **Y**, and for the positions, change **Left** to 2.5, **Pos Y** to 2.5, **Pos Z** to 0, **Right** to 2.5, and **Height** to 7.

4. In the **Image** component, click on the circle button on the right side of the **Source Image** field and search for the **Hp_line** sprite.

5. Right-click on the **HPBar** GameObject, select **UI | Image**, name it HPBarFrame, and change the Anchor Presets to **stretch** on the left and **stretch** at the top. For the positions, change **Left** to 0, **Top** to 0, **Pos Z** to 0, **Right** to 0, and **Bottom** to 0.

6. In the **Image** component, click on the circle button on the right side of the **Source Image** field and search for the **Hp_frame** sprite.

7. Right-click on the **SelectedUnit** GameObject, select **UI | Image**, name it **SelectedUnitFrame**, and change the Anchor Presets to **stretch** on the left and **stretch** at the top. For the positions, change **Left** to 0, **Top** to 0, **Pos Z** to 0, **Right** to 0, and **Bottom** to 0.

8. In the **Image** component, click on the circle button on the right side of the **Source Image** field and search for the **Mini_frame0** sprite.

9. Drag and drop the **SelectedUnit** GameObject from the **Hierarchy** view into the **Prefab | UI** folder on the **Project** view to create a Prefab.

10. Right-click on the **SelectedUnit** GameObject in the **Hierarchy** view and select **Duplicate** to create a copy. Do this four times to have a total of five objects. There's no need to change the GameObject names because they are temporary until we make them dynamically selected.

The preceding steps might seem like a lot of work, but now we are using settings and procedures that we are already familiar with. Once all the UI elements are set up, we will only need to bind the actual data into the UI elements, which is a bit less complicated than making sure the UI is properly configured. The following screenshot is the result of the **Details** panel so far:

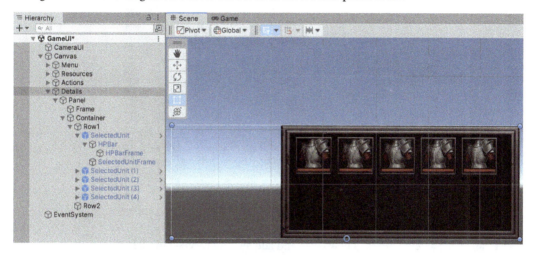

Figure 4.17 – The Details panel with the SelectedUnit Prefabs

In the resources counter, we added a temporary value of 9999 to fill the UI elements, and here we did the same using a temporary sprite for the **SelectedUnit** GameObject and duplicating it twice. As we can see, the objects are nicely sitting side by side. The HP bar is also temporarily full and will have dynamic data attached to reflect the real values of each selected unit.

Now we can proceed to the last part of the **Details** panel, which is the portrait of the selected unit using a **Render Texture** (a type of texture that Unity creates and updates at runtime) to display the 3D model with an idle animation. Using the Render Texture, we can render what is visible to one specific camera in a texture, and it is updated in real time. Let's add the Render Texture for the portrait:

1. Create a new folder, inside the **Assets** folder, and name it `Textures`. Right-click on the new **Textures** folder and select **Create | Render Texture**. You can name it `PortraitRenderTexture`. We will use the default values, so no change is required.

2. Right-click on the existing **Materials** folder, select **Create | Material**, and name it `PortraitMaterial`.

3. Left-click on **PortraitMaterial** to select it and, in the **Inspector** view, you will notice a property called **Albedo** with a small circle on the left side. Click on the small circle and it will display a window to select a texture. Then, search and select **PortraitRenderTexture**.

Once selected, you will see that the box on the left side of **Albedo** is no longer gray. Instead, a blue image is in that box to represent that a texture is set to that material, as shown in the following screenshot. There is no need to change any other material property.

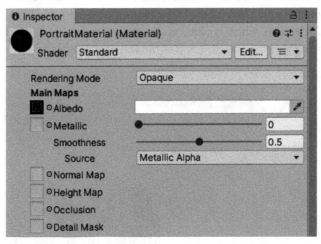

Figure 4.18 – The Render Texture set in the Albedo material property

Now we can create a new camera, a new layer, and add a temporary 3D model so we can test it. The idea here is that the camera will capture everything that is rendered in one specific layer, which we will call **Portrait**, and then send that information to the Render Texture. On the other side, we will have a **Raw Image** component that will render our **PortraitMaterial** material which will read the real-time information from the Render Texture linked to the **Albedo** property.

Let's first create the new layer. We are going to create a few more later, so get yourself familiarized with this process:

1. Click on **Edit | Project Settings…**.
2. In the new window, select **Tags and Layer** on the left panel.
3. After that, on the right side, expand the last option, called **Layers**.
4. In the **User Layer 6** (or any other empty field), type in Portrait.

The following screenshot shows the new layer created. After creating the new layer, you can close this window.

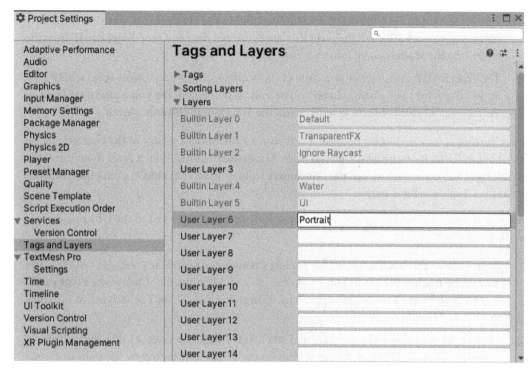

Figure 4.19 – The Tags and Layers settings with the new Portrait layer created

Now that we have the **Portrait** layer set up, we can move on with the new camera:

1. Left-click on the + button located in the top-left corner of the **Hierarchy** view, select **Create Empty**, and name it PortraitRenderer. This time, we are not creating a new GameObject as a child of another GameObject in the scene; it must be created on the same level as the Canvas. Set **Position X**, **Y**, and **Z** to (0, 0, 0).

2. Right-click on the **PortraitRenderer** GameObject, select **Camera**, and name it PortraitCamera.

3. Select **PortraitCamera** and, in the **Camera** component, select only the **Portrait** layer in the **Occlusion Culling** property. You can click on the other selected layers to remove them.

4. Change **Projection** to **Orthographic**, **Size** to 1, **Position X**, **Y**, and **Z** to (0, 1.2, 10), and **Rotation X**, **Y**, and **Z** to (0, -180, 0).

5. In the **Target Texture** property, click on the small circle button on the right side, and search and select **PortraitRenderTexture**.

6. Left-click on the **Audio Listener** component and select the **Remove Component** option. This component is already presented in the main camera on the level scene.

7. Right-click on the **PortraitRenderer** GameObject, select **Create Empty**, and name it Model. Select the new GameObject and, in the **Inspector** view, change **Layer** to **Portrait**.

8. Now we are going to add a Prefab that has a 3D model. In your **Project** view, go to the **ThirdParty | Mini Legion Lich PBR HP Polyart | Prefabs** folder and drag and drop **FreeLichHP** from the folder into the **Model** GameObject we just created.

9. The **FreeLichHP** Prefab must be a child of the **Model** GameObject. Select **FreeLichHP** and, in the **Inspector** view, change **Layer** to **Portrait**. When asked, **Do you want to set layer to Portrait for all child objects as well?**, select the **Yes, change children** object.

Target Texture is very important because it will redirect the camera render to the selected texture instead of rendering it on the whole screen. Also, **Occlusion Culling**, which is set to only cull the **Portrait** layer, will make sure we can use this camera without rendering objects from other cameras, such as the UI camera, for example.

Once that is done, the only thing left is to add the **Raw Image** component to the UI so we can see the 3D model there:

1. Right-click on the **Details** GameObject, select **Create Empty**, name it `PortraitModel`, and change the Anchor Presets to **bottom** on the left and **left** at the top. Change the **Pivot** values to 0 for **X** and 0 for **Y**, and for the positions, change **Pos X** to 0, **Pos Y** to 0, **Pos Z** to 0, **Width** to `150`, and **Height** to `150`.

2. Click on the **Add Component** button and add a **Raw Image** component to the **PortraitModel** GameObject. In the **Texture** property, click on the small circle button on the right side, then search and select **PortraitRenderTexture**.

3. Right-click on the **PortraitModel** GameObject, select **UI | Image**, name it `HPBar`, and change the Anchor Presets to **bottom** on the left and **stretch** at the top. Change the **Pivot** values to `0.5` for **X** and 0 for **Y**, and for the positions, change **Left** to 7, **Pos Y** to 7, **Pos Z** to 0, **Right** to 7, and **Height** to `10`.

4. In the **Image** component, click on the circle button on the right side of the **Source Image** field and search for the **Hp_line** sprite.

5. Right-click on the **HPBar** GameObject, select **UI | Image**, name it `HPBarFrame`, and change the Anchor Presets to **stretch** on the left and **stretch** at the top. For the positions, change **Left** to 0, **Top** to 0, **Pos Z** to 0, **Right** to 0, and **Bottom** to 0.

6. In the **Image** component, click on the circle button on the right side of the **Source Image** field and search for the **Hp_frame** sprite.

7. Right-click on the **PortraitModel** GameObject, select **UI | Image**, name it `PortraitFrame`, and change the Anchor Presets to **stretch** on the left and **stretch** at the top. For the positions, change **Left** to 0, **Top** to 0, **Pos Z** to 0, **Right** to 0, and **Bottom** to 0.

8. In the **Image** component, click on the circle button on the right side of the **Source Image** field and search for the **Mini_frame0** sprite.

These last steps are quite simple and have much in common with the **SelectedUnit** Prefab. The biggest difference here is that we are using the **Raw Image** component instead of the **Image** component because the latter cannot render textures and, in our case, we are using the Render Texture so we can have **PortraitCamera** rendering the 3D model in **PortraitRendererTexture**. With all the work done, we can finally see the details panel complete in the following screenshot.

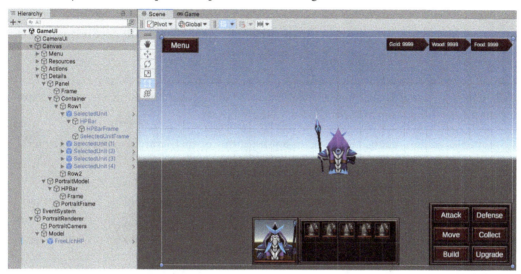

Figure 4.20 – The details panel set up with the selected units and a 3D model using Render Texture

It is worth noting that the 3D model at the center of the screen facing the opposite direction will not be rendered in the game. In fact, if we click on the **Game** view, that model is not onscreen. This can be a little bit confusing, but the **Scene** view will show everything in the scene and, when using multiple cameras, it is common to see things like this. It is perfectly normal if extra GameObjects are not rendered in the **Game** view.

Now we can move to the last UI element, which is the minimap in the bottom-left corner of the screen, which will also use a Render Texture to draw the level scene from the top view.

Rendering the minimap

The minimap is one of the most important resources for the player to master an RTS game. It shows a top view of the level and the discovered areas and it can track the enemy movements and highlight where the player is now. In this chapter, we are going to set up the minimap UI and the new camera to display the current level in the minimap. In later chapters, with more data available and game mechanics implemented, we will add more features to the minimap, so we need a solid foundation now.

So, we just learned how to use the Render Texture to display the camera render in a texture, and then add that texture to the UI. Now we are going to do the same process again but with a new Render Texture, Material, and Camera.

As we did previously, we can start creating the Render Texture and the Material that will be used by the minimap in the UI:

1. Right-click on the **Textures** folder and select **Create | Render Texture**. You can name it `MiniMapRenderTexture`.

2. Right-click on the **Materials** folder, select **Create | Material**, and name it `MiniMapMaterial`.

3. Select **MiniMapMaterial** and, in the **Inspector** view, add **MiniMapRenderTexture** to the **Albedo** property as we did for the portrait previously.

We do not need a new layer this time because we are going to render the whole game scene, except the **UI** (and the **Portrait**) layers. The next step is to create and set up a new camera for the top view, using the Render Texture and Material created for the minimap:

1. Left-click on the + button located in the top-left corner of the **Hierarchy** view, select **Camera**, and name it `MiniMapCamera`.

2. Select **MiniMapCamera** and, in the **Camera** component, locate the **Occlusion Culling** property and make sure to select only the following layers: **Default, TransparentFX, Ignore Raycast,** and **Water**. Do not select **UI** or **Portrait**.

3. Change **Projection** to **Orthographic, Size** to **60, Position X, Y,** and **Z** to (**-3, 120, 1**), and **Rotation X, Y,** and **Z** to (**90, 0, 0**).

4. In the **Target Texture** property, click on the small circle button on the right side, search and select **MiniMapRenderTexture**.

5. Left-click on the **Audio Listener** component and select the **Remove Component** option. This component is already attached to the main camera on the level scene.

6. Drag and drop **MiniMapCamera** from the **Hierarchy** view into the **Project** view inside the **Prefabs | UI** folder. After creating the Prefab, delete **MiniMapCamera** from the **Hierarchy** view by right-clicking on it and selecting **Delete**.

We created the Prefab for the minimap camera because it will be instantiated on the level scene, and not on the UI scene. The minimap camera needs to capture the 3D objects from the level, which is in a different scene and cannot be accessed easily from the UI scene.

To overcome this restriction and make it easier to access the camera from another scene without creating a dependency, the camera data will be written on **MiniMapRenderTexture** by **MiniMapCamera** and read from the same place by the UI element we are going to create next.

Finally, we will create the UI element that will display the minimap. It is quite simple, like the other UI elements that we covered in this chapter. However, now we are going to introduce and use a **Mask**, a component that restricts the child GameObjects to the shape of the parent GameObject, so the minimap can have a round shape in the UI:

1. Right-click on the **Canvas** GameObject, select **Create Empty**, name it MiniMap, and change the Anchor Presets to **bottom** on the left and **left** at the top. Change the **Pivot** values to 0 for **X** and 0 for **Y**, and for the positions, change **Pos X** to 0, **Pos Y** to 0, **Pos Z** to 0, **Width** to 200, and **Height** to 200.

2. Right-click on the **MiniMap** GameObject, select **Create Empty**, name it Mask, and change the Anchor Presets to **bottom** on the left and **left** at the top. Change the **Pivot** values to 0 for **X** and 0 for **Y**, and for the positions, change **Pos X** to 0, **Pos Y** to 0, **Pos Z** to 0, **Width** to 200, and **Height** to 200.

3. Click on **Add Component** to add an **Image** component and, in the **Source Image** property, find and select the **lil_roundbackground** sprite.

4. Click on **Add Component** to add a **Mask** component.

5. Right-click on the **Mask** GameObject, select **Create Empty**, name it Renderer, and change the Anchor Presets to **bottom** on the left and **left** at the top. Change the **Pivot** values to 0 for **X** and 0 for **Y**, and for the positions, change **Pos X** to 0, **Pos Y** to 0, **Pos Z** to 0, **Width** to 200, and **Height** to 200.

6. Click on **Add Component** to add a **Raw Image** component and, in the **Texture** property, find and select **MiniMapRenderTexture**.

7. Right-click on the **MiniMap** GameObject, select **Create Empty**, name it Frame, and change the Anchor Presets to **bottom** on the left and **left** at the top. Change the **Pivot** values to 0 for **X** and 0 for **Y**, and for the positions, change **Pos X** to 0, **Pos Y** to 0, **Pos Z** to 0, **Width** to 200, and **Height** to 200.

8. Click on **Add Component** to add an **Image** component and in the **Source Image** property, find and select the **big_roundframe** sprite.

The **Mask** component will use the sprite from the **Image** component, which is the parent GameObject, and apply the mask on the children GameObjects to restrict their shape to be the same as the parent GameObject. Any transparency on the sprite will be ignored by the mask – that is why the child GameObject, **Renderer**, which has the Render Texture applied to the **Raw Image** component, will only draw what is on the shape of the **Mask** component. Adding a sprite for the frame on top of the masked object is also a great way to hide any imperfections from the mask such as blurred pixels on the edges when using shapes such as circles or more complex forms.

The following screenshot shows all the visible UI elements that we created in this chapter, as well as the current structure of the GameObjects in the **GameUI** scene:

Figure 4.21 – The final structure of our game UI

We are done. We finished the UI structure, and all elements are ready to be wired with our game logic and can be updated dynamically in the next chapters. The only thing left now is loading the UI scene on top of the level scene, which we are going to do next.

Loading the UI scene additively

To see all the work we have done in this chapter, we will need to load the UI scene on top of the level scene, which we call loading additively, which is the opposite of the standard single scene load that replaces the current loaded scene with a new one. We want to use both scenes together, adding the UI to our level.

Before we start, we need to tell Unity what are the scenes we want to consider as part of the game. By default, no scene is added to the game build, and that is also a problem in the Editor because when we want to manually load a scene, Unity will need to know what that scene is.

To add scenes, click **File | Build Settings…**. At the top of the **Build Settings** window, in the **Scenes in Build** section, we will need to add our scenes. Drag and drop both the **Level01** and **GameUI** scenes from the **Scenes** folder in the **Project** view into the area below the **Scenes in Build** section. Make sure to have the **Level01** scene as the first on the list, otherwise, it will not be possible to load the **GameUI** additively. You can drag and drop the scenes on the list to change their order, as shown in this screenshot:

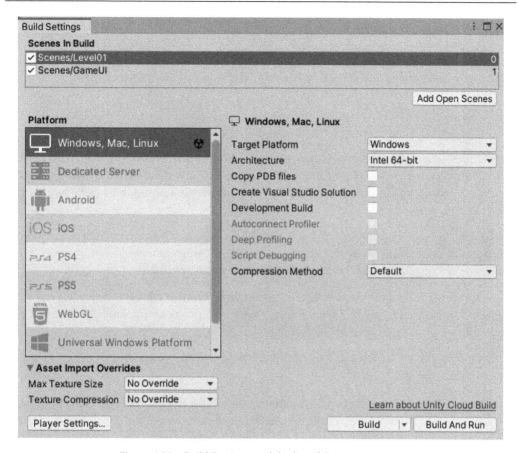

Figure 4.22 – Build Settings and the list of the game scenes

As you can see, we did not add the **Playground** scene to that list because it is a scene made for development and we do not want it in the game build; it is made only for the Editor. You can also add all scenes you created to the list – just make sure the first one is not **GameUI**.

Let's create our first and only script for this chapter, which is very short and simple. Create a new script inside the **Scripts | Level** folder and name it LevelManager. Double-click to open it in your IDE and add the following content:

```
using UnityEngine;
using UnityEngine.SceneManagement;
namespace Dragoncraft
{
    public class LevelManager : MonoBehaviour
    {
        [SerializeField] private GameObject _miniMapCameraPrefab;
```

```
    private void Start()
    {
        Instantiate(_miniMapCameraPrefab);
        SceneManager.LoadScene("GameUI", LoadSceneMode.Additive);
    }
  }
}
```

Seen in the preceding code, the LevelManager class will be used extensively throughout the book, and this is the very first version of it with very simple code. For now, the goal of this class is to hold a reference to _miniMapCameraPrefab and use it to instantiate a GameObject at the beginning of the Start method.

The following line uses the Unity API called SceneManager.LoadScene, which receives two parameters: the name of the scene we want to load and the loading mode, which is optional. The default mode is Single (unload the current scene and load the new one).

Then we are setting the loading mode to Additive so the level scene will not be unloaded, and the UI scene will be loaded on top of it. Do not forget to add UnityEngine.SceneManagement at the beginning of the class because this API is not part of the default Unity API.

Now we can add it to the level scene and create a Prefab so it will be easier to update it on all level scenes we have in the future:

1. Open the **Level01** scene, click on the + button in the top-left corner of the **Hierarchy** view, and select **Create Empty**. Change the new GameObject name to LevelManager.

2. Click on **Add Component** and find and add the LevelManager script.

3. Drag and drop **MiniMapCamera** from the **Prefab | UI** folder in the **Project** view into the **Mini Map Camera Prefab** property on the **LevelManager** component.

4. Create a new folder called Level inside the **Prefab** folder.

5. Drag and drop **LevelManager** from the **Hierarchy** view into the **Project** view inside the **Prefabs | Level** folder.

After the preceding steps, we should finally be able to click on the **Play** button and see our new UI loaded on top of the level and the minimap rendering the whole level from the top view, as we can see in the final screenshot. Remember to add the **LevelManager** Prefab to all levels you created, including the Playground.

Figure 4.23 – Level01 and GameUI running together

Most of the UI elements are static, but there are a few things already working, such as the **Menu** button, the portrait idle animation, and the minimap rendering the current level. Feel free to play with the UI elements and explore all options. You can always review the steps and revert to the project on the GitHub repository if something is not working properly.

Summary

Congratulations on reaching the end of this chapter. We now have our level and our UI with the basic structure and foundation we need to build our gameplay and core mechanics.

In this chapter, we learned how to create UI elements such as buttons and sprites, how to use a few available methods on button click without writing any lines of code, how to set up and position the UI elements using both horizontal and vertical layout groups, and about the anchor presets and pivot values.

We also learned how to use layers, Render Textures, Materials, and multiple cameras at the same time to render different parts of the world into UI elements, such as the portrait for the selected unit and the minimap.

Finally, we have our first version of the level manager class, which is currently loading our UI scene additively and instantiating the minimap on the level scene.

In *Chapter 5*, *Spawning an Army of Units*, we are going to start adding some units to the level, configure their basic status, and create some debug tools to make it easier to test. Starting in the next chapter, we will go more into the game code and mechanics, with a couple of design patterns introduced to spawn multiple objects with memory optimizations and update the UI using messages.

Further reading

We only scratched the surface of the camera and layouts. Here are some links to the official Unity documentation to learn more about the main APIs and components used in this chapter:

- *Cameras*: `https://docs.unity3d.com/2023.1/Documentation/Manual/CamerasOverview.html`

- *Horizontal Layout Group*: `https://docs.unity3d.com/2023.1/Documentation/Manual/script-HorizontalLayoutGroup.html`

- *Vertical Layout Group*: `https://docs.unity3d.com/2023.1/Documentation/Manual/script-VerticalLayoutGroup.html`

- *Render Texture*: `https://docs.unity3d.com/2023.1/Documentation/Manual/class-RenderTexture.html`

- *Tags and Layers*: `https://docs.unity3d.com/2023.1/Documentation/Manual/class-TagManager.html`

- *Mask*: `https://docs.unity3d.com/2023.1/Documentation/Manual/script-Mask.html`

- *Canvas Scaler*: `https://docs.unity3d.com/Packages/com.unity.ugui@1.0/manual/script-CanvasScaler.html`

- *Graphic Raycaster*: `https://docs.unity3d.com/Packages/com.unity.ugui@1.0/manual/script-GraphicRaycaster.html`

- *Render Texture*: `https://docs.unity3d.com/2023.1/Documentation/Manual/class-RenderTexture.html`

Part 2:
The Combat Units

In this second part of the book, you will learn everything that is needed to create combat units in RTS games, starting with how to create configurable unit characters, with different types, and how to spawn them on the map.

Then, you will learn how to select the spawned units on the map and give them commands to perform actions, such as moving to a position or attacking an enemy, as well as how to implement close combat and ranged attacks.

Finally, you will see how we can use the Pathfinder algorithm in the Unity Engine to help our units always find the best path to move around the map, avoiding other units and GameObjects.

This part includes the following chapters:

- *Chapter 5, Spawning an Army of Units*

- *Chapter 6, Commanding an Army of Units*

- *Chapter 7, Attacking and Defending Units*

- *Chapter 8, Implementing the Pathfinder*

Spawning an Army of Units

RTS games have a gameplay mechanic that allows you to spawn units and create large armies to battle enemies. But having so many objects in the same scene that can be easily created and destroyed requires good memory and CPU management, to avoid problems such as frame rate drops due to CPU spikes, out-of-memory issues, and memory leak issues due to inefficient memory usage. There is one programming pattern that works well in this scenario, which is called **Object Pooling**, and we are going to learn how to implement and use it to spawn our units.

In this chapter, we will create the configuration files for our units using the `ScriptableObject` class from Unity, which will have all the attributes we need to use a unit in our game. We are also going to see how to update the UI using a decoupled solution by learning how to implement and use the **Message Queue** pattern, which provides a great way for scripts or systems to communicate with each other without creating a hard dependency between them.

This chapter also introduces a debugging tool that will be used throughout the book to make it easier to test all the features developed. We are going to use this tool to update the resources in the UI and also to spawn units in the scene.

In this chapter, we will cover the following topics:

- Configuring the unit ScriptableObject
- Spawning units using the Object Pooling pattern
- Updating the UI using the Message Queue pattern
- Creating a debugging tool for the Editor

Technical requirements

The project setup in this chapter with the imported assets can be found on GitHub at `https://github.com/PacktPublishing/Creating-an-RTS-game-in-Unity-2023` in the `Chapter05` folder inside the project.

Configuring the unit ScriptableObject

In *Chapter 3, Getting Started with Our Level Design*, we were introduced to the `ScriptableObject` class, which we used to configure the level and map elements of our game, and saw how we can use it to create data files that are easily manipulated in the Unity Editor and with C# scripts. We are going to continue using this powerful yet lightweight class to configure the entire game, starting with the units that the player will generate and train before commanding them to gather resources, defend the settlement, attack enemies, and explore the map.

Even though we will have different units, such as warriors and mages with different stats, their properties (`Level`, `LevelMultiplier`, `Health`, `Attack`, `Defense`, `WalkSpeed`, and `AttackSpeed`) will be the same. Creating a solid foundation for the base stats will allow us to rapidly iterate the values for better game balancing and further customizations.

Let us start with a few scripts to define the `UnitType`, `UnitData`, and `UnitComponent` classes. These classes will expand as we add more content and features to our game, but for now, they will have the base for the unit set up. First, create a new folder inside **Scripts** and name it `Unit`. Inside that folder, add a new script called `UnitType`, which will have the following content:

```
using System;
namespace Dragoncraft
{
    [Serializable]
    public enum UnitType
    {
        Warrior,
        Mage
    }
}
```

This is a simple script that has only an `enum` class with the `Warrior` and `Mage` values, which will be used to determine the unit type. We do need to add the `Serializable` attribute to the `enum` class; the ScriptableObject will be able to serialize and render it in the Unity Editor.

Now, add a new script in the same **Unit** folder and name it `UnitData` – this is the ScriptableObject that will hold the data definition for our unit. We can have multiple variations of units, but they will always have these base properties, as you can see in the following script:

```
using UnityEngine;
namespace Dragoncraft
{
    [CreateAssetMenu(menuName = "Dragoncraft/New Unit")]
    public class UnitData : ScriptableObject
    {
        public UnitType Type;
```

```
    public int Level;
    public float LevelMultiplier;
    public float Health;
    public float Attack;
    public float Defense;
    public float WalkSpeed;
    public float AttackSpeed;
    }
}
```

Once again, we are using the handy `CreateAssetMenu` attribute, which will allow us to easily create a `ScriptableObject` file based on the previous class.

Let us see how to create and set up the warrior unit:

1. Right-click on the existing **Data** folder, select **Create | Folder**, and name it `Unit`.

2. Right-click on the new **Unit** folder, select **Create | Dragoncraft | New Unit**, and name it `BasicWarrior`.

3. Select the new asset (`BasicWarrior`) and add the values as shown in the following figure:

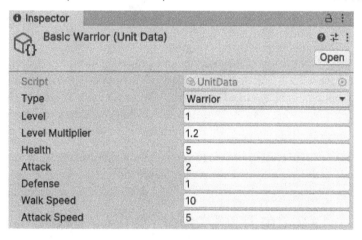

Figure 5.1 – The Basic Warrior configuration in the UnitData asset

The properties shown in the preceding figure will be used in later chapters to determine the amount of damage the unit can deal out and take, the attack and movement speed, the standard level and health points, and the multiplier that will increase the values when the unit is upgraded.

In this chapter, we are setting up only one warrior so, for now, we can focus on creating all the classes needed before starting to add unit variants. The next script we are going to create will be attached to the Prefab responsible for keeping the values up to date during the game. This means that the script will copy the base values from the ScriptableObject and then, during the game session, the values in the component will be updated depending on whether the unit takes damage or when it is upgraded.

Let us create the new script to manage the unit behavior (called `UnitComponent`) in the same **Scripts | Unit** folder as the previous scripts and add the following content:

```
using UnityEngine;
namespace Dragoncraft
{
    public class UnitComponent : MonoBehaviour
    {
        public string ID;
        public UnitType Type;
        public int Level;
        public float LevelMultiplier;
        public float Health;
        public float Attack;
        public float Defense;
        public float WalkSpeed;
        public float AttackSpeed;
    }
}
```

As we can see, the `UnitComponent` class has the same properties as the `UnitData` class, except for one extra property called `ID`. The `UnitData` class is a ScriptableObject, which is an asset that can store data values but cannot be modified and persisted while the game is running. For this reason, we have `UnitComponent`, which is a `MonoBehaviour` class that can be attached and modified during the game session, so the properties will reflect the actual values based on the game state.

The `ID` property will be used to create a unique identifier for the unit instance, so when we have more than one unit instance in the game, it will be easier to identify the one that was affected by the player's actions or the enemies' attacks. Now that we have the unit data, we can move on to the spawner.

Spawning units using the Object Pooling pattern

The Object Pooling pattern was created to optimize memory usage and CPU performance by reusing objects that are pre-allocated in a pool, instead of instantiating a new object when needed and then destroying it. When the `ObjectPool` script is created, a fixed number of instances will be allocated and added to a list. Then, when an object is required, the Object Pool will give you an existing copy. When you do not need the object anymore, for example when a unit is destroyed by the enemy, the object is returned to the Object Pool.

We can use the Object Pooling pattern when we frequently need to create and destroy objects that are all the same or very similar. A good use case is bullets that are fired from a weapon; they are all the same and created and destroyed a lot. In this case, the Object Pooling pattern will reduce the overhead on the CPU to create and destroy an object as well as making it faster to reuse an object instead of instantiating a new object. Memory is also optimized because all objects are allocated and reused, reducing memory leaks every time a new object is created or destroyed.

One of the cool things about the Object Pooling pattern is that its implementation is generic enough, meaning it will work with most objects in the game. In our case, the script accepts any Prefab as an object that can be reused and will be stored in a list as a GameObject. The script also allows us to set the pool size, which is the number of objects to be created and reused, and provides an option to create more objects when a new one is requested but none is available.

Implementing the Object Pooling pattern

Create a new **ObjectPool** folder inside the **Scripts** folder, and then add a new script called ObjectPoolComponent. Inside the script, add the following code:

```
using System.Collections.Generic;
using UnityEngine;
namespace Dragoncraft
{
  public class ObjectPoolComponent : MonoBehaviour
  {
    [SerializeField] private GameObject _prefab;
    [SerializeField] private int _poolSize;
    [SerializeField] private bool _allowCreation;
    [SerializeField] private List<GameObject> _gameObjects
      = new List<GameObject>();

    private void Awake()
    {
      for (int i = 0; i < _poolSize; i++)
      {
        _gameObjects.Add(CreateItem(false));
      }
    }
  }
}
```

The overall script is quite simple, but let us start with the properties and initialization. The Prefab object can be any Prefab or object from the scene since the script does not care about what is in the gameObjects list if it is a GameObject. The poolSize property will limit the number of copies available to use, which may vary depending on the object we want to use. For example, something like bullets could have a higher limit of 100, while units such as our warrior might be fine with a limit of 10. This kind of value must be tweaked according to the level of difficulty; for example, an easier map may require 10 warriors, while a harder map may require the spawning of 30 or more warriors to beat the enemies and bosses.

If we are not sure of the limit, or the limit is not something we would use to restrict the gameplay, the allowCreation property could be toggled to enable the script to keep creating new instances on demand and make them available in the list. This means that if the limit is set to 10 but 2 more instances are requested and created, the pool will then have 12 objects available. This option must be used wisely; otherwise, it would defeat the purpose of the Object Pool, since you are always requesting more than the limit. If that is the case, the limit should be a higher value – this is all about the game balance after all; unfortunately, there is no magic number that will fit all cases.

In the Awake method, which is the first method executed by MonoBehaviour when the GameObject with the attached script is loaded, we are initializing the gameObjects list using the poolSize property as the number of elements we need to create. We are going to use the following CreateItem method to add each element to the list:

```
private GameObject CreateItem(bool active)
{
    GameObject item = Instantiate(_prefab);
    item.transform.SetParent(transform);
    item.SetActive(active);
    return item;
}
```

The active parameter here is used to enable or disable the GameObject after the Instantiate method is called to create the object. In the Awake method, the parameter is false, so all objects will be created but disabled right after that, and they remain in the pool but are not visible. In the next piece of code, we will see the opposite: when the active parameter is true.

The SetParent method is also called in the newly created GameObject to make sure it stays as a child of the Object Pool, in the **Hierarchy** view, by setting the current transform as the parent. After the SetActive method is called, the new GameObject is returned.

We have finished the CreateItem method, which adds a new GameObject to the Object Pool, and now we are going to create a new GetObject method in the following code block to get one GameObject from the Object Pool:

```
public GameObject GetObject()
{
    foreach (GameObject item in _gameObjects)
    {
        if (!item.activeInHierarchy)
        {
            item.SetActive(true);
            return item;
        }
    }

    if (_allowCreation)
```

```
    {
        GameObject item = CreateItem(true);
        _gameObjects.Add(item);
        return item;
    }
    return null;
}
```

The preceding method is the main part of the Object Pooling pattern and contains the main logic behind it. In the first line, we are going to loop through the gameObjects list until we find the first item that is not active by checking the activeInHierarchy property. Since this is the only condition, the first object in the list that is not yet active becomes active and is then returned to be used by the script that requested the GameObject.

After the loop, if no object is available, the code will check the allowCreation property to create a new object on demand. If that is the case, the new object is created using the CreateItem method but, this time, the active parameter is set to true, which means the object will be created and enabled. Then, the object is added to the gameObjects list and returned to the script that requested it.

Keep in mind that the allowCreation property is a crucial part of the Object Pool configuration. The script using the Object Pool cannot decide whether a new object can be created or not. Hence, it is essential to set a good value for poolSize and only utilize allowCreation to prevent the Object Pool from blocking the gameplay by not allowing more instances of the object.

The last part is returning null, irrespective of whether there is no available object or no new object was created. The script calling this method must check whether the GameObject returned is null; otherwise, an error will be thrown, and the game might crash or stop – which is never a good outcome for the player.

Now that we have the Object Pooling pattern implemented, we can start to use it to create a spawner. While the ObjectPoolComponent script is responsible for handling the list of objects and the instantiation, the next script, called BaseSpawner, will use the Object Pool to get an object.

Creating a BaseSpawner class

Create a new folder inside the **Scripts** folder and name it Spawner. Then, create a new script inside the **Spawner** folder, name it BaseSpawner, and copy the following code into the new script:

```
using UnityEngine;
namespace Dragoncraft
{
    [RequireComponent(typeof(ObjectPoolComponent))]
    public class BaseSpawner : MonoBehaviour
    {
        private ObjectPoolComponent _objectPoolComponent;
```

```
    private void Awake()
    {
      _objectPoolComponent =
        GetComponent<ObjectPoolComponent>();
    }

    public GameObject SpawnObject()
    {
      return _objectPoolComponent.GetObject();
    }
  }
}
```

The preceding script is a base for other scripts, which can be expanded by adding more functionalities while retaining its basic functionality of getting an object from the pool; this functionality will always be available to those implementing this class.

The RequireComponent attribute is used here to automatically attach the ObjectPoolComponent script when we add the BaseSpawner script to a GameObject since it is a required component.

One of the advantages of adding the RequireComponent attribute is that the objectPoolComponent variable will never be null, so no validation or error handling is needed before using the variable. Therefore, we are using the GetComponent method without checking whether the variable is valid – it will always be valid due to the RequireComponent attribute.

This base class only has one method, SpawnObject, because the goal is that this class only has the code required to use the pool and is generic enough so that it can be easily expanded. Note that the Object Pool is hidden in this class, so the classes expanding and using it do need to know that the Object Pool exists – hiding the Object Pool is a good way to avoid exposing too many features to all classes; it makes development simple and reduces the misuse of the ObjectPool class.

Before moving forward with a class that will expand the BaseSpawner script, we first need to implement another very useful pattern: the Message Queue.

Updating the UI using the Message Queue pattern

The **Message Queue** pattern, also known as the **Event Queue** pattern, consists of a list of messages that are sent to all other scripts that are listening for a specific message. This is a powerful pattern that helps to decouple different scripts since you can just send a message with an update and the listener will receive and process it, without having shared variables across the whole project.

It also has good performance since the listeners are not checking for new messages on every loop. When a message is received, the message queue will trigger the listeners that are assigned to the type of message received.

We are going to implement the Message Queue pattern and use it to update the game UI and spawn new units. The idea is that when a new unit is required to be added to the game, a message will be sent by the script requesting it and received by the script that will then get a unit from the Object Pool. Both patterns work together, but without knowing about the existence of the other, you would have a completely decoupled solution.

Implementing the Message Queue pattern

Create a new folder inside the **Scripts** folder called MessageQueue and then add a new script named MessageQueueManager. Replace the content with the following code block, which defines the listeners and instance variables and the constructor for the class, with a new pattern called **Singleton** implementing a static instance:

```
using System;
using System.Collections.Generic;
namespace Dragoncraft
{
    public class MessageQueueManager
    {
        private readonly Dictionary<Type,
            List<Delegate>> _listeners;

        private static MessageQueueManager _instance;
        public static MessageQueueManager Instance
        {
            get
            {
                return _instance ?? (_instance =
                    new MessageQueueManager());
            }
        }

        private MessageQueueManager()
        {
            _listeners = new Dictionary<Type, List<Delegate>>();
        }
    }
}
```

This class is the Message Queue pattern implementation but the class itself also contains another pattern: a rather famous one called Singleton. This pattern gives global access to a class instance and ensures that only one single instance of that class will be created, so the state is persisted. The Singleton pattern is a bit controversial because, when used excessively, it may lead to a bad architecture that would have a lot of interdependencies, resulting in the opposite of a decoupled system.

That said, Singleton is still a very popular and useful design pattern when used wisely. Since this design pattern can ensure a single instance and global access, this is exactly what we need for the MessageQueueManager class. To implement it, we need a `private static` variable, which we named `instance`, and a `public static` property named `Instance`, which will return the `instance` variable if it exists or will create a new one if it is `null`. The `??` operator is a shortcut that executes the code on its left-hand side if the evaluated condition is `true` – in our case, if `instance` is not `null` – otherwise, it executes the right side, which initializes the `instance` variable.

Note that we only implement the `get` method for the property because the `set` method is optional and not required for the instance that will be created inside this class, and the only thing exposed is the `get` method to return its value. The last part of the Singleton pattern is the `private` constructor, which allows the instance creation only internally in this class, so no other class can create an instance of it. We have all the right parts to make sure the Singleton pattern works properly: a `static` variable to persist the value, a method to get the instance or create a new one if it is `null`, and the `private` constructor.

In this first piece of code, we have only one thing that is part of the Message Queue pattern, which is the `Dictionary` collection of `listeners` that will hold a collection of `Type` and `Delegate`. `Type` can be any variable type, but in our case, it will be a message type, and a list of `Delegate` – that is, a list of functions that will be invoked when that message is sent. `Dictionary` ensures that there is a unique key for our message type, and the corresponding value is a list containing all `listeners` of that type assigned by all scripts in the Message Queue pattern. These listeners in `Dictionary` listen to the specified message type and execute a callback when it is triggered.

To avoid changes in the `Dictionary` collection that has all the listeners, we added the `readonly` keyword to the variable declaration in the class. Due to this keyword, the `AddListener` and `RemoveListener` methods will only be able to add or remove items from the dictionary. You can add both methods right after the class constructor:

```
public void AddListener<T>(Action<T> listener)
{
  List<Delegate> listeners = null;
  if (_listeners.TryGetValue(typeof(T), out listeners))
  {
    listeners.Add(listener);
  }
  else
  {
    listeners = new List<Delegate> { listener };
    _listeners.Add(typeof(T), listeners);
  }
}
```

```
}

public void RemoveListener<T>(Action<T> listener)
{
  List<Delegate> listeners = null;
  if (_listeners.TryGetValue(typeof(T), out listeners))
  {
    listeners.Remove(listener);
  }
}
```

We have here the two operations that can be performed on the dictionary of listeners, which are AddListener and RemoveListener. They have the same method signature with a generic type parameter, <T>, a placeholder for a specific type that must be set when calling this method. The Action<T> parameter is a function that must have the same type in its callback parameter.

Since the listeners variable is Dictionary, we can use the built-in TryGetValue method, which will return true if the key we are looking for exists, which can be found using the typeof keyword. Then, if it is true, the value is set to the listeners variable by the out parameter. This is a very safe and fast way to get something from the Dictionary data type.

In the AddListener method, if the key we are looking for already exists in the dictionary, the new listener is added to the listeners list. Otherwise, a new list is created containing only the new listener and it is added to the dictionary with the T type as the key. The RemoveListener method is quite similar, with the main difference being that if a listener is found for the key that we are looking for, we remove the listener from the dictionary.

Now that we can add and remove listeners, the only missing method for the Message Queue pattern is the one that sends a new message, defined next. Add it as the last method in the class:

```
public void SendMessage(IMessage message)
{
  List<Delegate> listeners = null;
  if (_listeners.TryGetValue(message.GetType(),
    out listeners))
  {
    for (int i = 0; i < listeners.Count; i++)
    {
      listeners[i].DynamicInvoke(message);
    }
  }
}
```

The SendMessage method has an IMessage parameter, which is an interface that we are going to define next, and all messages sent must be inherited from that interface. As we did with the previous methods, we need to find the listeners by the message type in the dictionary. When we find them, we loop through the list of listeners using the for loop and execute the action that is in the listener by calling the DynamicInvoke method and providing the message parameter.

This is the core of the Message Queue pattern because it will forward the message to all other scripts that added listeners for the specific message type. DynamicInvoke will execute the Action<T> delegate that was added using the AddListener method, where the message type is the parameter sent by DynamicInvoke and received by Action<T> as a callback.

Creating the message interface

The previous piece of code completes the Message Queue pattern implementation, and the only thing missing now is the following interface. Add a new script called IMessage in the **Scripts | MessageQueue | Message** folder. Then, add the following code:

```
namespace Dragoncraft
{
    public interface IMessage
    {
    }
}
```

The interface keyword defines a contract where the class that implements it must have all the methods as well. In this case, IMessage does not have any method – we are only using this interface as a base type to group all classes that can be considered a message type.

Now, we can create the first real message type. Create a new folder named Messages inside the existing folder **Scripts | MessageQueue**. Then, create a new folder named **Unit** inside the folder **Messages** we just created. Finally, add a new script in the **Scripts | Message Queue | Messages | Unit** folder and name it BasicWarriorSpawnMessage. This class will not have any method yet, but it will implement the IMessage interface, as shown here:

```
namespace Dragoncraft
{
    public class BasicWarriorSpawnMessage : IMessage
    {
    }
}
```

The BasicWarriorSpawnMessage class will be used to spawn new warrior units, using the Object Pooling pattern that we implemented earlier in this chapter. Now, we can create a new class that implements BaseSpawner with custom code for spawning the basic warrior unit.

Implementing the warrior unit spawner

Add a new script called BasicWarriorSpawner inside the **Scripts | Spawner** folder. The following code is the first part of the BasicWarriorSpawner class that has a reference to the UnitData ScriptableObject. The OnEnable and OnDisable methods are from the MonoBehaviour class and are used to add and remove the listeners for the Message Queue pattern, respectively:

```
using System;
using UnityEngine;
namespace Dragoncraft
{
  public class BasicWarriorSpawner : BaseSpawner
  {
    [SerializeField] public UnitData _unitData;
    private void OnEnable()
    {
      MessageQueueManager.Instance
        .AddListener<BasicWarriorSpawnMessage>(
        OnBasicWarriorSpawned);
    }
    private void OnDisable()
    {
      MessageQueueManager.Instance
        .RemoveListener<BasicWarriorSpawnMessage>(
        OnBasicWarriorSpawned);
    }
  }
}
```

This class implements BaseSpawner, which means that the methods from that class are also available here. We need to keep in mind that BaseSpawner itself implements MonoBehaviour, so all Unity life cycle methods are at our disposal, and we can attach the script to a GameObject. Therefore, we are adding both the OnEnable and OnDisable methods here; they are from the MonoBehaviour class and can be used here since BasicWarriorSpawner is both BaseSpawner and MonoBehaviour.

Here, we can see the first usage of the Message Queue by adding and removing a listener when the script is either enabled or disabled. Note that both called methods, AddListener and RemoveListener, are accessed through the Singleton pattern using the public static Instance variable from the MenssageQueueManager class. We do not need to create an instance of this class to use it; we simply access the methods we need using the Instance object.

Both methods, AddListener and RemoveListener, have a generic parameter type in the signature, which is denoted as the <T> variable, here set as BasicWarriorSpawnerMessage. The method parameter type is Action<T>, which means that although it is a generic parameter type as well, it must have the same type as the previous <T>.

The function used as a parameter here is defined next (note that its parameter is of the BasicWarriorSpawnerMessage type). We can add this method after the last one in the same class:

```
private void OnBasicWarriorSpawned(BasicWarriorSpawnMessage message)
{
    GameObject warrior = SpawnObject();
    UnitComponent unit =
      warrior.GetComponent<UnitComponent>();
    if (unit == null)
    {
        unit = warrior.AddComponent<UnitComponent>();
    }
    unit.ID = System.Guid.NewGuid().ToString();
    unit.Type = _unitData.Type;
    unit.Level = _unitData.Level;
    unit.LevelMultiplier = _unitData.LevelMultiplier;
    unit.Health = _unitData.Health;
    unit.Attack = _unitData.Attack;
    unit.Defense = _unitData.Defense;
    unit.WalkSpeed = _unitData.WalkSpeed;
    unit.AttackSpeed = _unitData.AttackSpeed;
}
```

The SpawnObject method used here is defined in the BaseSpawner base class, which is why we do not need to define it and we can use it anywhere in the class. The method will return an available copy of the Warrior Prefab here and the first thing we do is to get the UnitComponent GameObject from it, which has all the values for the unit stats. If the UnitComponent GameObject is null, this means the Warrior copy was not used yet, so we add the UnitComponent script.

In the last part of this method, we reset the UnitComponent values to be the same as the base values defined in the ScriptableObject assigned to the unitData variable. This is required so that when we request a new warrior, the Warrior instance will not use the previous values, so it is a reused object but with brand new stats.

There is only one value that we do not set from unitData, which is ID. We want the ID value to be unique and reliable so we can identify the warrior later and, for example, apply damage to it or get its attack value to cause damage to an enemy. The ID value is generated every time a new object is spawned, and the value is a random **Global Unique Identifier (GUID)** that can be generated by the Guid.NewGuid API. In case you are wondering what a GUID looks like, it looks something like this: b0eb2812-ec8c-4fa0-91a9-adbb81476cc8.

That was the last bit of the unit spawning system; we can move forward to the UI update that will also use messages to update the amount of resources that the player holds, and any increase or decrease that can happen during the gameplay.

Creating the resource type

Now we are going to create a new message that will update the UI with the current values of the Gold, Wood, and Food resources, irrespective of whether the player receives or spends the game resources. To do that, we first need to create a new enumerator for the resource type available in the game.

Go ahead and create a new folder called Resource inside the **Scripts** folder, and add a new script called ResourceType. Then, add this code:

```
namespace Dragoncraft
{
    public enum ResourceType
    {
        Gold,
        Wood,
        Food
    }
}
```

The ResourceType enum has the three possible resources available in the game: Gold, Wood, and Food. This new enum will be used in many different scripts, and the first one that we are going to create is a new message type to update the game when a resource is added or removed from the game.

To create the new message type, add a new script in the **Scripts | MessageQueue | Messages | UI** folder and name it UpdateResourceMessage. Then, replace the content with the following code block:

```
namespace Dragoncraft
{
    public class UpdateResourceMessage : IMessage
    {
        public int Amount;
        public ResourceType Type;
    }
}
```

The UpdateResourceMessage class also implements the IMessage interface, but now it has two properties that need to be set for this message type: ResourceType and the amount of resources to be updated (Amount). The amount can be either positive or negative to make it simple to add or remove the resources gained or used by the player.

Updating the UI

Now that we have a message type for the resources update, we can create a new script to change the values in the UI when the message is received. This script is very similar to the `BasicWarriorSpawner` class but, instead of spawning things, we are just adding and removing the listener to update the UI. Therefore, the script implements `MonoBehaviour`, so we can use the `OnEnable` and `OnDisable` life cycle methods.

So, create a new folder named **UI** inside the existing folder **Scripts**, then add a new script to the **Scripts | UI** folder and name it `ResourceUpdater`. Then, enter the following code:

```
using TMPro;
using UnityEngine;
namespace Dragoncraft
{
  public class ResourceUpdater : MonoBehaviour
  {
    [SerializeField] private ResourceType _type;
    [SerializeField] private TMP_Text _value;
    private int _currentValue;

    private void OnEnable()
    {
      MessageQueueManager.Instance
        .AddListener<UpdateResourceMessage>(
        OnResourceUpdated);
    }

    private void OnDisable()
    {
      MessageQueueManager.Instance
        .RemoveListener<UpdateResourceMessage>(
        OnResourceUpdated);
    }
  }
}
```

In this script, we have a couple of properties exposed so we can set the values using the **Inspector** view in the Editor. The first is `ResourceType`, where we will be able to define the resource we are going to update in this script. The second property is a reference to the `TMP_Text` field that displays the text value in the UI. To use the TextMeshPro types in our code, such as `TMP_Text`, we need to include `TMPro` at the beginning of the script.

Note that we are using the Singleton instance of `MessageQueueManage` in `OnEnable` and `OnDisable`, as we did for `BasicWarriorSpawner`. However, this time, the message type is `UpdateResourceMessage`, and `Action<T>` is set to a new `OnResourceUpdated` function that has the same message type, `UpdateResourceMessage`. We are going to see this kind of implementation a lot moving forward, every time we need to add and remove listeners for a specific message type in our code.

We also have a class variable called `currentValue` that we are going to use so we know what the current value for the resource is. Since the value is converted into a string before setting it to the `value` variable, as we are going to see in the `UpdateValue` method, we do need to have a numeric representation; it is easier to add or remove the amount desired without converting the variable from a string to an integer:

```
private void OnResourceUpdated(UpdateResourceMessage message)
{
    if (_type == message.Type)
    {
        _currentValue += message.Amount;
        UpdateValue();
    }
}
```

The `OnResourceUpdated` method is quite simple but has important validation. The conditional checks whether the message received is for the resource type that this script is attached to and only updates the value if that conditional is `true`. The `+=` operation will add the value received in the `Amount` property, and even if that is negative, it is fine since we are adding a negative number, which results in a subtraction operation.

We are updating the text in the UI using the `UpdateValue` method so we can use it in a couple of places without duplicating code. `UpdateValue` is a small method, but it has a very nice way of creating strings that have both text and variables. To concatenate strings and variables, we can add the dollar symbol ($) at the beginning of the string and then include the variables we want to concatenate inside the double quote marks using curly brackets ({ }):

```
private void UpdateValue()
{
    _value.text = $"{_type}: {_currentValue}";
}
```

The preceding code will create a string such as `Gold: 10` by including the resource type and its current value. Note that the resulting string is set to the `text` property of the TMP_Text variable (not directly to the `value` property) because it is an object with many other properties and methods, and we want to update only the text content.

To ensure that the TMP_Text variable is set in the **Inspector** view, we will implement validation in the Awake method:

```
private void Awake()
{
  if(_value == null)
  {
    Debug.LogError("Missing TMP_Text variable in the script");
    return;
  }
  UpdateValue();
}
```

If the variable is null, we call the LogError method to display a red error message in the **Console** view. We will also use a return statement to prevent the execution of the UpdateValue method in case the variable is null, which will prevent a null reference error message.

At last, we have our Object Pooling and Message Queue patterns implemented and used on the unit spawn and UI update scripts. Now, we can add the scripts to the Prefabs and set them up before creating a simple debugging tool in the Editor to test everything we have done so far in this chapter.

Creating a debugging tool for the Editor

One of the easiest and fastest ways to test things while we are developing is by creating debugging tools to help us execute specific methods and trigger events so we can validate that our systems are working without having to play the game. This can be used for simple things such as adding more gold for the player or even spawning units and upgrading buildings.

Creating the Object Pool for the Warrior unit

Before we move to the actual debugging tool, let us first add the scripts we created so far to the Prefabs and GameObjects that will need them:

1. Open the Level01 scene.

2. Left-click on the + button in the top-left corner of the **Hierarchy** view, select **Create Empty**, and name it ObjectPools. In the **Inspector** view, in the **Transform** component, set **X**, **Y**, and **Z** to (0, 0, 0).

3. Right-click on the new **ObjectPools** GameObject, select **Create Empty**, and name it BasicWarriorObjectPool. In the **Inspector** view, in the **Transform** component, set **X**, **Y**, and **Z** to (0, 0, 0).

The new GameObjects will stay in the level scene so we can group all the Object Pools used together. For now, we only have the Object Pool for the `Warrior` unit. You can leave **ObjectPools** as the last object in the **Hierarchy** view, as shown here:

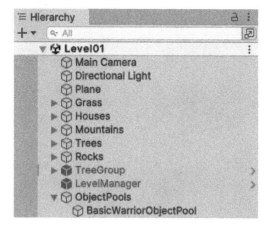

Figure 5.2 – The ObjectPools GameObject and Basic Warrior

Next, let us add the scripts we just created for the Object Pooling pattern and the spawner, which will be used to spawn `Warrior` units:

1. Left-click on the **BasicWarriorObjectPool** GameObject in the **Hierarchy** view to select it.

2. In the **Inspector** view, click on **Add Component** and search for the `BasicWarriorSpawner` script. You will see that once this script is added, **ObjectPoolComponent** is also added to the GameObject.

3. In **ObjectPoolComponent**, click on the small circle button on the right side of the **Prefab** property and type `FootmanHP` to search for the Prefab that is already included in the project from the packages we imported in *Chapter 2, Setting Up Unity and the Project*. Set **Pool Size** to `10` and click on the checkbox on the right side of the **Allow Creating** property to enable it.

4. In the **BasicWarriorSpawner** component, click on the small circle button on the right side of the **Unit Data** property and search for the **UnitData** ScriptableObject called **BasicWarrior** that was created earlier.

5. Save the scene.

This will be our setup for now with an Object Pool size of 10 and the option to create more objects if we request more and none are available. Each warrior unit and enemy will have a similar setup but with different **Prefab** and **Pool Size** settings. As we can see in the following figure, the **GameObjects** list is empty (it will be populated once the script is initialized in the game):

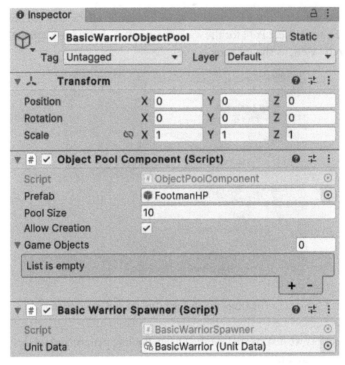

Figure 5.3 – Setting up the Object Pool for the Basic Warrior

Before moving on to the creation of the debugging tool, let us first add the `ResourceUpdater` component as well so we will be able to test all the scripts at once. The resources are in the `GameUI` scene; however, since the resources are using Prefabs, we can perform the changes in the Prefab so they will be applied to all copies of it on any scene. First, we need to open the Prefab in the edit mode:

1. In the **Project** view, double-click on the **ResourceButton** Prefab located in the **Prefabs | UI** folder to open it in the edit mode.

2. In the **Hierarchy** view, left-click on the **ResourceButton** GameObject to select it.

The following figure shows the Prefab opened in the edit mode:

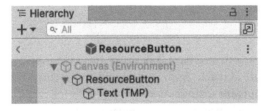

Figure 5.4 – Editing the ResourceButton Prefab

Note that there is no scene name in the **Hierarchy** view; instead, we can see the name of the Prefab and all the objects in the hierarchy. Any changes here will be applied to all copies of the Prefab on all scenes, but that is not the case if you change the Prefab or its values in the scene that is using it.

Now, we can add the new script to the Prefab and set it up:

1. In the **Inspector** view, click on **Add Component** and search for the `ResourceUpdater` script.

2. Drag and drop the **Text (TMP)** GameObject from the **Hierarchy** view into the **Value** property of the **ResourceUpdater** component.

Since the **Text (TMP)** GameObject is part of the same Prefab, this change will be applied to all copies of the Prefab. It would also work if we dragged and dropped assets from the **Project** view into a Prefab script component. However, if we do the same in the scene instead of doing it in the Prefab edit mode, the change is not applied to all Prefab copies. The following figure has the **ResourceUpdater** script added to the **ResourceButton** Prefab:

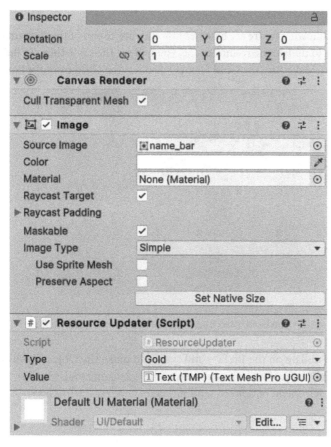

Figure 5.5 – Adding the ResourceUpdater script to the Prefab

Note that we did not change the value of the **Type** property, which has the default value of the first `ResourceType` enum, which is **Gold**. The actual resource type will be different in each copy of the Prefab, so we do not need to set it up in the Prefab edit mode. Instead, we are going to edit each resource in the `GameUI` scene to have the custom resource type set on their Prefab copies that the `ResourceUpdater` script added to all the copies:

1. Open the **GameUI** scene.

2. In the **Hierarchy** view, expand **Canvas** and then **Resources** to find the **Gold**, **Wood**, and **Food** GameObjects, which are Prefabs linked to **ResourceButton**.

3. Left-click on each one of them to select it and, in the **Inspector** view, we can select the **Type** value in the **ResourceUpdater** component to match the GameObject name – for example, **Gold**, **Wood**, or **Food**.

4. Repeat the last step so all three resources have the proper resource type set.

5. Save the scene.

The resources in the following figure are blue because they are copies of the **ResourceButton** Prefab. Once you select them to update the resource type, you will see the same thing in the **Inspector** view as shown in *Figure 5.6*. Make sure the **Gold** GameObject has the **Gold** resource type, and so on.

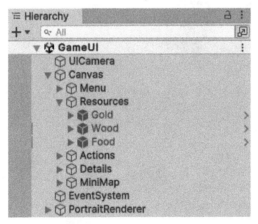

Figure 5.6 – The Gold, Wood, and Food resources in the GameUI scene

That is all we need to update on the Prefabs and scenes to make the Object Pooling, Spawner, and Message Queue systems work. Now, let us move on to the debugging tool, which is basically an editor script that will have a few menu options to send messages and test the listeners.

Creating the debug script

Create a new folder called Debug inside the **Scripts | Editor** folder, and add a new script in the newly created folder and name it UnitDebugger. This script must be inside the **Editor** folder; otherwise, we cannot use the UnityEditor API. Replace the script content with the following code:

```
using UnityEditor;
namespace Dragoncraft
{
    public static class UnitDebugger
    {
        [MenuItem("Dragoncraft/Debug/Unit/Spawn Warrior")]
        private static void SpawnWarrior()
        {
            MessageQueueManager.Instance.SendMessage(
                new BasicWarriorSpawnMessage());
        }
    }
}
```

When we add the MenuItem attribute, a menu option will be created in the Unity Editor that matches the path in the attribute. In this case, a menu called Dragoncraft will be displayed in the Editor with the respective submenus. The method with the attribute must be static, otherwise, it will not work since Unity cannot instantiate the class for us, it can only call static methods. We also make the class itself static to avoid having any method that is not static – a static class can only have static methods.

Here, we are using the only method from MessageQueueManager that has not been called yet, which is SendMessage. Before adding it to the gameplay mechanics in later chapters, we need to test whether our systems are working properly using this debugging tool, so for now, this is the only place we are going to call the SendMessage method. The only parameter is a message object, which, in this case, is a new BasicWarriorSpawnMessage object that we create to send a message to spawn a warrior unit.

Now, let us create another debug script, for updating the UI when adding or removing resources. Add a new ResourceDebugger script in the **Scripts | Editor | Debug** folder, and replace the content with the following code:

```
using UnityEditor;
namespace Dragoncraft
{
    public static class ResourceDebugger
    {
        [MenuItem("Dragoncraft/Debug/Resources/+10 Gold",
            priority = 0)]
```

```
private static void AddGold()
{
  MessageQueueManager.Instance.SendMessage(
    new UpdateResourceMessage {
      Type = ResourceType.Gold,
      Amount = 10 });
}

[MenuItem("Dragoncraft/Debug/Resources/-10 Gold",
  priority = 1)]
private static void SubtractGold()
{
  MessageQueueManager.Instance.SendMessage(
    new UpdateResourceMessage {
      Type = ResourceType.Gold,
      Amount = -10 });
}
  }
}
```

This script is very similar to the previous one, a `static` class with `static` methods, but now that we are adding more than one `MenuItem`, we can also use the `priority` parameter to have them in the order we want. One debug option will add 10 gold while the other subtracts 10 gold from the resources. To do that, we are creating a new `UpdateResourceMessage` as the `SendMessage` parameter and setting a couple of properties for `Type` and `Amount`.

You can duplicate both `AddGold` and `SubtractGold` a couple of times to create `AddWood`, `SubtractWood`, `AddFood`, and `SubtractFood`, changing `Type` to the equivalent value for each method. Once the scripts are saved and Unity recompiles, we will be able to see the following menu and options in the Unity Editor:

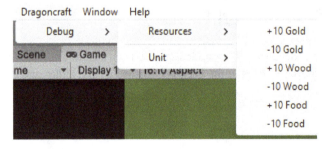

Figure 5.7 – The Debug menu with Resources and Unit options

If you click on any of those options without running the game, nothing will happen because the GameObject listening for `UpdateResourceMessage` is not active when the game is not running in the Editor. Now, open the **Level01** scene and hit the play button in Unity Editor to run the game. This time, you should be able to use all menu options and even spawn a warrior unit in the center of the screen since we did not set a position and the default is (`0, 0, 0`).

Note that the resources are updated in the top-right corner of the screen every time you use an option from the **Dragoncraft | Debug | Resources** menu. As for the Object Pool, we can see in the following figure that the Object Pool instantiates 10 copies of the warrior Prefab and shows them in the **GameObjects** list, and the unit **Debug** menu will spawn one of those copies in the scene.

Figure 5.8 – The Object Pool with 10 objects, 1 warrior spawned in the scene, and the resources UI updated

In this last part of the chapter, we put it all together and tested our new systems. Feel free to add more **Debug** menu options with different values or even create new spawners for different Prefabs, which are in the **ThirdParty** assets folder. Everything we have done so far using the warrior unit can now be easily replicated for other units and enemies, and we will do that in later chapters when we add more content to the game. We are still creating a few base systems and foundations to help build the *Dragoncraft* game, and as soon as we have all the pieces, it will be so much easier (and fun) to put them together and create the gameplay mechanics.

Summary

Well done on reaching the end of the chapter. We have built a brilliant foundation for the systems that we are yet to develop in the upcoming chapters of this book.

In this chapter, we learned how to implement the Object Pooling pattern, the Message Queue pattern, and even the Singleton pattern. All these patterns are widely used in professional games, and it is very important to learn about and implement them so we understand what game systems can benefit from them.

Even though we are building an RTS game, everything we saw in this chapter can be used and implemented in any kind of game. In fact, most of the scripts are generic enough that we could just copy them to a different project and use them out of the box. This is the beauty of well-designed classes and decoupled solutions.

In *Chapter 6*, *Commanding an Army of Units*, we will be getting more into the gameplay mechanics. We are going to see how to script actions that will spawn multiple units and how to select such units and move them together to different positions on the map. We will use most of the systems and components developed so far, and expand a few of them to have more functionality.

Further reading

Although we learned the basics of the three patterns introduced in this chapter, I recommend you check out the following links and learn more about them as well as see other examples and usages. We also saw a few new features from C# and .NET that are worth reviewing if you have any doubts about them, or just want to see more example usages. Here are the links:

- *Singleton*: `https://gameprogrammingpatterns.com/singleton.html`
- *Object Pool*: `https://gameprogrammingpatterns.com/object-pool.html`
- *Event Queue*: `https://gameprogrammingpatterns.com/event-queue.html`
- *MenuItem*: `https://docs.unity3d.com/ScriptReference/MenuItem.html`
- *Generic type parameters*: `https://learn.microsoft.com/en-us/dotnet/csharp/programming-guide/generics/generic-type-parameters`
- *Out parameter modifier*: `https://learn.microsoft.com/en-us/dotnet/csharp/language-reference/keywords/out-parameter-modifier`
- *Delegate class*: `https://learn.microsoft.com/en-us/dotnet/api/system.delegate`
- *Interface*: `https://learn.microsoft.com/en-us/dotnet/csharp/language-reference/keywords/interface`
- *MonoBehaviour life cycle*: `https://docs.unity3d.com/Manual/ExecutionOrder.html`

6

Commanding an Army of Units

One of the most exciting mechanics of an RTS game is preparing the strategy like a chess game and trying to picture what the opponent will do next. Preparing such a strategy involves carefully selecting and moving our units to the best possible places, or even splitting the army to cover different locations on the map.

So far, we have implemented the unit spawning feature, but now we are going to use it to select and move the units on the map. A new **Debug** menu shortcut will make it easier to instantiate more units in the Unity Editor to test the new features of this chapter.

This chapter will introduce the **Physics API** and **Raycast** to select the units on the map, and some vector manipulations to move the units and organize them into a nice grid formation. We are also going to expand some scripts and Prefabs created in previous chapters and have our first contact with materials and the default Unity Shader.

By the end of this chapter, you will know how to interact with the player input on the UI and the 3D world using Raycast, how to apply a material to a 3D model, and how to select 3D objects using the mouse position. Plus, you will have seen how we can select the units spawned in the map and how to select them and give a command to move to any location on the map, one of the core features of an RTS game.

In this chapter, we will cover the following topics:

- Preparing the Prefabs and UI
- Selecting the units
- Moving the units

Technical requirements

The project setup in this chapter with the imported assets can be found on GitHub at https://github.com/PacktPublishing/Creating-an-RTS-game-in-Unity-2023 in the Chapter06 folder inside the project.

Preparing the Prefabs and UI

In *Chapter 5, Spawning an Army of Units*, we were able to instantiate many units using the **Debug** menu we created. However, the units were spawned at the same place, and we were not able to do anything else with them. Now, we are going to develop a new script that will make it possible to select the units, which is fundamental to giving them an action to perform, such as moving or attacking a target.

We are going to add a few new debugging options to our game to make it possible to enable or disable the camera movement with the mouse, as well as add shortcuts to spawn new units. Then, we are going to start preparing the UI by adding a few new GameObjects to our **LevelManager** Prefab, and also create visual feedback in the UI to help the player see what is being selected using the mouse on the map.

Adding more debug options

Before creating the script responsible for selecting the units, we will need to make some adjustments to the project. The first couple of changes are going to make our lives easier moving forward; these include adding an option to enable or disable the camera movement and adding a shortcut to call the **Debug** menu option that spawns a new unit. Both modifications are quite simple, but they will help reduce the time taken to test our code changes in the game scene.

Let us start with the option to enable or disable the camera movement. Open the script located in `Scripts/Camera/CameraController.cs` in your IDE and add the following line before the `Start` method, the first method defined in the following class:

```
[SerializeField]private bool _disableCameraMovement = false;
```

The default value will be `false`, so we need to manually change it in the scene when we want to disable the camera movement.

Next, add the following code inside the existing `Update` method. We are going to ignore both methods, `UpdateZoom` and `UpdatePan`, if the `disableCameraMovement` variable is set to `true`:

```
private void Update()
{
#if UNITY_EDITOR
  if (_disableCameraMovement)
  {
    return;
  }
#endif
  UpdateZoom();
  UpdatePan();
}
```

Some new things are introduced here: the `#if` and `#endif` **conditional compilation** directives, which will allow us to selectively include a piece of code in the compilation, and the `UNITY_EDITOR` **scripting symbol**, which is a scripting symbol defined in the Unity Engine. Any piece of code that is between the `#if` and `#endif` conditional directives will only be included in the build, whether the scripting symbol is defined by our code or the Unity Engine. When using the Unity Editor, or running the game there, the `UNITY_EDITOR` scripting symbol is defined by the Unity Engine so any code between the conditional compilation directives will be executed.

Here, we are allowing the `disableCameraMovement` variable to be validated in the `Update` method only if we are running the script in the Unity Editor, which means that this piece of code will never be executed unless we are in the Unity Editor.

Remember to pay attention to the code inside the conditional compilation directives because it will be removed and not executed when the condition is `false`. This means that if a variable is created inside it but used outside the validation, for example, an error will be thrown for a missing variable when the condition is not `true`.

Now, let us enable the property to disable the camera movement when moving the mouse. To do so, open the **Level01** scene and select the **Main Camera** GameObject in the **Hierarchy** view. Notice that, in the **Inspector** view, the new **Disable Camera Movement** property is now displayed on the **Camera Controller** component. Enable it by clicking on the checkbox on the right side of the property name and save the scene to persist the changes.

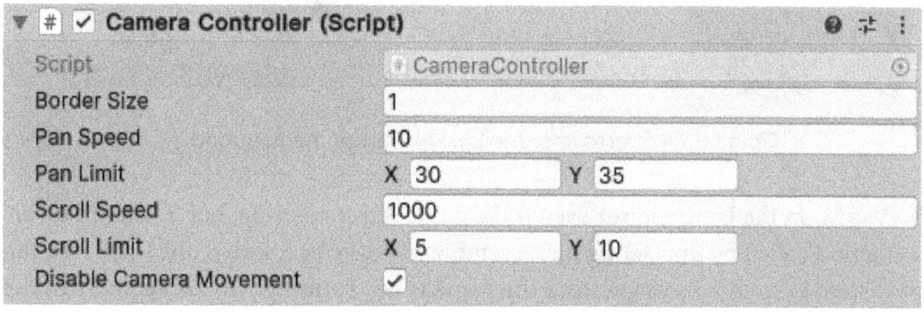

Figure 6.1 – Camera Controller now has the Disable Camera Movement property exposed

Now, if we play the **Level01** scene in the Editor, we will notice that the camera is not moving anymore when the mouse reaches the screen limits. This is very useful for testing new features and functionalities without worrying about the camera moving. We can always enable the camera movement again, even while the game is running in the Editor.

The next modification is adding a shortcut to the **Debug** menu option that spawns a new unit; this will make it easier to add more units without having to click on the menu option every time we want a new unit in the scene from the **Debug** menu.

Open the script located in `Scripts/Editor/Debug/UnitDebugger.cs` in your IDE and replace the `MenuItem` string with the following line:

```
[MenuItem("Dragoncraft/Debug/Unit/Spawn Warrior %g")]
```

The only actual change was adding the `%g` at the end of the string. This code will be parsed by Unity and a shortcut will be created when *Ctrl* + *G* is pressed on Windows or *Command* + *G* is pressed on macOS.

Note that the percent symbol (`%`) refers to the *Ctrl* key on Windows and the *Command* key on macOS. We can also use the ampersand symbol together with the percent symbol as `%&g` to add the *Alt* key to the shortcut, creating the combination *Ctrl* + *Alt* + *G*. It is also possible to add the hash symbol, `#`, which translates to the *Shift* key; therefore, using `%#&g` creates the shortcut *Ctrl* + *Shift* + *Alt* + *G*. This helps in creating a lot of different shortcuts.

It is good practice to create a shortcut for the custom menu items that we really want to use instead of adding one shortcut for every possible option. You can change the letter *G*, but be careful to not conflict with an existing shortcut in Unity Editor.

After saving the script, go back to the Unity Editor and check whether **Spawn Warrior** now has the shortcut shown on the right side of the menu option:

Figure 6.2 – Shortcut for the Spawn Warrior menu option

The shortcut works on the Unity Editor even if the game is not running, but it will not do anything because the method called by the **Debug** menu sends a message to spawn a unit; however, there will be no scripts listening to this message since the game is not running. Go ahead and run the game to test the new shortcut; it will instantiate the warrior unit on the center of the screen each time you press the key combination *Ctrl* + *G* (*Command* + *G* on macOS).

Next, we are going to prepare the Prefabs and UI before creating the scripts to interact with them.

Preparing the Prefabs and UI

We are going to start moving the Prefabs with scripts used in the level scenes into the same `LevelManager` Prefab to make it easier to add to a new scene, making sure we do not forget to include any script required for the game to work properly.

Let us first create a new Prefab for the `ObjectPools` GameObject and add it to the
`LevelManager` Prefab:

1. In the **Project** view, create a new folder called `ObjectPool` inside the existing **Prefabs** folder.

2. Open the **Level01** scene and, in the **Hierarchy** view, drag and drop the **ObjectPools** GameObject
    into the new **Prefabs | ObjectPool** folder to create the Prefab.

3. In the **Hierarchy** view, drag and drop the **ObjectPools** GameObject into the **LevelManager**
    GameObject. **ObjectPools** will be a child of **LevelManager** in the hierarchy.

ObjectPools will display a white icon with a green plus symbol underneath **LevelManager**; this means
that a new GameObject that is not part of the Prefab object was added to the Prefab. Do not save the
scene now because it will save this change only in `Level01`, and we want that to be applied to every
scene that already has (or will have) the `LevelManager` Prefab. Instead, select the **LevelManager**
GameObject in the **Hierarchy** view and, in the **Inspector** view, click on the **Overrides** dropdown,
then the **Apply All** button:

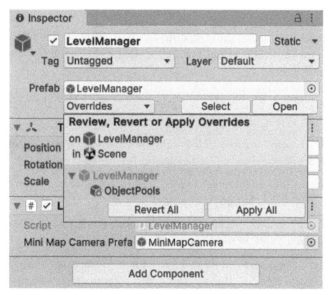

Figure 6.3 – Applying the changes to the LevelManager Prefab

Once **Apply All** is clicked, the **ObjectPools** icon will change from white to blue, without the green
plus symbol, indicating that the `ObjectPools` Prefab is now a child of the `LevelManager` Prefab.

We have covered two different ways to edit a Prefab: double-clicking on the Prefab in the **Project** view
to edit it, and modifying it in the current scene and then applying the changes to the Prefab. Both
approaches have the same result, the only difference is that sometimes it is easier to edit in the scene
than in the isolated Prefab. Adding `ObjectPools` inside `LevelManager` works well because the
scripts on both Prefabs do not have dependencies with other objects from the scene, or to each other.

This is possible because we are using a few techniques to decouple the script dependencies, such as sending messages, for example.

Moving on, the next changes will be in the UI, to draw the selected area when dragging the mouse cursor to select the units.

Drawing the selected area in the UI

It is always important to give visual and sound feedback to the player's actions. Without feedback, the player might get lost or confused when trying to do something in the game. The main feature that we are going to develop in this chapter is the functionality to select units and give them one action to perform.

In this section, we will create the visual feedback, which is the feedback the player sees when the mouse is pressed and dragged across the screen to select the units before giving them an action. We will not be working on the functionality itself but it will be useful to see what we are selecting before highlighting the selected units and giving them a command to execute.

We are also going to split this feature into two different parts. Here, in the UI scene, we will draw the selected area while the player is dragging the mouse. Next, in the game scene, we will use Unity's Physics API to select the units.

Let us first create the image that will be manipulated using a script to change its size when the player moves the mouse cursor:

1. Open the **GameUI** scene and, in the **Hierarchy** view, right-click on the **Canvas** GameObject, select **UI | Image**, and name the new GameObject `UnitSelector`.

2. Select the **UnitSelector** GameObject and, in the **Inspector** view, change **Anchor Preset** to **bottom** on the left side and **left** on the top side.

3. Change both **Pivot X** and **Y** to 1 and then change both **Pos X** and **Pos Y** to 0. Then, change the **Width** to 1 and the **Height** to 1.

4. In the **Image** component, click on the color selector and change the values as follows: **R** to 0, **G** to 255, **B** to 0, and **A** to 55. This will change the color to green with transparency.

5. Create a new script inside the **Scripts | UI** folder and name it `UnitSelectorUI`.

6. Add the newly created **UnitSelectorUI** script to the **UnitSelector** GameObject.

Once the **UnitSelector** GameObject is set up, you should have the settings as shown in the following screenshot, with the **UnitSelectorUI** script attached to the GameObject at the bottom:

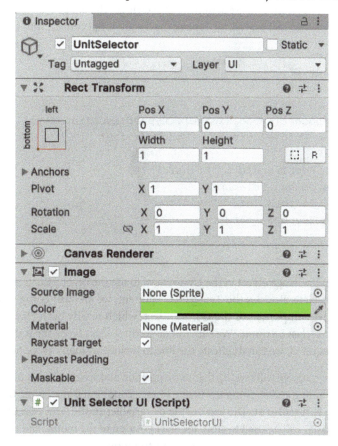

Figure 6.4 – UnitSelector GameObject configured

The green color has a transparency setting so it will not look like a green block on top of the screen, but instead, it will be possible to see through the selection box.

Moving to the script now, double-click on the **UnitSelectorUI** script to open it in your IDE for editing. This is quite a simple script that is responsible for hiding and showing the selected image based on the left mouse button press, hold, and release actions. It also changes the image scale to follow the mouse while the left button is pressed and moved. So, replace the default script content with the following code:

```
using UnityEngine;
using UnityEngine.UI;
namespace Dragoncraft
{
    [RequireComponent(typeof(Image), typeof(RectTransform))]
    public class UnitSelectorUI : MonoBehaviour
    {

        private RectTransform _transform;
        private Image _selector;
        private Vector3 _startPosition;
        private Vector3 _offset;

    }
}
```

Since we are modifying the image object in the UI, we need to use the `UnityEngine.UI` namespace to include the `Image` class. Besides the `Image` component, we are also going to need access to the `RectTransform` component of the GameObject, which is why both class types are in the `RequireComponent` class attribute. We can add up to three class types separated by commas. If more class types are required, we can duplicate the line to include as many as we need.

The `RectTransform` variable will modify the image position and scale, whereas the `Image` variable will be responsible for hiding and showing the image when a condition is met in the `Update` method, as we are going to see in a moment. The last couple of variables in the `UnitSelectorUI` class, which are both `Vector3`, will be used to store the start position of the mouse and the offset to adjust the position of the image related to the mouse position and movement.

In the following code block, which is the `Awake` method, we are going to initialize the variables we just added to the `UnitySelectorUI` class:

```
private void Awake()
{
  _transform = GetComponent<RectTransform>();
  _selector = GetComponent<Image>();
  _selector.enabled = false;
  _offset = new Vector3(Screen.width / 2, Screen.height / 2);
}
```

Here, we are initializing both the `RectTransform` and `Image` variables using the `GetComponent` method from the Unity API. Since we are using the `RequireComponent` class attribute, it is not necessary to verify whether they are `null` or not before using them – they will always have a valid value.

Next, we are hiding the image using the `enabled` property set as `false`. Note that this property will hide the image, but the GameObject will still be active. The last line is initializing the `offset` variable, setting the value as half of the screen size for **X** and **Y**, and leaving **Z** as **0** by omitting the value in the object constructor.

The second and last method in this class is `Update`, which is responsible for getting the left mouse button events and updating the image position, scale, and visibility:

```
public void Update()
{
    if (Input.GetKeyDown(KeyCode.Mouse0))
    {
        _selector.enabled = true;
        _startPosition = Input.mousePosition;
        _transform.localPosition = Input.mousePosition - _offset;
    }
    if (Input.GetKey(KeyCode.Mouse0))
    {
        _transform.localScale = _startPosition - Input.mousePosition;
    }
    if (Input.GetKeyUp(KeyCode.Mouse0))
    {
        _selector.enabled = false;
    }
}
```

We are using three different APIs from Unity to get the input key event:

- `GetKeyDown` will be `true` when one button is pressed
- `GetKeyUp` is valid when the pressed button is released
- `GetKey` will tell us whether the button is still pressed after the `KeyDown` event

`KeyCode` is an enum from Unity that has all the possible keyboard keys and mouse buttons, and `Mouse0` corresponds to the left mouse button. The right mouse button, as you might have guessed, is `Mouse1`.

In the first conditional check, when the left mouse button is pressed, the image is shown, and the current `mousePosition` is stored as `startPosition`. The image is also positioned where the mouse cursor is on the screen minus the offset that was defined in the `Awake` method, with the resulting vector set to the `localPosition` property.

The second validation, when the left mouse button is still pressed, considers that the image is already displayed and positioned by the mouse-pressed event, and then only resizes `localScale` based on `startPosition` minus the current `mousePosition`.

When the player releases the left mouse button, the last conditional will be `true`, and the only thing left here is to hide the image. Nothing else is required here since the image will be positioned and resized properly the next time the left mouse button is pressed and dragged across the screen.

Once the script changes are saved in the IDE, we can return to the Unity Editor and open the **Level01** scene to run it. The `LevelManager` class also loads the **GameUI** scene, so we will be able to see the preceding script in action by simply clicking with the left mouse button and dragging the cursor on the screen; a transparent green image will follow the mouse movement, as shown in the following screenshot:

Figure 6.5 – The green selector feedback when clicking and dragging the mouse cursor

Now that we have visual feedback for the unit selection action in the UI scene, we will need to prepare the level scene before moving to the script that is going to select the units and give them a command to change the position to somewhere else in the map.

Preparing the level scene

The `LevelManager` Prefab will grow once more with a new GameObject that will have the component to select and move the units. However, to keep this script self-contained without direct references from the scene, we are going to add a tag to a GameObject that will be required in the script.

There are a few methods to find one specific GameObject in the scene, such as the `GetComponent` method, which can also be used to search for a component in the parent GameObject or in the child

GameObjects. However, sometimes it is not enough or not possible to use the `GetComponent` method in the scene, especially if the scene has a lot of different GameObjects, which can cause some performance issues. We could also use the `GameObject.Find` method to search by the object name, but that is also a problem because we could have duplicate names.

One good solution is to use the `GameObject.FindGameObjectWithTag` method, which will return the first object with one specific tag. In this case, the tag is set manually by us in the **Inspector** view and the GameObject is free to have any name. We are going to use this approach and tag the **Plane** GameObject with a new custom tag that will have the same name so we can find it in the upcoming script because we are going to need the `MeshCollider` component attached to it.

First, let us create the custom tag and attach it to the **Plane** GameObject:

1. Open the **Level01** scene and select the **Plane** GameObject in the **Hierarchy** view.

2. In the **Inspector** view, at the top left, click on the drop-down field named **Tag**, which has the default value of **Untagged**. Select the last option, **Add Tag…**, which will show the **Tags & Layers** settings in the **Inspector** view.

3. In the first item named **Tags**, there is an empty list. Click on the + button on the right side, type the name `Plane`, and click on the **Save** button.

4. In the **Hierarchy** view, left-click on the **Plane** GameObject again to select it in the **Inspector** view. Now, click on the **Tag** dropdown one more time, select the newly created tag named **Plane**, and save the scene.

The following screenshot shows the **Plane** GameObject with the tag that has the same name:

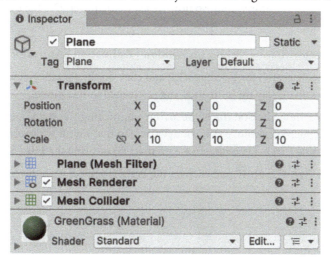

Figure 6.6 – The Plane GameObject with the Plane tag

That is the only change we will need in the scene to make our next script able to find the GameObject and component that are needed. Now, we can create a new GameObject inside **LevelManager** and the new script:

1. Right-click on the **LevelManager** GameObject in the **Hierarchy** view, select the **Create Empty** option, and name the new GameObject UnitSelector.

2. Create a new script inside the **Scripts | Unit** folder and name it UnitSelectorComponent.

3. Select the **UnitSelector** GameObject and add the newly created script to **UnitSelectorComponent** as a component.

4. Select the **LevelManager** GameObject and, in the **Inspector** view, click on the **Overrides** dropdown and click on the **Apply All** button to save the changes in the Prefab; then, save the scene.

Now that we have visual feedback for the unit selection action and all the changes required in the level scene, we can move to the next script, which will indeed select all the units and move them to the desired location.

Selecting the units

There is one more change we need to apply to the project, but this time, we are going to edit an asset from the **ThirdParty** folder. You probably noticed that both the warrior spawned and the wizard in the UI have an animation that cycles through different states continuously. This is because the **Animator** of both 3D models was set up like that to showcase the possible animation states. We want to have better control over the animations, so now we are going to remove the transitions between the animation states:

1. Open the **Animator** view by clicking on **Window | Animation | Animator**.

2. In the **Project** view, search for FootmanHP and click on the Prefab, you will notice that the **Animator** view will display the animation states attached to this model.

3. In the **Animator** view, click on the white arrows connecting the states and press the *Delete* key on your keyboard to remove that connection. Do that for all white arrows and save the changes by pressing *Ctrl + S* (*Command + S* on macOS) or using the **File | Save** menu option.

4. Next, search for FreeLichHP and click on the Prefab to select it on the **Animator** view.

5.   Delete all the white arrow connections as well and save the changes.

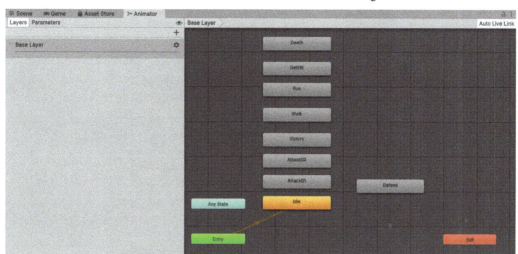

Figure 6.7 – The FootmanHP Animator states after removing the white arrow connections

Be careful not to delete the animation states, which are the gray boxes. The orange arrow connection cannot be deleted because the Animator needs to know at least what the entry point to the animation is; this is why the **Entry** animation state is connected to the **Idle** animation state, which is also an orange box. The preceding screenshot shows the Animator states for the **FootmanHP** Prefab after the white arrows are removed. Although they are organized a bit differently, the animation states are the same for the **FreeLichHP** Prefab as well.

Now, both models will cycle through only the **Idle** animation state instead of going through all the animation states and back to the beginning. Without the animation connections, it is possible to play the **Run** animation state, for example, and the model will keep running until we change to a different animation state. The animation control will be done via scripts later in this chapter when we start to move the selected units.

Setting a custom color for the selected units

Before selecting the units, we will need to add visual feedback to highlight what were the selected units, and we are going to do that by tinting the model color to a custom color. For the warrior units, the color will be a light blue; this color will be applied with a bit of transparency using a Shader property.

Open the script located at **Scripts | Unit |** UnitData.cs in your IDE and add the following property as the last one in the class; it will hold the color information:

```
public Color SelectedColor;
```

Now that we have the new property, we need to set a value to it. Select the ScriptableObject located at **Data | Unit |** `BasicWarrior.asset` and change the color to the **Hexadecimal** value of `00A9FF`. Once the color is set, the ScriptableObject will be as shown here:

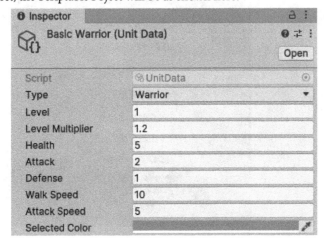

Figure 6.8 – Basic Warrior with the selected color set as light blue

With the color set in the ScriptableObject, we need to pass the value to the `UnitComponent` class, but before that, there is one more modification that we will need to do in another script. Open the script located at **Scripts | Spawner |** `BasicWarriorSpawner.cs` and replace the content of the current `OnBasicWarriorSpawned` method with the following code:

```
private void OnBasicWarriorSpawned(BasicWarriorSpawnMessage message)
{
  GameObject warrior = SpawnObject();
  UnitComponent unit = warrior.GetComponent<UnitComponent>();
  if (unit == null)
  {
    unit = warrior.AddComponent<UnitComponent>();
  }
  unit.CopyData(_unitData);
}
```

The only change in the preceding code is that we removed all the variables from here and will pass the `UnitData` object directly to the `UnitComponent` class instead.

Once we save the script, an error will be thrown saying that the `CopyData` method is not defined in the `UnitComponent` class. So, let us open the script located at **Scripts | Unit |** `UnitComponent` to add it, as shown in the following code:

```
public Color SelectedColor;
```

```
public void CopyData(UnitData unitData)
{
  ID = System.Guid.NewGuid().ToString();
  Type = unitData.Type;
  Level = unitData.Level;
  LevelMultiplier = unitData.LevelMultiplier;
  Health = unitData.Health;
  Attack = unitData.Attack;
  Defense = unitData.Defense;
  WalkSpeed = unitData.WalkSpeed;
  AttackSpeed = unitData.AttackSpeed;
  SelectedColor = unitData.SelectedColor;
}
```

Moving the variables to the new CopyData method helps to reduce the places we need to modify when adding or changing a new variable. The BasicWarriorSpawner class does not need to know what properties the basic warrior has, the class is only responsible for receiving the message and adding a new unit to the level. Since SelectedColor is only used in UnitComponent, it makes much more sense to add it here, so we created a new public class variable for it and set the value from the UnitData object.

There are a few more changes that we need to make in the UnitComponent class; let us start by adding a couple of classes as required components to this class. Add RequireComponent, as shown here, on top of the class declaration:

```
[RequireComponent(typeof(BoxCollider), typeof(Animator))]
```

The Animator component is already attached to the GameObject so, in this case, it will only return the reference to it. However, BoxCollider is a new component, and it will be attached to the GameObject when the UnitComponent script is attached to it, as a dependency component.

Now, we can add the two variables that will hold the references for the Animator and Renderer components and get them in the Awake method:

```
private Animator _animator;
private Renderer _renderer;

private void Awake()
{
  _renderer = GetComponentInChildren<Renderer>();
  _animator = GetComponent<Animator>();
  _animator.Play("Idle");
}
```

We are going to use GetComponentInChildren to find the Renderer component, and we cannot add it to the list of required components, since it is not part of the current GameObject, but it is attached to a child of it. For this reason, we are going to check whether the renderer variable has a valid value before using it.

We do not need to check whether the Animator component is null before using it because it is part of the required components we just added to the class. The last line in the Awake method calls the Play method to start an animation named Idle. As we just saw in *Figure 6.7*, this animation state is present in the Animator. We can play any of those animations using this method.

For now, the only method we need to add to this class is the following one called Selected, which will enable or disable the EMISSION keyword from the Material default Shader and set EmissionColor to the SelectedColor value:

```
public void Selected(bool selected)
{
  if (_renderer == null)
  {
    Debug.LogError("Renderer component is missing!");
    return;

  }
  Material[] materials = _renderer.materials;
  foreach (Material material in materials)
  {
    if (selected)
    {
      material.EnableKeyword("_EMISSION");
      material.SetColor("_EmissionColor", SelectedColor * 0.5f);
    }
    else
    {
      material.DisableKeyword("_EMISSION");
    }
  }
}
```

The first validation we do in this method is to check whether the renderer variable is not null. We know that the renderer variable will not be null because the Prefab we are using does have the renderer component, but it is always good to check whether an object is not null to avoid an exception thrown in our code. If renderer is null, we exit the method using the return keyword and print an error to let us know what is wrong.

The `renderer` object has a list of materials of the model we are using; therefore, we need to loop through all the materials, using `foreach`, and change the color of each material if the `selected` parameter is true. In this case, we also use the `EnableKeyword` method to enable the `EMISSION` keyword, otherwise we use the `DisableKeyword` method to disable the `EMISSION` keyword and we do not change the color of the material.

The `EnableKeyword` and `DisableKeyword` methods change keywords from the default Shader that is in the material, and enabling or disabling `EMISSION` makes it change the `material` emission to one specific color, which is white by default. For this reason, we are also calling the `SetColor` method to change `EmissionColor` to `SelectedColor` multiplied by `0.5f`, which reduces the intensity of the color by 50% to not make it so strong that we cannot see the model details.

That was the last change for now in the `UnitComponent` class, and we are ready to start writing the code to select the units.

Defining the unit selector component

We already created the `UnitSelectorComponent` class when it was added to the **UnitSelector** GameObject inside the **LevelManager** Prefab, so now we need to define its content.

The first thing is to replace the content with the following code, which has the class variables that we are going to use in this class:

```
using System.Collections.Generic;
using UnityEngine;
namespace Dragoncraft
{
  public class UnitSelectorComponent : MonoBehaviour
  {
    private MeshCollider _meshCollider = null;
    private Vector3 _startPosition;
    private List<UnitComponent> _units = new List<UnitComponent>();
  }
}
```

Here, we have the `MeshCollider` component reference, which will be used to find the mouse click location in the 3D coordinates. The `startPosition` property is very similar to the variable with the same name that we created in the `UnitSelectorUI` class, but here we are going to use a different method to get its value, and finally, a list of all the selected units to keep track of them.

The first method in this class is Awake, which is responsible for getting the MeshCollider component in the **Plane** GameObject, finding it by the tag named Plane that we created previously in this chapter. Here, we can see the FindGameObjectWithTag method in action, which returns the first GameObject with that tag if found:

```
private void Awake()
{
    GameObject plane = GameObject.FindGameObjectWithTag("Plane");
    if (plane != null)
    {
        _meshCollider = plane.GetComponent<MeshCollider>();
    }
    if (_meshCollider == null)
    {
        Debug.LogError("Missing tag and/or MeshCollider reference!");
    }
}
```

Both the plane and the meshCollider variables can be null if the current level scene does not have the Plane tag set to the **Plane** GameObject, so both validations are here to check whether that is the case and print an error message to let us know if something is wrong and to double-check it.

The next method defined is Update, which will constantly check whether the player pressed and released the left mouse button. We are using the same KeyCode.Mouse0 here as used in the UnitSelectorUI class, but this time, we only verify whether the player pressed or released the button (we do not need to check for mouse dragging here):

```
private void Update()
{
    if (Input.GetKeyDown(KeyCode.Mouse0))
    {
        _startPosition = GetMousePosition();
    }
    if (Input.GetKeyUp(KeyCode.Mouse0))
    {
        Vector3 endPosition = GetMousePosition();
        SelectUnits(_startPosition, endPosition);
    }
}
```

Both the startPosition and endPosition variables are found using the GetMousePosition method, which will be defined in the following code block. Once the player presses the left mouse button, startPosition is retrieved and, once the left mouse button is released, endPosition is retrieved.

In the second `if` validation, when the left mouse button is released, we also execute the `SelectUnits` method, which will be defined soon, with both variables, `startPosition` and `endPosition`.

Now, let us add the following `GetMousePosition` method to our `UnitSelectorComponent` class:

```
private Vector3 GetMousePosition()
{
  Ray ray = Camera.main.ScreenPointToRay(Input.mousePosition);
  RaycastHit hitData;
  if (_meshCollider.Raycast(ray, out hitData, 1000))
  {
    return hitData.point;
  }
  return Vector3.zero;
}
```

Since we are not in the 3D environment here, different from the 2D coordinates in the UI, we need to use the **Raycast** to literally cast a ray from `mousePosition` to `meshCollider` and check whether that was a hit. Then, we can find the position in the 3D world that corresponds to `mousePosition`.

Although we have multiple cameras in our game, there is only one that is considered the main camera, which is the one in the scene that is called **Main Camera** and has the tag with the same name. This camera is our top view of the map, and by using the `ScreenPointToRay` method, we can convert the current `mousePosition` into a `Ray` object that can be used to check the `Raycast` hit.

The `meshCollider` variable inherits the `Raycast` method from the `Collider` class and has `Ray` as an input parameter and `RaycastHit` as the output parameter in the method signature. That is why the `Raycast` method needs the variable defined before and uses the `out` keyword on the left side of the `hitData` parameter. The last parameter is the max distance to cast the ray, which we can leave as `1000`. The `Raycast` method will return `true` with something hit by the ray, and we can safely return the `point` property. Otherwise, we return the `zero` value.

Finally, here is the method for selecting the units using everything we set up so far in this chapter. `SelectUnits` takes the start and end position of the mouse pressed and released buttons, respectively, and tries to find whether there is any collider between the two points:

```
private void SelectUnits(Vector3 startPosition, Vector3 endPosition)
{
  foreach (UnitComponent unit in _units)
  {
    unit.Selected(false);
  }
  _units.Clear();

  Vector3 center = (startPosition + endPosition) / 2;
```

```
float distance = Vector3.Distance(center, endPosition);
Vector3 halfExtents = new Vector3(distance, distance, distance);

Collider[] colliders = Physics.OverlapBox(center, halfExtents);
foreach (Collider collider in colliders)
{
    UnitComponent unit = collider.GetComponent<UnitComponent>();
    if (unit != null)
    {
        unit.Selected(true);
        _units.Add(unit);
    }
}
}
```

The player can also unselect the units by simply clicking anywhere in the map – that is why the first thing we do here is to loop through all the units in the list calling the Selected method with the false value on each one, and then call the Clear method to remove all units from the list before starting to check whether there are any selected units again.

The main logic of this method comes next with a few mathematical operations:

- The first one is to find the center of the two points, startPosition and endPosition, which can be done by summing the two vectors and dividing it by 2.

- Next, we need to find the distance between the center position and endPosition so we know the size of the halfway position, and for that, the Vector3 class has a handy method called Distance.

- Then, we create a new Vector3 where all axes have the distance value.

We had to do all those calculations because we are going to use the Physics API from Unity, which has a method called OverlapBox that takes the center point and a Vector3 position with half of the size of the area we have. Imagine that this method will create a 3D cube, using the center and halfExtents variables to determine the dimension and position of the cube, and then the method will return all the colliders found within the cube. This is how we are going to find whether any unit is between the two points we get when the player presses and releases the left mouse button.

The last bit of the method is quite simple; we're just looping through all the colliders found and checking whether they have a component called UnitComponent attached to them. This is necessary because the OverlapBox method will return all the possible colliders within the box area, and that includes **Plane** and any other object. One way to find out whether the selected collider is a unit is by checking whether the GameObject has the UnitComponent component, as we know a unit should have it. Then, if the component does exist, we call the Selected method with the true value to highlight the unit and add it to the list.

Running the game in the current state is possible to add units using the shortcut *Ctrl + G* and using the mouse to select the unit in the center of the screen. By doing that, you will notice that the selected unit changes its tint color to light blue when selected and back to the regular color if you click anywhere else on the map.

Now that the selection is done, we can go ahead to the last part of this chapter, which is moving the selected units to where we want and in formation.

Moving the units

Now that we have all the unit selection logic in place, we only need a couple more changes in the UnitSelectorComponent class to have it ready to move the units where we want on the map. The Update method, which is already defined in the UnitSelectorComponent class, needs one more validation that we are going to add at the end of the method, as we can see in the following code block, which will be used to check whether the right mouse button was released, using KeyCode. Mouse1:

```
private void Update()
{
    ...
    if (Input.GetKeyUp(KeyCode.Mouse1))
    {
        Vector3 movePosition = GetMousePosition();
        MoveSelectedUnits(movePosition);
    }
}
```

Once we have the desired move position, which is where the player clicked on the map using the right mouse button, we can call the MoveSelectedUnits method, which is defined next, as well as a new class variable named distanceBetweenUnits:

```
private float _distanceBetweenUnits = 2.0f;

private void MoveSelectedUnits(Vector3 movePosition)
{
    int rows = Mathf.RoundToInt(Mathf.Sqrt(_units.Count));
    int counter = 0;
    for (int i = 0; i < _units.Count; i++)
    {
        if (i > 0 && (i % rows) == 0)
        {
            counter++;
        }
```

```
        float offsetX = (i % rows) * _distanceBetweenUnits;
        float offsetZ = counter * _distanceBetweenUnits;
        Vector3 offset = new Vector3(offsetX, 0, offsetZ);

        _units[i].MoveTo(movePosition + offset);
    }
}
```

The `MoveSelectedUnit` method does one more thing besides moving the units: it also organizes the selected units to create a formation like a grid, where all units are side-by-side with an equal distance between them. For example, if we have four units, they will be placed in four positions split between two rows and two columns.

In the first line, we are determining the number of rows in which we will have to organize the units. This is calculated by the square root of the total number of units, using the `Sqrt` method, and then converting this floating point value to an integer by using the `RoundToInt` method. Both methods are part of the `Mathf` API. With that value, we can start looping through all the selected units in the list.

Inside the `for` loop, we are checking whether a new row is required by using the `%` mod operator, which will return the remainder after dividing the `i` variable by the `rows` variable. If that is zero, this means a new row is required; then, the `counter` variable is increased. There is also a validation to avoid increasing `counter` in the very first row if the value of `i` is 0.

Next, we calculate both `offsetX` and `offsetZ`, which will be used to create a `Vector3` variable with an offset to be added to `movePosition` so we can move each one of the selected units to a different position near to each other in the formation. The `offsetX` value is the distance between the units multiplied by the remainder of the division of `i` by `rows`, which will change `offsetX` based on the column of each unit in our imaginary grid. The `offsetZ` value only changes if a new row is required, which increases the `counter` value and is multiplied by the distance between the units.

The last line calls the `MoveTo` method, sending `movePosition` plus the calculated `offset` as the parameter. Note that the `MoveTo` method is not defined in the `UnitComponent` class yet, but this is what we are going to do next.

To add the new `MoveTo` method in the `UnitComponent` class, open the script located at `Scripts/Unit/UnitComponent.cs` in your IDE and add the content in the following code block. Here, we are adding a couple of new private class variables that we are going to need, `movePosition` and `shouldMove`, as well as the `MoveTo` method used in the preceding script:

```
private Vector3 _movePosition;
private bool _shouldMove;

public void MoveTo(Vector3 position)
{
    transform.LookAt(position);
```

```
   _movePosition = position;
   _animator.Play("Run");
   _shouldMove = true;
}
```

The MoveTo method will update a few variables, but it is not going to move the unit; instead, here we are setting up the value of the movePosition and shouldMove variables so the Update method will take care of moving the GameObject on each frame toward movePosition.

Before moving the GameObject, we need to rotate it to face the direction in which we want to move, which is the position parameter, using the LookAt method. Then, we also store movePosition to be used by the Update method and call the Play method to start the Run animation state. The last variable set is shouldMove, which will be used to start or stop the unit movement and animation.

Next is the core of the unit's movement, which is the Update method. If we set the shouldMove variable to false, the method does nothing as the return keyword is executed at the beginning:

```
private void Update()
{
  if (!_shouldMove)
  {
    return;
  }

  if (Vector3.Distance(transform.position, _movePosition) < 0.5f)
  {
    _animator.Play("Idle");
    _shouldMove = false;
    return;
  }

  Vector3 pos = (_movePosition - transform.position).normalized;
  transform.position += pos * Time.deltaTime * WalkSpeed;
}
```

If the unit can be moved, we then check whether the distance between the current position and movePosition is lower than 0.5f to stop it from moving any further. Because of the animation and the floating point nature, both positions will rarely be the same, so instead, we check whether they are close enough to consider that the unit reached the desired destination using the Distance method from the Vector3 class. If we need to stop moving, the Idle animation state is played, the shouldMove variable changes to false, and we exit the Update method with the return keyword.

At the end of the method, if we still need to move the unit, we first find the position by subtracting the current position from movePosition, and then use the normalized vector to get the final position value. Finally, we can increase the current position by the normalized difference calculated in the line above it, multiplied by both the current `deltaTime` and `WalkSpeed`. The `WalkSpeed` value is defined in the ScriptableObject that we created for the basic warrior.

Now, we are good to go ahead and save the script, return to the Unity Editor, and run the game. To test the changes we made in this chapter, press the shortcut *Ctrl + G* a few times, select the units in the center of the screen (they spawn in the same position), and then click somewhere on the map to see them move and organize themselves in a nice grid formation:

Figure 6.9 – The units selected and moving together in formation

We can also select the units, move the camera to a different position on the map, and see the army of units marching through the lands in the mini map. As we progress in the book, we are beginning to see the *Dragoncraft* game taking shape as we are adding more and more features, and they will soon start to interact with each other, building up the mechanics of an RTS game.

Summary

Congratulations on reaching the end of this chapter; our project is starting to become a game and soon we will have *Dragoncraft* as a playable RTS experience. Being able to select and move the units to a specific place on the map is one of the core mechanics of an RTS game and an important part of the strategy when playing this type of game.

In this chapter, we learned how to interact with both the UI and the 3D world using the mouse buttons and position. We also saw how to change the Material Shader property to highlight a model, how to select a 3D object using Raycast and the Physics API, and the logic to place all the units into a grid formation as they move together to a new position.

This chapter was all about the mechanics of selecting and moving the units, with visual feedback to the player and decoupled scripts that isolate the features from each other. We are just starting to implement the unit actions, and soon we will have all the core mechanics working in our game.

In *Chapter 7, Attacking and Defending Units*, we are going to implement the Command pattern to change the unit action after selecting them to attack or defend. We will also introduce a new unit, the wizard, which uses a ranged attack to cast damage spells. The damage and health logic will also be introduced as we progress to have more features implemented in the units.

Further reading

In this chapter, we saw quite a few new APIs from Unity that we might use again in later chapters. The following list includes resources for some of them, including more examples and use cases:

- *Physics.OverlapBox*: `https://docs.unity3d.com/2023.1/Documentation/ScriptReference/Physics.OverlapBox.html`

- *GameObjects.FindGameObjectsWithTag*: `https://docs.unity3d.com/2023.1/Documentation/ScriptReference/GameObject.FindGameObjectsWithTag.html`

- *MenuItem*: `https://docs.unity3d.com/2023.1/Documentation/ScriptReference/MenuItem.html`

- *Collider.Raycast*: `https://docs.unity3d.com/2023.1/Documentation/ScriptReference/Collider.Raycast.html`

- *Camera.ScreenPointToRay*: `https://docs.unity3d.com/2023.1/Documentation/ScriptReference/Camera.ScreenPointToRay.html`

7

Attacking and Defending Units

The core mechanic of an RTS game is the strategy used by the player. One of the most important decisions to be made is how to move and position the army of units, as well as when to defend from an enemy invasion and when to attack and raid. In this chapter, we are going to implement the main actions that the player can use to command the selected units: move, attack, defend, and collect.

We will also see how to update the details panel and the actions panel in the UI to reflect the units selected, displaying a dynamic 3D mode in the portrait UI and how many units were selected, as well as what the actions available for the selected units are and which buttons should be enabled in the UI based on that.

By the end of this chapter, you will know how to use the Command pattern to create the actions and how to update the UI elements using our existing message queue system. You will also have learned how to add a new unit to the game, the mighty Mage, and how to implement both a melee and ranged attack by throwing a fireball.

In this chapter, we will cover the following topics:

- Updating the UI with the selected units
- Attacking and defending with units
- Adding the new Mage unit and ranged attack

Technical requirements

The project setup in this chapter with the imported assets can be found on GitHub at https://github.com/PacktPublishing/Creating-an-RTS-game-in-Unity-2023 in the Chapter07 folder inside the project.

Updating the UI with the selected units

In *Chapter 6, Commanding an Army of Units*, we developed a component that can select the units spawned in the map and give them an action to move as a group to where we click on the screen. With that feature in place, we can now update the bottom of the UI to reflect how many units were selected as well as change the 3D model in the portrait UI.

Besides the selected units at the bottom center of the UI, we will also update the action buttons at the bottom right of the screen to display only the actions that the selected units can perform and use the Command pattern on each button.

But first, let us start with the unit details.

Setting up the selected unit details

The **GameUI** scene, which only has UI elements and is loaded additively on top of any level scene, has most of the features already implemented, such as the mini-map display in the bottom-left corner, the resource counters at the top right, and the menu on the top left side of the screen. Now, we are going to implement the details panel feature at the bottom center of the screen, which will display the 3D model of the first unit selected and the list of units selected.

We can start by removing all the placeholders we added when the UI was created:

1. Open the **GameUI** scene and, in the **Hierarchy** view, expand the **Canvas | Details | Panel | Container** GameObjects, until you find the **Row1** GameObject.

2. Delete the five GameObjects named **SelectedUnit** by right-clicking on each one and selecting the **Delete** menu option (you can also select all of them at once instead of one by one).

3. Still in the **Hierarchy** view, expand **Portrait Renderer | Model** and delete the **FreeLichHP** GameObject.

4. Save the changes to the scene **GameUI**.

After removing the placeholders, we can start adding the real objects that will be used to update the details. For each unit selected, we are going to display one **SelectedUnit** Prefab in the details panel. Since we will need to show and hide the units in the details panel, we can use our `ObjectPoolComponent` script to speed up our development process and make sure we have pre-allocated all the units that can be displayed in the UI, which will be up to 10 units due to the space in the UI. This is very common in RTS games since there is not much space in the UI to display large armies and it is very likely that the player will select groups of units to perform actions instead of selecting all to perform the same thing.

Since we are in a different scene, with different needs, we are not going to use the Prefab named **ObjectPools**. Instead, we are going to create a new GameObject in the scene that will hold the `ObjectPoolComponent` script and the **SelectedUnit** Prefab:

1. In the **Hierarchy** view, click on the + button, located in the top-left corner of the window, and select **Create Empty**. Name the new GameObject `ObjectPoolsUI`.

2. Right-click on the new **ObjectPoolsUI** GameObject and select **Create Empty**. Name the new GameObject **SelectedUnitObjectPool**.

3. Left-click on the **SelectedUnitObjectPool** GameObject and, in the **Inspector** view, click on **Add Component** and add the **ObjectPoolComponent** script.

4. In the **ObjectPoolComponent** parameters, left-click on the circle button on the right side of the **Prefab** property. In the new window, click on the **Assets** tab and search for the **SelectedUnit** Prefab. Double-click to add it and close the window. Alternatively, you can drag and drop the same Prefab located in the **Prefabs | UI** folder.

5. Set **Pool Size** to `10` and do not enable the **Allow Creation** option.

6. Save the **GameUI** scene.

The object pool will ensure that we have up to 10 GameObjects pre-allocated for us, which is the limit that will fit the details panel UI. We are also not enabling the **Allow Creation** option, which will not allow the component to create more instances on demand, so it will be important that we check that the object returned by the object pool is not `null` before attempting to use it.

Now that we have finished adding the object pool for the selected units in **SceneUI**, we can set up the message to update the UI and extend the `GameObject` class to have a new, handy method for helping with changing layers of a GameObject.

Extending the GameObject class

We are going to create a couple of scripts: the `UpdateDetailsMessage` script, which will be the message we are going to send when we need to update the details panel with all the data needed, and the `DetailsUpdate` script, which will be the code to receive the message and update the details panel accordingly.

However, we will add another script first, the `GameObjectExtension` class, which will contain an **extension method**. This is a very handy feature from C# that allows us to create new static methods to extend an existing class, even from code that is not in our project, such as from the Unity Engine itself. The extension method must be `static`, with the first parameter being the class we want to extend, as well as the `this` keyword before the type.

So, create a new folder inside the **Scripts** folder and name it Extension. Then, create a new script with the name GameObjectExtension, double-click to open the script, and add the following content (it is important to note that we don't need a folder named Extension or a script named Extension to create an extension method; we are using Extension in both the folder and script names to keep it organized):

```
using UnityEngine;
namespace Dragoncraft
{
  public static class GameObjectExtension
  {
    public static void SetLayerMaskToAllChildren(
      this GameObject item, string layerName)
    {
      int layer = LayerMask.NameToLayer(layerName);
      item.layer = layer;

      foreach (Transform child in
        item.GetComponentsInChildren<Transform>())
      {
        child.gameObject.layer = layer;
      }
    }
  }
}
```

The GameObjectExtension script has an extension method that extends the GameObject class from the Unity Engine by adding a new method to set LayerMask on all child GameObjects in the Hierarchy. This will be used to ensure that the GameObject and all child GameObjects are in the same layer mask, which is used to render the GameObject in a specific camera, as we have done for the 3D model unit portrait.

The first requirement for an extension method is that the method must be static, and for that reason, we are also creating class as a static class so every method declared inside it must be static as well.

The second requirement for an extension method is that the first parameter of the method must be of the type that we want to extend, preceded by the this keyword. In this case, we want to extend the GameObject class from the Unity Engine, so the first parameter is a GameObject object. All other parameters, after the first one, are optional in extension methods, but in this case, we have one string parameter, layerName, that we are going to use to find the internal Unity ID for the layer we want to set.

In both the Unity Editor and our code, we are going to refer to the layers by their names, which is fine and much better than having an integer that has no meaning. For example, calling it the **Portrait** layer gives us an idea of what it is used for, while `layer 6` has no real meaning to us. However, in a GameObject, we can only set the layer name as an integer value, so we first need to find and convert our `layerName` string into the corresponding integer value using the `NameToLayer` method from the `LayerMask` class. Then, we can set the layer name of the item to the integer value.

> **Note**
>
> A **layer** is a configuration that we can add to any GameObject to separate it in the scene and can be used in the UI and script. For example, we can use layers to configure what a GameObject can or cannot collide with, making it easy to control collisions without having to write code for this case.

Once we have the layer value, we can loop through all the children using the `GetComponentsInChildren` method, which will return a list of GameObjects of the type we want – in this case, it is the `Transform` type. Using `foreach`, we can access each item of the list of `Transform` returned by the `GetComponentsInChildren` method, using the `child` variable. We are getting the list of `Transform` objects because we cannot get a `GameObject` from a `GameObject`; instead, we need to access the `gameObject` variable, which is in the `child` variable.

The last bit of this method is setting the layer, using the integer value, to all the children objects.

At this point, we also need one more script with the message type and parameters that we are going to listen to. So, in the **Scripts** | **MessageQueue** | **Messages** | **UI** folder, create a new script named `UpdateDetailsMessage` and add the following content to it:

```
using System.Collections.Generic;
using UnityEngine;
namespace Dragoncraft
{
    public class UpdateDetailsMessage : IMessage
    {
        public List<UnitComponent> Units;
        public GameObject Model;
    }
}
```

We are going to use our message queue component, which we developed in *Chapter 5*, *Spawning an Army of Units*, so we need a message type that will reflect what we want to do – and what we want to do is update the details panel. The message of the `UpdateDetailsMessage` type will carry two parameters – the `Units` list, which is a list of the selected units, so we can update the details panel, and the main 3D unit `Model` that we are going to display in the portrait UI.

The next script is the one we were waiting for – the one to update the details panel. We will do this by using the two scripts just defined. So, create a new script in the **Scripts | UI** folder and name it DetailsUpdater. This script is a little longer, but we will go through it bit by bit, starting with the properties that we are going to expose in the **Inspector** view to set up the script:

```
using System.Collections.Generic;
using UnityEngine;
namespace Dragoncraft
{
    public class DetailsUpdater : MonoBehaviour
    {
        [SerializeField] private ObjectPoolComponent _objectPool;
        [SerializeField] private int _maxObjectsPerRow = 5;
        [SerializeField] private GameObject _row1;
        [SerializeField] private GameObject _row2;
        [SerializeField] private GameObject _portraitModel;
    }
}
```

Here, we have all the properties needed to update the details panel:

- objectPool is a reference to the SelectedUnitObjectPool GameObject that we created earlier in this chapter.

- This time, we are not going to use a spawner class; instead, we want to populate the details panel with the number of units selected, which will first fill row1 and then row2.

- The maxObjectsPerRow variable monitors when a row is full and checks whether we need to move to the next one or stop. The default value is 5 because it is the max size that will fit the row, but we use it as an exposed property in the **Inspector** view so we can easily tweak the value later if needed.

These four properties, objectPool, row1, row2, and maxObjectsPerRow, will be used for the selected units, but the last one, portraitModel, holds a reference to the **Model** GameObject in the **Hierarchy** view, which will receive the 3D model to be displayed as a portrait. As soon as we finish this script and add it to the **Details** GameObject, we are going to set it up with the proper values and references.

Updating the selected units

We are going to continue working on the DetailsUpdater class, now moving on to the logic that will update the details panel by adding the selected units in the UI.

The message queue component will be the trigger to update the details panel – for that reason, we are going to listen out for when the UpdateDetailsMessage message type is received. We need to add the listener to the OnEnable method and remove the listener in the OnDisable method, as shown here:

```
private void OnEnable()
{
  MessageQueueManager.Instance.
    AddListener<UpdateDetailsMessage>(OnDetailsUpdated);
}

private void OnDisable()
{
  MessageQueueManager.Instance.
    RemoveListener<UpdateDetailsMessage>(
      OnDetailsUpdated);
}

private void OnDetailsUpdated(UpdateDetailsMessage message)
{
  CleanUpRows();
  UpdateRows(message.Units);
  UpdateModel(message.Model);
}
```

When the message is received, the OnDetailsUpdated method will be called so the script knows that the details panel must be updated based on the new units and models received in the message parameters. The details panel will be updated a lot during the gameplay, and we need to make sure that the previous units were removed from the UI before setting up the new ones that we received in the message parameter. The first method, CleanUpRows, will take care of that, followed by the UpdateRows method, which will update and display the newly selected Units. The last method, UpdateModel, will change the Model GameObject in the portrait.

These three methods also have an extra attribution: if no unit is selected, they will use this information to clean up the details panel and the portrait UI. This means that the UpdateDetailsMessage message must be sent even if no units are selected. Let us define the first of these methods, doing the cleanup twice since we have two rows:

```
private void CleanUpRows()
{
  RemoveObjectsFromRow(_row1);
  RemoveObjectsFromRow(_row2);
}
```

```csharp
private void RemoveObjectsFromRow(GameObject row)
{
    for (int i = 0; i < row.transform.childCount; i++)
    {
        row.transform.GetChild(i).gameObject.SetActive(false);
    }
}
```

Since we have two identical rows, and we need to clean up both, we can create the generic RemoveObjectsFromRow method, which receives a GameObject and loops through all the children GameObjects using childCount to get the number of children, and then we hide each GameObject using the SetActive method.

We are using the object pool component to pre-allocate the selected units in the details panel – that is why we need to disable all the children GameObjects so they will be hidden and returned to the object pool for the next time an object is requested.

After cleaning up both rows, we are ready to update them with newly selected units if there are any. The next method, UpdateRows, takes the list of units and loops through each element (note that if the list is empty, with size 0, the loop is ignored; in this case, the details panel will be cleaned up, but no new units will be added):

```csharp
private void UpdateRows(List<UnitComponent> units)
{
    for (int i = 0; i < units.Count; i++)
    {
        if (i < _maxObjectsPerRow)
        {
            AddObjectToRow(_row1, _objectPool.GetObject());
        }
        else if (i < _maxObjectsPerRow * 2)
        {
            AddObjectToRow(_row2, _objectPool.GetObject());
        }
        else
        {
            Debug.Log(
                $"More than {_maxObjectsPerRow*2} units selected");
            break;
        }
    }
}

private void AddObjectToRow(GameObject row,
```

```
   GameObject selectedUnit)
{
   if (selectedUnit != null)
   {
     selectedUnit.transform.SetParent(row.transform, false);
   }
}
```

Updating the rows is quite simple; we only need to request a new GameObject from the object pool and add it to one of the rows. However, we need to consider the limit per row, which is defined in the class variable called maxObjectsPerRow. The first row gets new GameObjects until the loop counter reaches the value of maxObjectsPerRow. Then, if the loop continues, we check whether the counter is less than maxObjectsPerRow multiplied by the number 2. Once both rows are full, with the limit reached on each of them, we can print a debug message to let us know in the Console view and use the break keyword to stop the loop and exit the method.

The AddObjectToRow method is used to set the new GameObject parent to be the desired row. We need to check that selectedUnit is not null because the object pool for this object does not have the **Allow Creation** property enabled, so it is better to check whether the object is valid before using it. Since we are doing the same thing for both rows, it makes sense that this snippet of code is in a different method so it can be reused.

Updating the portrait model

The code to display the selected units in the details panel is all done now, and we can move to the last bit of this class, which is the method to update the 3D model that will represent the first unit selected of the group. In *Chapter 4, Creating the User Interface and HUD*, we manually added a 3D mode to the portrait frame in the details panel to represent a selected unit. Now, we are going to make it change dynamically, based on the GameObject we received in the message.

The following UpdateModel method is responsible for cleaning up the UI by removing the previous 3D model set in the portraitModel property. When the player clicks anywhere on the map without selecting a unit, we also need to remove the previous 3D model from the portraitModel property; however, if there is no new 3D model, the model parameter will be null, and we can use the return keyword to exit the method and skip the last lines:

```
private void UpdateModel(GameObject model)
{
   for (int i = 0; i < _portraitModel.transform.childCount; i++)
   {
     Destroy(
       _portraitModel.transform.GetChild(i).gameObject);
   }
```

```
    if (model == null)
    {
        return;
    }

    GameObject newPortrait = Instantiate(model);
    ResetTransform(newPortrait);
    RemoveUnitComponent(newPortrait);
    newPortrait.SetLayerMaskToAllChildren("Portrait");
}
```

The 3D model in the portraitModel property is not loaded using the object pool, so in this case, we can use the MonoBehaviour method Destroy to remove the old 3D model from the portraitModel property. Although we added only one 3D model at a time in the portraitModel property, it is a good practice to loop through all the children GameObjects and remove them one by one to make sure no 3D model is left in the portraitModel property when cleaning it up.

After cleaning up the previous model from the UI, we need to check whether there is a new model to instantiate in the UI, which will be the case them the model parameter is not null. The last four lines of the method will adjust the GameObject to be used as a portrait model. We start with the Instantiate method, which will create a copy of the GameObject. This is required, otherwise we would manipulate the original GameObject, but instead we need to create a copy so we can freely change it here.

Once we have a fresh copy of the GameObject, we need to adjust the position and rotation of the new object, so it will be placed in the center of the portrait frame and facing the camera. We also remove a few components from the GameObject that are not required since we are using this GameObject as a representation, with no intention of updating it after displaying it in the portrait frame. The last line is calling the SetLayerMaskToAllChildren extension method that we created earlier in this chapter, which will change the GameObject layer so it will be rendered only by the portrait camera.

With the new GameObject created, we can adjust the transform and remove the components that we do not need. The last couple of methods in this class, defined here, will access and modify the properties of this new portrait model, getting it ready to be displayed in the correct position:

```
private void ResetTransform(GameObject portrait)
{
    portrait.transform.position = Vector3.zero;
    portrait.transform.rotation = Quaternion.identity;
    portrait.transform.SetParent(
        _portraitModel.transform, false);
}

private void RemoveUnitComponent(GameObject portrait)
```

```
{
    UnitComponent unit =
        portrait.GetComponent<UnitComponent>();
    unit.Selected(false);
    Destroy(unit);

    Rigidbody rigidbody = portrait.GetComponent<Rigidbody>();
    Destroy(rigidbody);

    BoxCollider boxCollider =
        portrait.GetComponent<BoxCollider>();
    Destroy(boxCollider);
}
```

In the first method, ResetTransform, we are setting position to the zero value, which has the value of (0, 0, 0). We are also setting rotation to the default value of Quaternion, using the identity value, which has the value of (0, 0, 0, 1).

After adjusting the position and rotation, we need to set the portraitModel GameObject as a parent of the portrait GameObject, using the SetParent method, so the 3D model will appear in the frame in the details panel.

At this point, the script is ready to be used, but we need to modify the **UnitSelectorComponent** script to send the messages that will update the details panel.

However, before doing that, let's prepare the unit actions in the UI as well, because those UI elements will also be affected by the selected units and will also require a message from the **UnitSelectorComponent** script.

Setting up the selected unit actions

The actions panel, which is the bottom-right element of the UI, contains up to six buttons that represent the actions the selected units might perform. In *Chapter 4, Creating the User Interface and HUD*, we added six buttons in the UI, but now we are going to display only the actions the units can execute, hiding the other buttons. Before updating the UI, we first need to define what these actions are and add them to the unit configuration.

Although the actions are enums, as we are going to see in the following code, the selected units can have and perform more than one action, so it is not possible to just have a simple variable with one action. We could declare one variable for each action or even have a list of actions but, instead, we are going to use something more elegant and with better performance, called **bitwise**.

Bitwise operation for the actions

Bitwise is an operation that manipulates bits that make up a number. We are going to use the number 2 to the power of X, where X can be any number greater than 0 and in a sequence. For example, we can have the calculations $2^0=1$, $2^1=2$, and $2^2=4$. When we do not have a calculation that results in 3, the bitwise operation will allow us to consider that 3 corresponds to 1+2, which is the result of 2^0 and 2^1. To give a practical example, if 1 is blue and 2 is red, the value 3 is like a group that has both blue and red. This is a very fast operation for the processor, and we will take advantage of that by using it to define our actions.

Now we are going to create a script that will have all the possible actions a unit can perform and the corresponding value so we can use the bitwise operation to group more than one action. Create a new script in the **Scripts | Configuration** folder, name it `ActionType`, and replace the content with the following code:

```
using System;
namespace Dragoncraft
{
    [Flags]
    [Serializable]
    public enum ActionType
    {
        None = 0,
        Attack = 1,
        Defense = 2,
        Move = 4,
        Collect = 8,
        Build = 16,
        Upgrade = 32
    }
}
```

This script is quite simple, just an `enum` definition, but there are three details that are required to make the bitwise operation work properly:

- First, the `Flags` attribute will let our code know that a variable of this enum type can hold multiple values using bitwise operations.

- Second, the `Serializable` attribute allows these values to be serialized in the ScriptableObject or GameObjects.

- Third, we have the `enum` values, which follow the power of 2 pattern instead of having a sequential value. This is how we will understand that a value of 3 means that both `Attack` and `Defense` are included in the value.

Now that we have `ActionType` defined, we can start using it to define the actions that a unit can perform. Open the existing script located at **Scripts** | **Unit** | `UnitData.cs` and add the following property as the last public variable declared in the `UnitData` class:

```
public ActionType Actions;
```

Save the script and search for the ScriptableObject called **BasicWarrior** in the **Project** view. When you left-click to select it, in the **Inspector** view, you will see a new property called **Actions**. Because of the `Flag` attribute we added to the `ActionType` enum, here we can select multiple values, as we can see in the following screenshot:

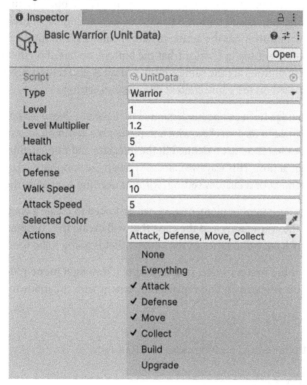

Figure 7.1: The actions for BasicWarrior

BasicWarrior will have the **Attack**, **Defense**, **Move**, and **Collect** actions. The value **Everything** was not defined in our enum class, but Unity adds that one for us to select all values that are not 0. Plus, the value **None** was added to the enum class as a value of 0 – because of that, if we select this option, all the others will be unselected (this is also done by Unity automatically).

Now we know all the actions the selected unit can perform; the next step is to define what happens on each action and implement each behavior. To define each action behavior, we will be introduced to a new programming pattern called **Command**.

The Command pattern

Command is another great design pattern that we can use on any game project if we can identify a feature or functionality that can benefit from it. This pattern is basically an interchangeable object that implements a common interface, making it easier to swap implementations without changing the logic.

One great example of the Command pattern's usage in games is the buttons on a joystick: imagine that button A has the jump action and button B has the crouch action. Now imagine that, when the player is in a game menu, button A has the action to accept changes and button B to close the menu. We could also have a custom setting to change what the buttons do based on the player's preference. In all these scenarios, both buttons are the same, always button A and button B, but the action linked to each button changes based on the game context or a custom setting.

This is where the Command pattern can be implemented to make our lives easier. Each action a button might perform can be defined in its own class, which implements a common interface. In this example, we could have a method called `Execute` defined in the interface and implemented on each class that corresponds to a different action. This way, we can easily swap the variable declaration to something else depending on the context or a custom setting, without needing to change the button logic.

Before starting our own Command implementation, we first need to create a couple of messages that are going to be used by the commands and the class that will create the commands. Both messages are quite similar, but they have a difference in the context that is going to be used.

Create the first script, for the first message, in the **Scripts** | **MessageQueue** | **Messages** | **UI** folder. Name it `UpdateActionsMessage` and replace the content with the following:

```
namespace Dragoncraft
{
    public class UpdateActionsMessage : IMessage
    {
        public ActionType Actions;
    }
}
```

This message will be sent when units are selected, so the UI updates accordingly. This is why the only property in this class is called `Actions`, in plural form, because all the actions defined in the **UnitData** ScriptableObject will be sent in this message. This property could have, for example, both Attack and Defense actions.

Now create the second script, for the second message, in the **Scripts | MessageQueue | Messages | Unit** folder, and name it `ActionCommandMessage`. Replace the content with the following code:

```
namespace Dragoncraft
{
  public class ActionCommandMessage : IMessage
  {
    public ActionType Action;
  }
}
```

While the first message had the context of updating the UI based on the actions the selected units have, `ActionCommandMessage` will be sent when a button is clicked by the player. Therefore, we have the `Action` property, now in singular form, because it will only have a single value. For example, this property can have the value of `Move`, which will indicate the action of movement for the selected units.

With these two message scripts created, we can move on to the command itself by first creating the interface and then four classes that will implement the interface and send a message of the type `ActionCommandMessage` when the command is executed.

First, create a new folder called `Command` inside **Scripts**. Then, add a new script and name it `ICommand`. This interface class has only one method defined:

```
namespace Dragoncraft
{
  public interface ICommand
  {
    void Execute();
  }
}
```

The `Execute` method defined in the `ICommand` interface will be implemented on the following scripts.

Now add a new script to the same folder and name it `AttackCommand`. Replace the content with the following code:

```
namespace Dragoncraft
{
  public class AttackCommand : ICommand
  {
    public void Execute()
    {
      MessageQueueManager.Instance.SendMessage(
        new ActionCommandMessage { Action =
```

```
            ActionType.Attack });
    }
  }
}
```

The `AttackCommand` class implements the `Execute` method from the `ICommand` interface. When called, the method will use the message queue pattern to send a message of the `ActionCommandMessage` type with the `Action` property set to `Attack`. It is quite a simple script that basically broadcasts `ActionType` when the button that has the command assigned is clicked.

The previous script takes care of the `Attack` command, but we also need to create the other three scripts responsible for the `Collect`, `Defense`, and `Move` actions. To do this, take the following steps:

1. Add a new script in the same folder and name it `CollectCommand`.

2. Copy the content from the **AttackCommand** script into the new **CollectCommand** script.

3. Rename the `AttackCommand` class to `CollectCommand` and change `ActionType.Attack` to `ActionType.Collect` in the new script.

4. Add a new script to the same folder and name it `DefenseCommand`.

5. Copy the content from the `AttackCommand` script into the new `DefenseCommand` script.

6. Rename the `AttackCommand` class to `DefenseCommand` and change `ActionType.Attack` to `ActionType.Defense` in the new script.

7. Add a new script in the same folder and name it `MoveCommand`.

8. Copy the content from the `AttackCommand` script into the new `MoveCommand` script.

9. Rename the `AttackCommand` class to `MoveCommand` and change `ActionType.Attack` to `ActionType.Move` in the new script.

The four scripts we just created are very similar, but for the Command pattern, we need each of the actions to be in separate classes. And, since they are different classes, they have different messages to be sent when the `Execute` method is called.

Now that we have all the messages and commands that we need created, we can start to use them in the UI.

Dynamic buttons and actions

We are going to add a couple more scripts to help use the commands we created for the button clicks and to dynamically set up the buttons in the actions panel. The actions panel UI is already using the **DefaultButton** Prefab, but now we will attach a new script to call a method when the button is clicked, as well as setting up the button text and action based on the command assigned to it.

In the **Scripts | UI** folder, add a new script and name it `ButtonComponent`. Then, replace the content with the following code:

```
using UnityEngine;
namespace Dragoncraft
{
    public class ButtonComponent : MonoBehaviour
    {
        public ActionType Action;
        private ICommand _command;
    }
}
```

The `ButtonComponent` class has two properties: one is `ActionType`, which will be set by the script responsible for setting up the actions panel, and the other is `command`, which will be executed by the button click event. Note that the variable is of the type `ICommand`, which means that it can hold the value of any class that implements the `ICommand` interface. This is very useful when creating a generic variable that can have different values depending on the classes that implement the interface.

Besides the command property, we also have the `Action` property, which will be set up before the button is enabled. That is why we can use the `switch` expression on that property in the `Start` method defined next, which is called by `MonoBehaviour` internally when the `GameObject` is enabled. Add the following method after the property definitions:

```
private void Start()
{
    switch (Action)
    {
        case ActionType.Attack:
            _command = new AttackCommand();
            break;
        case ActionType.Defense:
            _command = new DefenseCommand();
            break;
        case ActionType.Move:
            _command = new MoveCommand();
            break;
        case ActionType.Collect:
            _command = new CollectCommand();
            break;
        case ActionType.Build:
        case ActionType.Upgrade:
        case ActionType.None:
        default:
```

```
        break;
    }
}
```

Depending on `ActionType`, the command property will be a different object but still `ICommand`. We are only checking the types of `Attack`, `Defense`, `Move`, and `Collect` because we've created command scripts for these actions. All the other actions and the `default` keyword will just break the `switch` expression, leaving the command property as `null`, since it is not defined.

The last bit we are going to add to this class is the `OnClick` method, which will call the `Execute` method from the command. Add it after the `Start` method:

```
public void OnClick()
{
    _command?.Execute();
}
```

The `OnClick` method has only one line because it does not matter what the command attached to the button is: it will call the `Execute` method on any command.

The question mark after the `command` variable is a very cool feature from the C# language that works like an `if` conditional, which checks whether the object is `null` before executing the method after the dot. In this case, we do need to add it because, as we can see in the `Start` method, the not-implemented commands will keep the command property as `null` and would result in a no reference exception when calling the method from the `null` object.

This is the whole Command pattern implementation, which is quite simple and useful when we have a use case that could easily use it. Now, we can update the **DefaultButton** Prefab with a few more changes to attach the script and make it work properly:

1. In the **Project** view, navigate to the **Prefabs | UI** folder and double-click to open the **DefaultButton** Prefab in edit mode.

2. In the **Inspector** view, click on **Add Component** and search for **ButtonComponent**, then add it to the GameObject.

3. In the **Button** component that was already there, in the **On Click ()** box, click on the + button to add an item to the list.

4. Drag the **DefaultButton** GameObject from the **Hierarchy** view and drop it into the item that was added to the list, which has the value of **None (Object)**.

5. After dropping the GameObject there, the dropdown that has the value **No Function** will be enabled. Select the **ButtonComponent | OnClick** method on the menu.

6. Change **Selected Color** to red by clicking on the color field and dragging the cursor to the red color or setting the **Hexadecimal** value of **FF0000**.

Now, the **DefaultButton** Prefab is set up and ready to be used by our next script. Since we edited the Prefab, all six copies of it that are in **SceneUI** were also modified due to being linked to the Prefab, and we do not need to update each button individually. The Prefab for **DefaultButton,** with the default settings, is shown in the following screenshot. We can see that **ButtonComponent** is added, and **Selected Color** has been changed to red, so the button will have a red highlight when clicked:

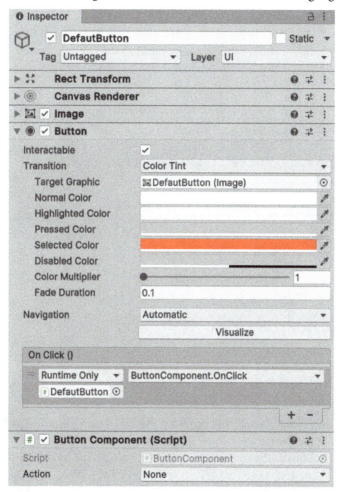

Figure 7.2: The DefaultButton Prefab after attaching the ButtonComponent script

We can leave the **Action** property as **None** because it will be set by the next script, which is the class responsible for configuring all the buttons based on the selected units' actions.

The new `ButtonManager` class, which we are going to create now, will be responsible for listening to `UpdateActionsMessage` and displaying or hiding the buttons based on the actions received, as well as updating `ActionType`, which will define the command used and the button text.

In the **Scripts | UI** folder, create a new script and name it `ButtonManager`. Open the script and replace its content with the following code:

```
using System;
using System.Collections.Generic;
using TMPro;
using UnityEngine;
namespace Dragoncraft
{
  public class ButtonManager : MonoBehaviour
  {
    [SerializeField] private List<GameObject> _buttons;
    private void Awake()
    {
      DisableAllButtons();
    }

    private void DisableAllButtons()
    {
      foreach (GameObject button in _buttons)
      {
        button.SetActive(false);
      }
    }
  }
}
```

The `ButtonManager` class has only one property, the list of buttons, which will be configured with a list of buttons in the GameObject that is going to have this script attached. This list will have a reference to the six buttons in the actions panel, and they need to be hidden by deactivating each button as soon as the script is initialized in the Awake method. The `DisableAllButtons` method executed inside the Awake method will loop through each button and call the `SetActive` method on each one to deactivate and hide the button.

This is done because we will enable only the buttons that are going to be used, which are based on the actions that the selected units have. The actions will be received in the UpdateActionsMessage message, which will be sent when the units are selected. In the following methods, we are going to add and remove the listeners for the UpdateActionsMessage message when the GameObject is enabled or disabled, respectively:

```
private void OnEnable()
{
    MessageQueueManager.Instance.
      AddListener<UpdateActionsMessage>(OnActionsUpdated);
}

private void OnDisable()
{
    MessageQueueManager.Instance.
      RemoveListener<UpdateActionsMessage>(
        OnActionsUpdated);
}
```

As we did in other classes that use the message queue, we need to add and remove the listeners for the message type we want on the OnEnable and OnDisable methods, respectively. When the message is sent by MessageQueueManager, the OnActionsUpdated method will be called. We do this like so:

```
private void OnActionsUpdated(UpdateActionsMessage message)
{
    DisableAllButtons();

    int counter = 0;
    foreach (ActionType action in
      Enum.GetValues(typeof(ActionType)))
    {
        if (action == ActionType.None)
        {
            continue;
        }
        if (message.Actions.HasFlag(action))
        {
            SetButtonType(action, _buttons[counter]);
            counter++;
        }
    }
}
```

When the message is received, we also hide all buttons to configure and display only the ones we want. This works when no unit is selected, meaning that no button will be visible in the actions panel UI.

In the previous code, we are looping through each action defined in the ActionType enum and checking whether the value of the Actions property included in the message has the action using the HasFlag method, and then enabling the corresponding button. The loop will compare each enum value with the values in the Actions flag, ignoring the None value.

If we find one of the ActionType enums in Actions, then we configure the next available button. This is controlled by the counter variable, so each time a flag is found, we set up the button from the list in the position of the counter value, and then increase the counter to get the next button from the list when another flag is found. The OnActionsUpdated method just defined will enable the buttons for each action in the Actions property, and to enable the button, we are going to add a new method, SetButtonType.

The SetButtonType method, defined here, will configure the current button based on the ActionType enum found:

```
private void SetButtonType(ActionType action, GameObject button)
{
    ButtonComponent component =
        button.GetComponent<ButtonComponent>();
    component.Action = action;

    TMP_Text text =
        button.GetComponentInChildren<TMP_Text>();
    text.text = action.ToString();

    button.SetActive(true);
}
```

This method configures the button by getting the ButtonComponent component and setting the value of the Action property. Then, after locating the text of the TextMeshPro button, which is in the child GameObject of the button, we set its value to be the current action name. Finally, by calling the SetActive method, we enable the GameObject that will create the correct command based on the action.

Now that we have the Command pattern implemented and the scripts to make it work properly, we can configure the actions panel in the UI to be dynamically set up using the **ButtonManager** script. Let us open the **GameUI** scene and add the script:

1. In the **Hierarchy** view, expand the **Canvas, Actions, Container, Row1, Row2,** and **Row3** GameObjects.

2. Left-click on the **Container** GameObject and, in the **Inspector** view, click on **Add Component**. There, select **ButtonManager**.

3. Then, in the **Buttons** list, change the value from 0 to 6. This will create the elements for the button references.

4. Drag and drop each button from the three rows to each corresponding element:

    - **AttackButton** is **Element 0**

    - **DefenseButton** is **Element 1**

    - **MoveButton** is **Element 2**

    - **CollectButton** is **Element 3**

    - **BuildButton** is **Element 4**

    - **UpgradeButton** is **Element 5**

5. Save the scene.

Although we wrote quite a lot of code to update the actions panel, the configuration in the UI elements is simple; as we have just done, we only needed to drag and drop the buttons to the script. The following figure shows how the **ButtonManager** script is set up:

Figure 7.3: The ButtonManager script configured with the six buttons

Now we have everything we need to update both the details panel and the actions panel based on the selected units. However, if we run the game after all the changes so far in this chapter, we will not see any of it working. This is because we still need to send the messages that will update the buttons in the UI, which is what we are going to do next, as well as setting up the unit animation based on each action we command to the units.

Attacking and defending with units

The UI is almost finished, and all elements now have scripts attached to them that will update them to reflect what is happening during the gameplay. Now, we need to adjust the scripts responsible for selecting and moving the units and add the new messages and actions. Each action also has a different animation – this is why we must first define what the animation states are that we are going to play in each action or context.

Fortunately, both Warrior and Mage have the same animation states with slight variations on the animation state names. To see the animation states, open the **Animator** view using the **Window | Animation | Animator** menu option and, while this window is open, search in the **Project** view for the **FootmanHP** Prefab and right-click on the asset to select it. All the animation states will be displayed in the **Animator** view, as we can see in the following figure:

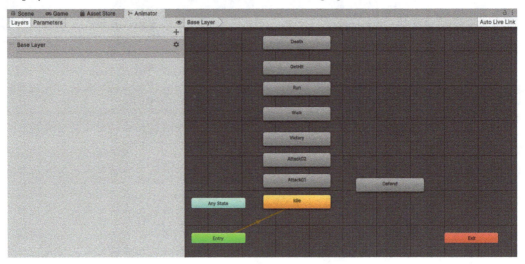

Figure 7.4: The animation states in the Warrior and Mage Prefabs

We are going to use most of the animation states that we can see in the preceding figure, including the two different attacks that, when randomly chosen, will cause our selected units to have different animations while attacking the same target.

To make it easier to check and play the animation states, we are going to add them to the **UnitData** Scriptable Object. This way, each unit will have the same animation states, even if they have different names in the animator. Each animation will be matched to an enum value, so in the code that will play the animation, we do not need to know that each unit has a different name.

To do this, create a new script inside the **Scripts | Unit** folder, name it UnitAnimationState, and replace the content with the following code:

```
namespace Dragoncraft
{
  public enum UnitAnimationState
  {
    Attack01,
    Attack02,
    Defense,
    Move,
    Idle,
    Collect,
    Death
  }
}
```

In this enum, we mapped each animation state that we are going to implement for the units, so it will be easier to play them by the type of the animation, including the two variations for the attack. Each animation state is a string, which is the name we can see on each node in *Figure 7.4*, and that is the information we need to play the animation itself from the code.

Even though we will use the UnitAnimationState enum to play one specific animation, we are going to need the string name of the animation state. Each animation state will be added to the **UnitData** Scriptable Object and a new method will translate the UnitAnimationState enum into the corresponding animation state name, returning the correct string for each animation state. So, open the **UnitData** script in your IDE and add the following properties to it:

```
public string AnimationStateAttack01;
public string AnimationStateAttack02;
public string AnimationStateDefense;
public string AnimationStateMove;
public string AnimationStateIdle;
public string AnimationStateCollect;
public string AnimationStateDeath;
public Color OriginalColor;
public float AttackRange;
```

In the code block, we have one property for each animation state – seven in total. The last two properties, `OriginalColor` and `AttackRange`, are not animation states. These two new properties have the original color of the material and the range that the unit will start to attack from a distance, respectively.

Once we have the properties for each animation state name, we can add the method responsible for returning the correct string based on the enum:

```
public string GetAnimationState(UnitAnimationState animationState)
{
  switch (animationState)
  {
    case UnitAnimationState.Attack01:
      return AnimationStateAttack01;
    case UnitAnimationState.Attack02:
      return AnimationStateAttack02;
    case UnitAnimationState.Defense:
      return AnimationStateDefense;
    case UnitAnimationState.Move:
      return AnimationStateMove;
    case UnitAnimationState.Idle:
      return AnimationStateIdle;
    case UnitAnimationState.Collect:
      return AnimationStateCollect;
    case UnitAnimationState.Death:
      return AnimationStateDeath;
    default:
      return AnimationStateIdle;
  }
}
```

The `GetAnimationState` method is very simple, mapping each `enum` to a `string` value. We are adding this method here instead of directly accessing the property of each animation state because the enum will make it easier to handle conditionals and validations.

Now that the Scriptable Object has the new properties, we need to add the corresponding values for the **BasicWarrior**. In the **Project** view, search for the **BasicWarrior** asset and select it to show the properties in the **Inspector** view. Set the value of each animation state to the values in the following screenshot:

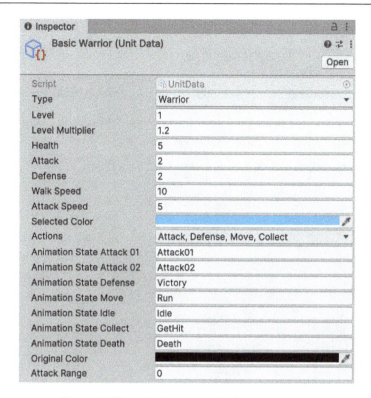

Figure 7.5: The animation states for the Basic Warrior

The screenshot shows all the new properties with the correct values for **BasicWarrior** to work, including each animation state.

After updating the Scriptable Object values, we have everything we need to start playing different animations based on the player's actions. However, before implementing each animation state in our code, we first need to make a few tweaks to the project.

Setting up layers and collisions

There are a few adjustments to the project that need to be made now before jumping into the unit animations. Earlier in this chapter, we changed the layer mask of the portrait GameObject to a custom layer so the portrait camera could render the 3D model in the UI portrait frame. Now we are going to add another custom layer, but this time it will be used for object collision:

1. Click on the **Edit | Project Settings** menu option to open a new window with the settings.

2. In the **Project Settings** window, select **Tags and Layers** from the left menu.

3. On the right side, expand the **Layers** list and add a new **Unit** value to **User Layer 7**, which should be empty since the previous value, **Portrait**, was added to the field above it.

4. Now, from the left menu, click on **Physics**.

5. On the right side, scroll down and expand **Layer Collision Matrix**, and uncheck all the rows and columns for the **Portrait** layer.

6. Uncheck the box where the **Unit** layer intersects on both rows and columns.

The following screenshot shows **Layer Collision Matrix** updated, which you can use as a reference for the matrix values:

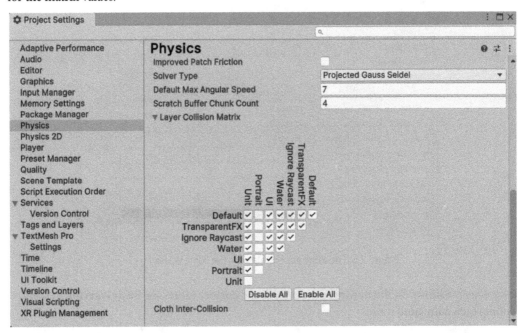

Figure 7.6: Layer Collision Matrix with the values updated

This is a very powerful setting in Unity because, after adding a GameObject to one specific layer, we can easily configure here what collisions must be ignored and Unity will take care of that for us, without needing any code to ignore collisions. It is good practice to use **Layer Collision Matrix** instead of trying to do the same with lots of conditionals in our code.

Since we are going to configure our units to be in the **Unit** layer, we need to tell the main camera to render this layer as well; otherwise, our units will not be visible on the map:

1. Open the **Playground** scene and, in the **Hierarchy** view, select **Main Camera**.

2. In the **Inspector** view, in the **Camera** component, click on the **Occlusion Culling** dropdown and select the **Unit** layer.

3. Repeat these two steps on the **Level01** scene and any other scene you may have created.

We do not need to change the **Occlusion Culling** setting on any other camera because the main cameras are the only ones that render the units on the map. Now that we have set up the layer, physics, and cameras, the only thing left is to set the new layer to the unit. We are going to do that in the **BasicWarriorSpawner** script, using the extension method that sets the layer masks to all the children GameObjects. Open the **BasicWarriorSpawner** script and add just the line highlighted in the OnBasicWarriorSpawned method:

```
private void OnBasicWarriorSpawned(BasicWarriorSpawnMessage message)
{
    GameObject warrior = SpawnObject();
    warrior.SetLayerMaskToAllChildren("Unit");
    ...
}
```

Using the SetLayerMaskToAllChildren extension method, we will ensure that every spawned Basic Warrior, even reused from the object pool, will have the correct layer.

Now that we have the layer set up in every place that is required, we can update the script that selects the units to also update the details panel and the actions panel on the UI.

Updating the UI

Earlier in this chapter, we created a couple of messages that both the details panel and the actions panel are listening to. The two messages, UpdateDetailsMessage and UpdateActionsMessage, need to be sent when the player selects the units, so the UI will reflect the units and their available actions. To do this, open the **UnitSelectorComponent** script and add the following code snippet at the beginning of the Update method:

```
private void Update()
{
    if (EventSystem.current != null &&
      EventSystem.current.IsPointerOverGameObject())
    {
      return;
    }
    ...
}
```

The code uses EventSystem to determine whether the mouse cursor is over a GameObject from the UI – if it is true, we exit the Update method since no action is required. When it is not null, EventSystem handles the mouse input in the UI, and it is currently only used in the **GameUI** scene.

EventSystem is unique, like a Singleton, and cannot exist more than once at a time. Because of this characteristic, we can access the current instance anywhere in the code. However, we need to check whether EventSystem is initialized and ready to be used by checking that it is not null before accessing its methods.

We are checking whether the mouse cursor is on top of a UI element to avoid the same event click being sent to both the UI and the level. When the player clicks on a button, we do not want the select units moving to the same position, because this is an interaction in the UI only.

This validation is required because the UI uses a **Graphic Raycaster**, which is different from the Physics Raycaster we are using to find where the player clicked on the map. Since they are different systems, the validation helps to ignore the units when the player clicks on the UI elements.

We also need to update one GameObject in the **GameUI** scene to disable the Raycast in the GameObject used as the feedback image for the selected units; otherwise, it will be ignored as well. Open the **GameUI** scene and, in the **Hierarchy** view, expand the **Canvas** GameObject and select the **UnitSelector** GameObject. In the **Inspector** view, uncheck **Raycast Target** in the **Image** component and save the scene.

Continuing with the changes in the UnitSelectorComponent class, we are going to update the existing SelectUnits method to also send both messages, UpdateDetailsMessage and UpdateActionsMessage, when the units are selected. Here, we only need to add the code that is highlighted:

```
private void SelectUnits(Vector3 startPostion, Vector3 endPosition)
{
    ...
    GameObject model = null;
    ActionType actions = ActionType.None;
    Collider[] colliders =
        Physics.OverlapBox(center, halfExtents);

    foreach (Collider collider in colliders)
    {
        UnitComponent unit =
            collider.GetComponent<UnitComponent>();
        if (unit != null)
        {
            unit.Selected(true);
            _units.Add(unit);
            if (model == null)
            {
                model = collider.gameObject;
                actions = unit.Actions;
            }
        }
    }
```

```
    }
    MessageQueueManager.Instance.SendMessage(
      new UpdateDetailsMessage {
        Units = _units, Model = model });

    MessageQueueManager.Instance.SendMessage(
      new UpdateActionsMessage { Actions = actions });
  }
```

Since we can only add one 3D model in the portrait UI, we are going to consider that model to be the first unit that was selected in case there is more than one. We create the `model` and `actions` variables and, inside the loop, we get the `gameObject` and actions that this unit can perform. We ensure that only the first unit is considered for the portrait by ignoring that code if the model is not `null` anymore.

At the end of the method, we send both messages – `UpdateDetailsMessage` requires the list of units, to display all of them in the details panel, and the `model` variable that is going to be copied and displayed in the portrait frame; `UpdateActionsMessage` only requires the actions that the unit can perform to build the list of buttons, and those are the actions that the selected unit in the portrait UI can execute. However, they will be applied to all the selected units.

The last thing to be done in this script is to update the existing `MoveSelectedUnits` method by adding the following code at the beginning of it:

```
private void MoveSelectedUnits(Vector3 movePosition)
{
  if (_units.Count == 0)
  {
    MessageQueueManager.Instance.SendMessage(
      new UpdateDetailsMessage {
        Units = _units, Model = null });

    MessageQueueManager.Instance.SendMessage(
      new UpdateActionsMessage {
        Actions = ActionType.None });
    return;
  }
  ...
}
```

This code is used to send a message when no unit is selected to clear the details panel, the portrait UI, and the actions panel. If the list of units is empty, we can also use the `return` keyword to skip the rest of the method execution after sending the two messages, `UpdateDetailsMessage` and `UpdateActionsMessage`.

When the list of units is empty, we can safely send both messages, `UpdateDetailsMessage` and `UpdateActionsMessage`, but with the empty list of units, a `null` model, and the `ActionType None`. This will remove the elements from the UI and make sure the UI is ready for the next time units are selected.

The next step is making the actions change the animation state of the selected units and the melee attack that will move the units to where the player clicks on the map, and then change the animation to attack.

Attacking and playing other animations

We need to expand the **UnitComponent** script further by adding the logic to accept the command issued by the action button pressed by the player and changing the animation state to the corresponding one. Although we are adding quite a lot of code to this existing class, most of the new methods are straightforward.

Open the **UnitComponent** script in your IDE and add the following properties at the beginning of the class. It is a good practice to group the variables by their visibility, as we can see in this code snippet, organized by `public` first and then by `private`:

```
public Color OriginalColor;
public float AttackRange;
public ActionType Actions;
...
private bool _shouldAttack;
private float _attackCooldown;
private ActionType _action;
private UnitData _unitData;
private float _minDistance = 0.5f;
```

The `public` variables are the new properties that the `UnitData` class has, and their values will be copied from the `unitData` Scriptable Object inside the `CopyData` method, so they are initialized with the corresponding property values from the unit configuration file. The `private` variables are support variables in this class, and we are going to use each of them in the next code blocks.

The selected units are passive, and they will react based on the action that is commanded by the player. For that reason, we are going to listen to the `ActionCommandMessage` message so we know what the command issued is and what this class needs to do to each selected unit. The following methods can be declared anywhere in the class since they are all new:

```
private void OnEnable()
{
    MessageQueueManager.Instance.
      AddListener<ActionCommandMessage>(
        OnActionCommandReceived);
```

```
}

private void OnDisable()
{
  MessageQueueManager.Instance.
    RemoveListener<ActionCommandMessage>(
      OnActionCommandReceived);
}

private void OnActionCommandReceived(
  ActionCommandMessage message)
{
  _action = message.Action;
  _shouldAttack = false;
}
```

We have learned already that when we need to listen to one specific message, the listener must be added to the message queue in the OnEnable method and removed in the OnDisable method. The purpose of each method is different, but both have the same generic type, which in this case is ActionCommandMessage, and the callback method, which is defined at the end of the code snippet as OnActionCommandReceived. This callback method does a couple of things: it stores the action received in the message, so it can be used in the class, and it changes the value of shouldAttack to false and stops any current attack when a new command is issued by the player. This is not a problem when the new command is also an attack; it will start again.

The existing CopyData method will have more properties to be copied from the UnitData Scriptable Object. Besides copying the values to be updated during the gameplay, we are also copying unitData itself to a private property so we can use it later to access the method that is responsible for retrieving the animation state name from the Scriptable Object. Keep the current content of this method and add the five lines at the end of it:

```
public void CopyData(UnitData unitData)
{
  ...

  OriginalColor = unitData.OriginalColor;
  AttackRange = unitData.AttackRange;
  Actions = unitData.Actions;
  _unitData = unitData;
  EnableMovement(false);
}
```

The new lines added to the existing `CopyData` method are used to set up the `OriginalColor`, `AttackRange`, `Actions`, and `unitData` properties, with the corresponding values from the `unitData` parameter received in this method. In the last line, we are calling the `EnableMovement` method with the value `false`, which indicates that we are not moving, so the animation state will change to idle.

Now we are going to define the new `EnableMovement` method in the next code block, which will play the animation state for the movement or idle actions, as well as updating the value of the `shouldMove` property:

```
private void EnableMovement(bool enabled)
{
    if (enabled)
    {
        _animator.Play(
            _unitData.GetAnimationState(
                UnitAnimationState.Move));
    }
    else
    {
        _animator.Play(
            _unitData.GetAnimationState(
                UnitAnimationState.Idle));
    }
    _shouldMove = enabled;
}
```

When the value of the parameter enabled is `false`, which is what we can see in the last line of the `CopyData` method, the unit movement is disabled and the `Idle` animation state will be played.

`EnableMovement` is a new method that is responsible for changing the value of the `shouldMove` variable, which is already used by the class for the unit movement. This method is also playing the animation by getting the animation state name from `UnitData` using the `UnitAnimationState` enum. Instead of having the name of the animation state hardcoded here, we are reading it from the unit configuration. This is good practice because, in our case, different units will have different animation state names for the same enum type.

The next method is an existing one and we are going to change one part of its code. The highlighted line is new, which resets the original color when the unit is not selected anymore (note that we also removed both lines that had `EnableKeyword` and `DisableKeyword`):

```
public void Selected(bool selected)
{
    ...
    if (selected)
```

```
{
  material.SetColor("_EmissionColor",
    SelectedColor * 0.5f);
}
else
{
  material.SetColor("_EmissionColor", OriginalColor);
}
...
}
```

We no longer need to enable or disable the emission of the material, only EmissionColor. We keep setting SelectedColor with half of the intensity, and when not selected, reset to the OriginalColor value defined in the Scriptable Object.

The following Update method changes the logic and expands to accommodate the different actions that the unit can perform. We can replace the code from the previous version of the Update method with the code in the following snippet instead:

```
private void Update()
{
  switch (_action)
  {
    case ActionType.Attack:
      UpdateAttack();
      break;
    case ActionType.Defense:
      UpdateDefense();
      break;
    case ActionType.Move:
      UpdateMovement();
      break;
    case ActionType.Collect:
      UpdateCollect();
      break;
    case ActionType.Build:
    case ActionType.Upgrade:
    case ActionType.None:
    default:
      EnableMovement(false);
      break;
  }
}
```

The previous `Update` method was made for the unit movement only. It is still valid, and we are going to use the unit movement code, but now it is extracted into a different method named `UpdatePosition`, which can be called from different methods. Since we have only four actions supported by the unit, which are `Attack`, `Defense`, `Move`, and `Collect`, those are the ones that we are going to implement in this class. All the other actions, which are not supported, will disable the unit movement to make sure we are not doing anything for not-implemented behaviors.

The `switch` case in the `Update` method is a simplified **state** pattern that will change the class behavior based on the `enum`, which indicates the state of the unit in this class. We ensure that the correct method is called based on the action type.

All of these four actions have one common characteristic: they all change the animation state to something else after the unit moves to the position that the player clicks on with the mouse. This means that after the player selects the units and clicks on the map, the action selected before the click will play the animation state once the selected units reach the destination. The selected units will walk until the desired position is reached, playing the walking animation, and then changing the animation to attack, defense, collect, or idle if the action was a simple movement.

The first action implemented is attack, as shown in the following code snippet. Since we have two different animation states for attack, `Attack01` and `Attack02`, we defined which one will be played by getting a random number:

```
private void UpdateAttack()
{
    UnitAnimationState attackState =
        (UnityEngine.Random.value < 0.5f) ?
            UnitAnimationState.Attack01 :
                UnitAnimationState.Attack02;

    UpdatePosition(_minDistance + AttackRange, attackState);
}
```

The `UpdateAttack` method uses the Unity API, which implements the `Random` class to get a random value between 0 and 1. If that value is less than `0.5`, we will play `Attack01` once the unit stops moving; otherwise, we will play `Attack02`. Instead of using traditional `if` and `else` statements, we are using a **ternary conditional operator**, which is an inline operation that checks a condition and returns the value after the question mark if `true`; otherwise, it will return the value after the colon.

Once the animation is defined, we call the `UpdatePosition` method, which has two parameters: the first is the minimum distance we consider that the unit reached the end position, plus `AttackRange`, which is 0 for melee attackers and greater than 0 for ranged attackers; the second parameter is `UnitAnimationState`, which should be played once the unit stops moving.

The next three methods that we are going to add to the class are quite similar, with the only difference between them being the animation state that should be played once the unit reaches the end destination set by the player clicking on the map:

```
private void UpdateDefense()
{
  UpdatePosition(_minDistance, UnitAnimationState.Defense);
}

private void UpdateMovement()
{
  UpdatePosition(_minDistance, UnitAnimationState.Idle);
}

private void UpdateCollect()
{
  UpdatePosition(_minDistance, UnitAnimationState.Collect);
}
```

Different from the previous method, UpdateAttack, these three methods have only one possible animation state once the unit stops moving and the minimum distance is used without adding any extra offset to the value.

The UpdatePosition method defined next is basically what we had in the previous version of the Update method that was replaced in this chapter by the new implementation. The main difference here is the usage of the range and state parameters instead of the previous hardcoded values:

```
private void UpdatePosition(float range,
  UnitAnimationState state)
{
  if (!_shouldMove)
  {
    return;
  }

  if (Vector3.Distance(
    transform.position, _movePosition) < range)
  {
    _animator.Play(_unitData.GetAnimationState(state));
    _shouldMove = false;
    _shouldAttack = true;
    return;
```

```
    }

    Vector3 direction =
      (_movePosition - transform.position).normalized;
    transform.position +=
      direction * Time.deltaTime * WalkSpeed;
    }
```

The shouldAttack property was also added here, with the value true, meaning the unit should move, but we are still going to check whether that is possible in the next method. The OnCollisionEnter method is called automatically by the MonoBehaviour class when the collider or Rigidbody components attached to the current GameObject start to touch another GameObject that also has a collider or Rigidbody component:

```
    private void OnCollisionEnter(Collision collision)
    {
      if (!collision.gameObject.CompareTag("Plane"))
      {
        _animator.Play(
          _unitData.GetAnimationState(
            UnitAnimationState.Idle));

        _shouldMove = false;
      }
    }
```

Because of the updates to **Layer Collision Matrix** that we made in the **Physics** settings, two GameObjects within the **Unit** layer will not collide with each other. This means that this OnCollisionEnter method will never be called when two units touch each other's collider or Rigidbody.

However, that is not true for GameObjects such as **Plane**, which is constantly in touch with the unit on top of it, so in this case, we ignore the collision if the other GameObject has the tag named Plane. If we are not touching another GameObject that is a Plane object, we can change the animation state to Idle and the shouldMove value to false so the unit will stay in the place that the collision happened, which could be a wall or a rock, for example, blocking the unit path.

Later, in *Chapter 8, Implementing the Pathfinder*, we are going to see how to make the units avoid collisions with obstacles and move around them when it is possible. But for now, we can keep it simple and make the units stop moving once they collide with something on their way.

We now have another feature implemented, which is the actions affecting the selected units' animation state once they finish their movement. Go ahead and play the **Playground** or **Level01** scene to see the animations for each action. You will need to spawn a few warriors using the debug menu, then select the units and click on one action button before clicking somewhere on the map. The units will move and, once they reach the point you checked, they will play the animation that corresponds to the action button clicked.

We've reached the point in this chapter where we can start testing the changes we've made so far by playing the **Playground** or **Level01** scene. Now, it is possible to select the units and see the same number of units in the details panel, as well as the first selected unit in the portrait frame and the buttons enabled based on what actions the unit can perform.

In the last section of this chapter, we are going to add a new unit type that has a ranged attack, which is the Mage.

Adding the new Mage unit and ranged attack

It is time to start adding more characters to our RTS game. We already have our Warrior character, and now we are going to use everything we have made so far to add a Mage character. We are going to expand the **UnitComponent** script to add the ranged attack as well, which will be a fireball cast by the Mage.

Now that we have built many support systems in our RTS game, we can define a clear path to use when we need to include and configure new units in the project. This is what we need to do:

1. Create a new **UnitData** Scriptable Object and configure all the base attributes and animation states. We can always use another unit as a reference for the value and tweak it.
2. Create a new message that will be sent when we need to spawn a unit.
3. Create a new spawner script and a new object pool, adding the reference to **UnitData** and the Prefab.

That covers the basics of adding a new unit. However, in the case of the Mage, we are also going to implement the ranged attack. Let us first set up the Mage using what we currently have before moving on to the new attack implementation.

Creating the data, object pool, and spawner

The first thing we are going to do is create the data for the Mage configuration. Right-click on the **Data | Unit** folder, select **Create | Dragoncraft | New Unit**, and name it `BasicMage`.

The following shows the values that **BasicMage** should have.

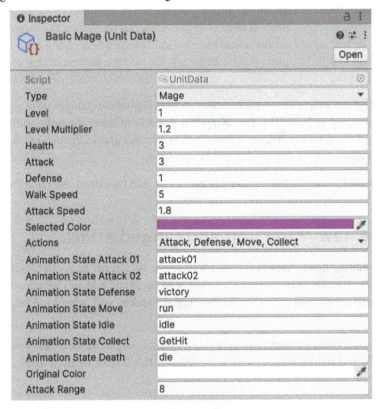

Figure 7.7: The Basic Mage configuration

We already created a type called **Mage**, which is why the value is available in the **Type** property to be selected. If we were adding a new **Type**, we would need to first add it in the **UnitType** script.

The value for **Selected Color**, which is purple, is **RGBA 191, 0, 255, 255**, and for **Original Color**, which is white, is **RGBA 255, 255, 255, 255**.

The next step is creating a new message that will be sent when we want to spawn a new Mage. In the **Scripts | Message Queue | Messages | Unit** folder, create a new script called `BasicMageSpawnMessage` and add the following content:

```
namespace Dragoncraft
{
  public class BasicMageSpawnMessage : IMessage
  {
  }
}
```

This class is very similar to the message we have for the Warrior, just an empty class because the most important part is the class itself. `BasicMageSpawnMessage` will be used by the spawner class that we are going to create next.

In the **Scripts | Spawner** folder, create a new script called `BasicMageSpawner`. This new class, as shown here, extends the `BaseSpawner` class, and it is like the spawner class that we made for the Warrior, with the difference being the message type we are listening to:

```
using UnityEngine;
namespace Dragoncraft
{
  public class BasicMageSpawner : BaseSpawner
  {
    [SerializeField] private UnitData _unitData;
    private void OnEnable()
    {
      MessageQueueManager.Instance.AddListener<
        BasicMageSpawnMessage>(OnBasicMageSpawned);
    }

    private void OnDisable()
    {
      MessageQueueManager.Instance.RemoveListener<
        BasicMageSpawnMessage>(OnBasicMageSpawned);
    }

    private void OnBasicMageSpawned(
      BasicMageSpawnMessage message)
    {
      GameObject mage = SpawnObject();
      mage.SetLayerMaskToAllChildren("Unit");
      UnitComponent unit =
        mage.GetComponent<UnitComponent>();

      if (unit == null)
      {
        unit = mage.AddComponent<UnitComponent>();
      }
      unit.CopyData(_unitData);
    }
  }
}
```

As usual with classes that wait for a message, we add the listener to the `OnEnable` method and remove the listener in the `OnDisable` method. In both methods, we are listing to the `BasicMageSpawnMessage` message and executing the `OnBasicMageSpawned` method when a new message is received. The `OnBasicMageSpawned` method has the same code as the `OnBasicWarriorSpawned` method in the `BasicWarriorSpawner` class. However, the difference here is the message type in the `BasicMageSpawnMessage` parameter.

Now that we have the spawner class created, we can add it to a new object pool in the **ObjectPools** Prefab:

1. In the **Project** view, search for the **ObjectPools** Prefab and double-click on it to open it in edit mode.

2. Once opened in edit mode, in the **Hierarchy** view, right-click on **ObjectPools** and select **Create Empty** to add a new GameObject. Name the new GameObject `BasicMageObjectPool`.

3. Click on the **BasicMageObjectPool** GameObject and, in the **Inspector** view, click on **Add Component**. Search for the **BasicMageSpawner** script.

4. In the **Object Pool Component**, click on the circle button on the right side of the **Prefab** property. In the new window, click on the **Assets** tab and search for **FreeLichHP**. Double-click on the Prefab to add it.

5. Change **Pool Size** to **5** and select the checkbox on the right side of the **Allow Creation** property.

6. In **Basic Mage Spawner**, click on the circle button on the right side of the **Unit Data** property. In the new window, click on the **Assets** tab, search for **BasicMage**, and double-click on the Scriptable Object to add it.

After following the steps, we will have our Prefab **ObjectPools** prepared to handle both the Warrior and Mage spawners. Since it is a Prefab, after editing it, all changes will be automatically applied to the scenes that are using the Prefab.

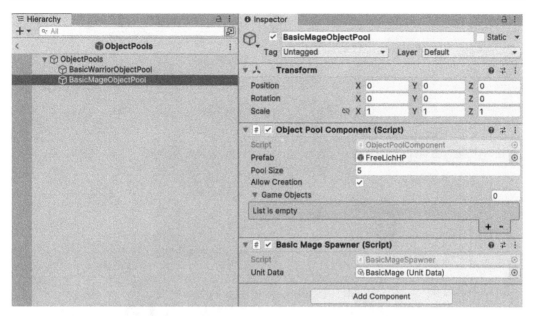

Figure 7.8: The basic Mage object pool after adding and setting up the components

We have done everything to spawn some Mages in the scene. Let us add a debug menu item to make it easier to spawn a Mage unit, as we did for the Warrior as well. Open **Scripts | Editor | Debug | UnitDebugger.cs** and add the following method to it:

```
[MenuItem("Dragoncraft/Debug/Unit/Spawn Mage %h")]
private static void SpawnMage()
{
  MessageQueueManager.Instance.SendMessage(
    new BasicMageSpawnMessage());
}
```

MenuItem will be added in the same place as **Spawn Warrior**. And, as we did with the other debug option, we are adding a keyboard shortcut for %h to execute the method when running the game in the Editor, which is bound to the *Ctrl + H* keys (or *Command + H* on macOS). When the method is executed, the BasicMageSpawnMessage message will be sent.

Now, we can open the **Playground** or **Level01** scene and play it in the Editor, using the debug menu or the shortcut to spawn Mage units. We can also spawn Warrior units and select all of them to execute actions. Note that their movement speed is different because we configured the Mage with a lower movement speed than the Warrior, so when selecting both units, they will reach the desired position at different moments.

Figure 7.9: Both Mage and Warrior units spawned on the map

When selecting a Mage unit, clicking on the **Attack** action, and then clicking on the map, we can see that the Mage will move closer to the position but stop far from it and start to play the animation state for the attack. This happens because the ranged attack is partially implemented, so the Mage knows that it cannot get closer to the position when starting to attack. Now we are going to add a projectile that will be thrown from a distance.

Setting up ranged attacks with fireballs

To attack from a distance, that is, a ranged attack, we need to throw something in the direction that the selected unit, in this case the Mage, is facing. We are going to create a projectile that, once created, will travel in a specific direction and live for a limited time. We can imagine here that the Mage will cast a fireball that, after a few seconds, will disappear.

The beauty of our implementation is that we are going to use two systems that we have in place: the spawner, to create the fireball, and the object pool, to keep a list of objects and avoid creating and deleting the instances by reusing them. Both systems are generic enough to be used for the projectiles. This is also good practice when developing games that need to throw an object repeatedly, such as bullets from a gun in a first-person shooter game, for example.

Create a new folder named `Battle` and add a new script named `ProjectileComponent`. Then, replace the content with the following code. It has very basic logic for a projectile, with a life cycle control that will set up the position before the GameObject is activated, and then deactivate the GameObject once the time to live reaches 0:

```
using UnityEngine;
namespace Dragoncraft
{

    [RequireComponent(typeof(Rigidbody))]
    public class ProjectileComponent : MonoBehaviour
    {

        [SerializeField] private float _timeToLive;
        [SerializeField] private float _speed;
        private float _countdown;

    }

}
```

The already familiar `RequireComponent` attribute is in this class to ensure we will have a valid `Ridigbody` attached to the GameObject. `timeToLive` is the time value, in seconds, that the projectile should be visible after the GameObject is activated, which is controlled by the `countdown` variable. `speed` will control the velocity of the Rigidbody, and both properties are exposed in the GameObject, so we can tweak the values easily in the **Inspector** view.

Next, we have a couple of methods that are part of this class: the `Update` method, from `MonoBehaviour`, which will control the projectile life cycle, and the `Setup` method, which will configure the initial values of the projectile:

```
private void Update()
{

    _countdown -= Time.deltaTime;
    if (_countdown <= 0)
    {

        gameObject.SetActive(false);
    }

}

public void Setup(Vector3 position, Quaternion rotation)
{

    transform.position = position;
    transform.rotation = rotation;
    _countdown = _timeToLive;
```

```
GetComponent<Rigidbody>().velocity =
  transform.rotation * Vector3.forward * _speed;
}
```

In the `Update` method, the `countdown` variable is decremented by the value of `deltaTime` on each frame and, when the `countdown` value reaches 0 or less, the GameObject is deactivated using the `SetActive` method. This will return the GameObject to the object pool once we have that implemented.

The `Setup` method is responsible for setting the initial values of `position` and `rotation`, which will depend on the Mage position when the projectile is created. The `countdown` variable is reset to the `timeToLive` value, so the `Update` method can decrease its value, and the `velocity` property of the `Rigidbody` is calculated considering the rotation, the `forward` vector, and the speed configured. Once the GameObject is activated, it will start to move based on the `velocity` property until the countdown reaches 0.

Now that we've added the script to control the projectile, we can create the Prefab that will have this component attached to it. The first step is to create a material, then attach it to a sphere, add the **ProjectileComponent** script to control the projectile, and save it as a Prefab to be easily used by the object pool:

1. In the **Project** view, right-click on the **Materials** folder, select the **Create | Material** option, and name it `RedFireball`. Click on the color field on the right side of the **Albedo** property and change the color to **RGBA 255, 0, 0, 255**, which is a standard red.

2. Open the **Playground** scene and, in the **Hierarchy** view, right-click anywhere and select the **3D Objects | Sphere** option to add a new GameObject with a sphere model. Name it `Fireball`.

3. Select the **Fireball** GameObject and, in the **Inspector** view, change **Scale** to **0.5, 0.5, 0.5**.

4. In the **Mesh Renderer** component, change **Element 0** in the **Materials** list to the **RedFireball** material. Click on the circle button on the right side of the **Element 0** property. In the new window, click on the **Assets** tab and search for **RedFireball**, and double-click on the material to add it.

5. Click on the **Add Component** button and add a **Rigidbody**. Uncheck the **Use Gravity** property.

6. Click on the **Add Component** button one more time and add the **ProjectileComponent** script that we just created. Change the **Time To Life** property to the value 2 and the **Speed** property to the value `15`.

7. In the **Project** view, create a new folder named `Projectile` inside the **Prefabs** folder.

8. Drag the **Fireball** GameObject from the **Hierarchy** view and, in the **Project** view, drop it into the **Prefabs | Projectile** folder to create a Prefab.

9. Delete the **Fireball** GameObject from the **Hierarchy** view.

We used the **Playground** scene just to create the GameObject sphere and configure it as our projectile. Once it was saved as a Prefab, it was no longer needed to keep the GameObject in the scene, so we deleted it in the last step.

The following screenshot shows the Prefab after all the steps are done:

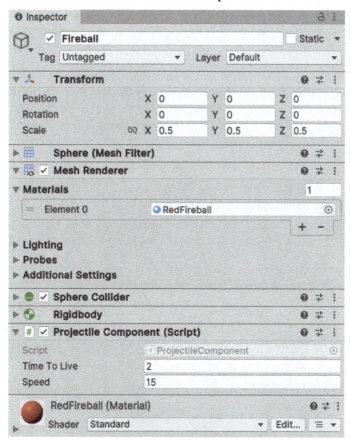

Figure 7.10: The Fireball Prefab set up to be used as a projectile

This Prefab will be used by the new spawner and object pool components that we are going to create and set up. As we are already familiar, any new spawner will need a message to be triggered, so create a script in the **Scripts | Message Queue | Messages | Battle** folder and name it `FireballSpawnMessage`. Then, add the following content:

```
using UnityEngine;
namespace Dragoncraft
{
  public class FireballSpawnMessage : IMessage
  {
```

```
    public Vector3 Position;
    public Quaternion Rotation;
  }
}
```

This message has a couple of properties, `Position` and `Rotation`, that are required to spawn the Fireball object in the correct position and rotation, so it can be moved to the right direction by the `ProjectileComponent` script.

Once we have the message class created, we move to the spawner script that will listen to the `FireballSpawnMessage` message and get one projectile from the object pool. This script is quite like the spawners we created for both Warrior and Mage, but there are a few differences. Create a new script in the **Scripts | Spawner** folder and name it `FireballSpawner`. Then, enter the following code:

```
using UnityEngine;
namespace Dragoncraft
{
  public class FireballSpawner : BaseSpawner
  {
    private void OnEnable()
    {
      MessageQueueManager.Instance.AddListener<
        FireballSpawnMessage>(OnFireballSpawned);
    }

    private void OnDisable()
    {
      MessageQueueManager.Instance.RemoveListener<
        FireballSpawnMessage>(OnFireballSpawned);
    }

    private void OnFireballSpawned(
      FireballSpawnMessage message)
    {
      GameObject fireball = SpawnObject();
      fireball.SetLayerMaskToAllChildren("Unit");

      ProjectileComponent projectile =
        fireball.GetComponent<ProjectileComponent>();
      projectile.Setup(message.Position, message.Rotation);
    }
  }
}
```

In the OnEnable and OnDisable methods, we are adding and removing, respectively, the callback to the OnFireballSpawner method from the message queue.

In the OnFireballSpawned method, we start by getting the new or reused GameObject from the object pool using the SpawnObject method and, with that object returned, we set the correct layer mask and get the ProjectileComponent component from it. In this case, we do not need to check whether the GameObject has this component because we were the ones that added it to the **Fireball** Prefab.

Once we have a ProjectileComponent variable, we can set it up by passing the position and rotation. The Setup method updates the fireball properties, and the fireball will start the movement right away.

Before moving to the Prefab configuration, we need to update one existing script. So far, we've created the message and the script that will listen to the message, but now we will update the script that sends the message when a ranged attack happens. This script is UnitComponent, and we need to add the following code at the end of the UpdateAttack method (keeping the current code too):

```
private void UpdateAttack()
{
    ...

    if (!_shouldAttack || AttackRange <= 0)
    {
        return;
    }

    _attackCooldown -= Time.deltaTime;
    if (_attackCooldown < 0)
    {
        MessageQueueManager.Instance.SendMessage(
            new FireballSpawnMessage {
                Position = transform.position,
                Rotation = transform.rotation });

        _attackCooldown = AttackSpeed;
    }
}
```

This code has three main parts:

- First, we are checking whether the `shouldAttack` variable is `true` or whether the `AttackRange` is equal to or less than 0. In this case, we can safely ignore the rest of the method using the `return` keywork, because it means that we should not attack or we cannot attack from a distance.

- Second, we are using the `attackCooldown` variable to control the time between the fireball fired and the next one so we can have, for example, one attack every two seconds while the selected unit action is attack. We decrease the value of `attackCooldown` using `deltaTime` from the Unity API, and then, when this value is less than 0, we send a message of the type `FireballSpawnMessage` with the position and rotation of the Mage at this moment.

- Finally, we reset `attackCooldown` to the `AttackSpeed` value so we can start the decrement process again until `attackCooldown` reaches 0 once more. This cooldown or countdown is a nice way to implement this kind of feature, using `deltaTime` from Unity to calculate how much time has passed in the game.

With all the code changes done, let us move to the Prefab configuration. We are editing the **ObjectPools** Prefab again, adding a new object pool there for the fireball:

1. In the **Project** view, search for the **ObjectPools** Prefab and double-click on it to open it in edit mode.

2. Once opened in edit mode, in the **Hierarchy** view, right-click on **ObjectPools** and select **Create Empty** to add a new GameObject. Name the new GameObject `FireballObjectPool`.

3. Click on the **FireballObjectPool** GameObject and, in the **Inspector** view, click on **Add Component** and search for the **FireballSpawner** script.

4. In the **Object Pool Component**, click on the circle button on the right side of the **Prefab** property. In the new window, click on the **Assets** tab and search for **Fireball**, then double-click on the Prefab to add it.

5. Change **Pool Size** to 5 and select the checkbox on the right side of the **Allow Creation** property.

After updating the **ObjectPools** Prefab, these changes will be reflected on all references of this object in the scenes that are using it.

As we can see in the following screenshot, we now have three object pools in this Prefab:

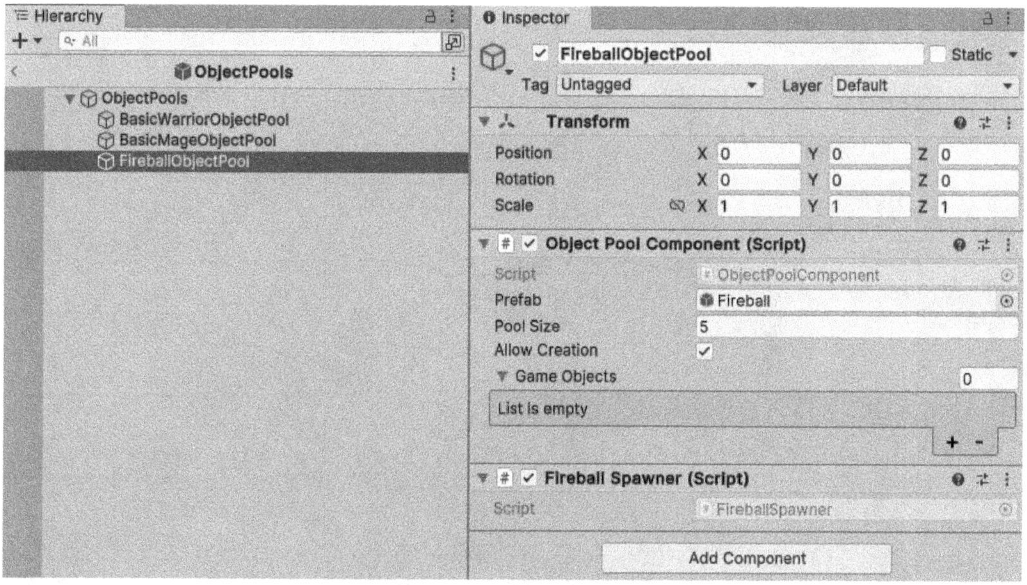

Figure 7.11: The Fireball object pool after setup

We are now all set to test everything we did in this chapter. Now that we have most of the basic systems in place, such as message queue and object pool, it is very easy and fast to add more units or even things such as projectiles or UI elements reusing what we have developed so far.

Go ahead and play in the Unity Editor with the **Playground** or **Level01** scene. Since both have the **ObjectPools** Prefab, we do not need to update them. Add a Mage to the map using **Dragoncraft | Debug | Unit | Spawn Mage**, click on the **Attack** button, and click anywhere in the map to see the Mage start to perform a ranged attack using the fireballs we created.

Figure 7.12: The Mage throwing a fireball projectile

Feel free to add more Mages and Warriors to the map and select all of them to attack. While the game is running and the Mage is attacking using Fireballs, you can look at the **ObjectPools** GameObject in the **Hierarchy** view to inspect how the object pool is handling the spawning of fireballs.

Summary

Congratulations on reaching the end of this chapter! It was quite a long one, but we now have two different units, the Warrior and the Mage, which use two different attacks, melee and ranged. Our game also now sends messages based on the actions that the player is using to update the UI and reflect the selected units in the details panel at the bottom of the UI, as well as updating the action buttons to have only what the selected units can perform.

Throughout the chapter, we learned how to implement the Command pattern and the State pattern, and how to reuse our message queue and object pool to add a new unit, projectiles, and even UI elements that are added in the details panel for each selected unit. We also saw how to play different animation states based on the context and how to expand our unit Scriptable Object to hold more data.

Our RTS game, Dragoncraft, is taking shape and progressing nicely on every feature or content we add to the gameplay – well done!

In *Chapter 8*, *Implementing the Pathfinder*, we are going to learn the concept behind one of the most used artificial intelligence techniques, the pathfinder. Then, we will learn how to implement it in our game using NavMesh from Unity, which is an amazing API to add an AI-controlled navigation and avoid obstacles. The pathfinder is widely used in RTS games to move the selected units to the destination point, avoiding the elements in the way.

Further reading

In this chapter, we saw quite a lot of new techniques, patterns, and Unity features. The following links will help you to understand each topic better, as well as giving you more examples that can make it easier to learn:

- *Command*: `https://gameprogrammingpatterns.com/command.html`
- *Extension methods*: `https://learn.microsoft.com/en-us/dotnet/csharp/programming-guide/classes-and-structs/extension-methods`
- *Bitwise operation*: `https://en.wikipedia.org/wiki/Bitwise_operation`
- *Bitwise and shift operations*: `https://learn.microsoft.com/en-us/dotnet/csharp/language-reference/operators/bitwise-and-shift-operators`
- *Uses of layers in Unity*: `https://docs.unity3d.com/2023.1/Documentation/Manual/use-layers.html`
- *Layer-based collision*: `https://docs.unity3d.com/Manual/LayerBasedCollision.html`
- *Graphics Raycaster*: `https://docs.unity3d.com/Packages/com.unity.ugui@1.0/manual/script-GraphicRaycaster.html`
- *State*: `https://gameprogrammingpatterns.com/state.html`
- *Ternary conditional operator*: `https://learn.microsoft.com/en-us/dotnet/csharp/language-reference/operators/conditional-operator`
- *OnCollisionEnter*: `https://docs.unity3d.com/ScriptReference/Collider.OnCollisionEnter.html`

8

Implementing the Pathfinder

One of the coolest features of an RTS game is that the selected units always find the best path to reach their destination, avoiding rocks, trees, buildings, and everything else on their way. There is an AI algorithm behind this feature, called a pathfinder, and it has a few different implementations. In this chapter, we will take a look at the most used algorithms and will focus on the **Navigation Mesh (NavMesh)**.

We will implement the **NavMesh** on Unity by using the AI Navigation package, setting up our units as agents and the level items as obstacles, and letting the **NavMesh** find the best path and move the selected units to the destination, avoiding or moving around the obstacles. We are also going to learn how to use the visual debugging tool from the AI Navigation package to see in real time how the **NavMesh** finds the best path for the agents.

By the end of this chapter, you will learn what a pathfinder algorithm is and the difference between the most common implementations. You will also learn how to implement the **NavMesh**, the industry standard for the pathfinder, using the Unity Engine and the AI Navigation package; this will expand our RTS game code to include new features but will keep everything we have done so far in this book working as expected.

In this chapter, we will cover the following topics:

- Understanding the pathfinder
- Implementing the pathfinder using the Navmesh
- Visual debugging of the NavMesh

Technical requirements

The project setup in this chapter, with the imported assets, can be found on GitHub at `https://github.com/PacktPublishing/Creating-an-RTS-game-in-Unity-2023` in the `Chapter08` folder inside the project.

Understanding the pathfinder

The **pathfinder** is an **Artificial Intelligence** (**AI**) algorithm that has the objective of finding the best path between two points in a 2D or 3D world. The algorithm has a graph data structure that uses a set of nodes, and their adjacent connected nodes, to search for the best path.

The pathfinder is commonly used in RTS games, so the selected units and even the enemies can find the best path while avoiding blocks in the way. We are going to get an overview of the pathfinder algorithm and understand how it works.

The Greedy Best-First and A* algorithms

Game worlds can be represented, for example, as a grid of squares or hexes, with every cell being a node. The cell neighbors are the adjacent nodes used to determine the path. In the following figure, we can see the best path that the pathfinder can find from the blue cell to the orange cell, which is just a straight line:

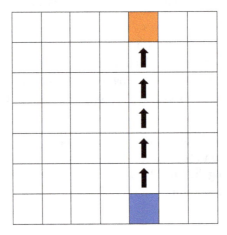

Figure 8.1 – Pathfinding in a direct line

This is the easiest route for the pathfinder. However, when we have obstacles in the path, such as walls or anything that blocks the straight path, the algorithm needs to calculate the best path. One common implementation is called **Greedy Best-First**, which tries to find the best path, trying to get closer to the objective.

However, this is not optimal, as we can see in the next figure – the algorithm adds unnecessary nodes to the path because, based on its calculations, it tries to get closer until it is blocked, then tries again in every direction until it finds the objective in the orange cell:

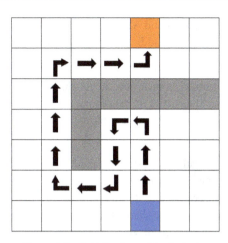

Figure 8.2 – Pathfinding with obstacles

Although this is a valid path and solution, most games cannot afford to lose time or CPU power to perform extra calculations in a suboptimal algorithm. The game engine is already quite busy when it comes to drawing the world and its elements, handling the physics, the input system, the game loop, the UI, and many internal systems.

The AI algorithms need to be extremely optimized to be used in a game – this is why Greedy Best-First is not commonly used in video games or other applications that require pathfinding, due to the high CPU usage. Although it is not an optimal solution for games, the Greedy Best-First algorithm is the foundation for more advanced algorithms such as the A* (A star), which is a more elegant solution and is better optimized to not waste time and CPU cycles to find the best path. The following figure shows the A* algorithm finding the best possible path from the blue cell to the orange cell, without wasting time on non-optimal nodes:

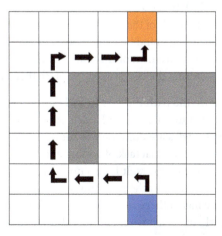

Figure 8.3 – Pathfinding with obstacles and the optimal path

The A* algorithm is widely used on games and AI projects, due to its optimal way of finding the best possible path without wasting resources or adding more nodes than required to the route.

The difference from the Greedy Best-First algorithm is that each node has a path cost, which is the cost from the start node to the current node. At each step, the A* algorithm selects the neighbor node with the lowest estimated cost, updates the cost of its neighbors, and continues until the goal is reached. It is also possible that no path is found to the target destination, and this is worth considering in all pathfinder implementations.

Although the A* algorithm is a great pathfinding strategy, and I highly recommend the books mentioned in the *Further reading* section at the end of this chapter to learn more about it, it is beyond the scope of this book to implement it from scratch. Our RTS game is not tile-based, which would increase the complexity of the pathfinder or require more changes to the project to make it tile-based. That would make the A* algorithm quite slow if we considered how many tiles our map would have and the time for the pathfinder to calculate the path and value on each node, also considering the obstacles and all moving parts in the scene.

We will keep the gameplay mechanic for the selected units to move freely on our map to any position, without requiring a tile to assign to each unit on the map. What we need is a pathfinder algorithm that can be added to our RTS game, without having to change what was developed so far, and with good performance, since we will have lots of units, enemies, and objects on the map once the game is completed. There is another pathfinder algorithm, used even more frequently in the game industry than the A* algorithm, and it is a great solution for 3D RTS games: the **NavMesh**.

The NavMesh

The previous pathfinder implementations were all based on the tiles, which for games with large maps would not be an optimal solution to find the best path, especially in games such as RTS. Imagine that you selected six units and then clicked on the map to move them. When that happens, the pathfinder will start to calculate the best path for each of the selected units and provide a path to follow. This is already quite a difficult operation, and it could be even more difficult if we moved the camera to a distant position or kept changing the movement of the units, which is quite common in RTS games, since the player is constantly considering the best strategy to place their units.

One common solution to optimize the pathfinder is to add path nodes, also known as **Points of Visibility** (**POVs**), manually added to the map, with each path node having a straight line to at least another path node. The path nodes are quite easy to set up and find the best path faster than the previous algorithms; however, the manual task of adding and tweaking them on the map could take a lot of time and might be prone to errors. It is also not possible to auto-generate path nodes, and even the manual path nodes could limit the positions available since it is likely that only a few path nodes would be added to the map; otherwise, it becomes difficult to calculate and find the best path. And, even after all that work, the result would be an inflexible route that is limited to only the pre-defined path nodes.

There is another option, a rather fancy and efficient one, called the **NavMesh**. The **NavMesh** is a network of nodes represented as a convex polygon. Using this approach, we can have large regions of the map represented by a few convex polygons. In the following figure, we are using the same map as in *Figure 8.3*, but this time, we divided the regions into four convex polygons labeled A, B, C, and D:

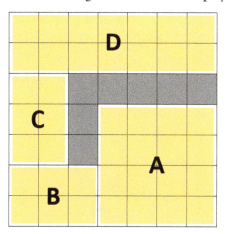

Figure 8.4 – The NavMesh convex polygons

Instead of having path nodes or POVs as fixed points on the map, we consider the entire convex polygon as a walkable area for the pathfinder algorithm. Each convex polygon is a node in the graph, which makes the search for a path very efficient since the data structure is quite compact, compared to all previous examples. The **NavMesh** also has a bonus in that the convex polygon nodes can be easily auto-generated, which results in a more natural path since we can consider any location contained in the convex polygon as a walkable position.

NavMesh boasts great performance for games and AI applications, and it is widely used in the games industry as the standard solution for the pathfinder algorithm. In fact, most of the modern game engines, such as the Unity Engine, already have the **NavMesh** implemented and available for us to use in our games. We are going to use the Unity AI Navigation package to implement our pathfinder behavior.

The Unity AI Navigation package

The **NavMesh** implementation in the Unity Engine has existed pretty much since the engine was released for the game developers to use. It was a core feature of the engine until the release of Unity 2022.2 when this feature was extracted from the engine and published as a package called AI Navigation. Since we are using Unity 2023 for our RTS game, we are going to focus on the latest **NavMesh** implementation on Unity, which is a bit different from the earlier versions before the AI Navigation package.

Let us first add the package to our project:

1. Click on the **Window | Package Manager** menu option to open the **Package Manager** view.

2. In the **Packages** dropdown, select **Unity Registry** and, in the search field, type in `navigation` to search for the package.

3. Right-click on the **AI Navigation** package on the left panel to select it and, on the right panel, click on **Install**.

The latest version of the AI Navigation package is **1.1.1**, but any version released after that will also work since they are all backward compatible. However, do not use any version lower than that recommended, as it might not work properly or have the latest features and fixes.

The following figure shows the package after being installed:

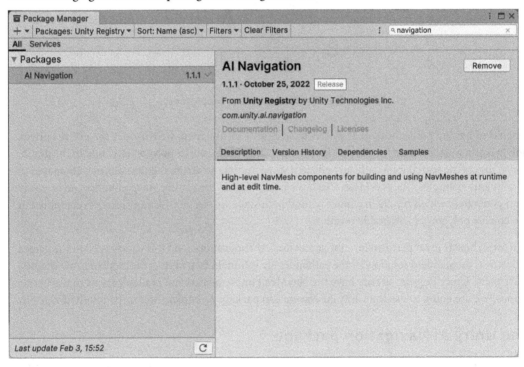

Figure 8.5 – The AI Navigation package installed using Package Manager

The AI Navigation package has four main features with many properties and customizations that we can adjust to tweak our settings. Luckily, most of the default settings are already good enough to use this package out of the box. Here is a short description of the four components:

- **NavMesh**: A data structure containing the walkable surfaces of the game world.

- **NavMesh Agent**: A character that can move toward the destination and avoid other agents.

- **Off-Mesh Link**: A navigation shortcut when the path cannot be represented as a walkable surface, such as when jumping between two areas.

- **NavMesh Obstacle:** A moving or stationary obstacle that the agents should avoid while navigating the game world. When the obstacle is not moving, as in the case of a wall, it creates a hole in the **NavMesh**, so the agents can move around the object or find a different route.

In our RTS game, we are going to use the **NavMesh** component in the **Plane** GameObject, which all scene levels have. The **NavMesh Agent** component will be added to all units that can be selected by the player, and the **NavMesh Obstacle** component is going to be used by all the level items that are placed on the map manually or dynamically, using our **LevelManager** script. We do not need the **Off-Mesh Link** component because our level is flat and continuous, without separated blocks.

Now that we understand how the pathfinder algorithm works, and we have the AI Navigation package imported into our project, we can move on to the implementation.

Implementing the pathfinder using the NavMesh

To implement the pathfinder using the **NavMesh** on Unity, we are going to add three components to the game from the AI Navigation package. The **NavMesh**, **NavMesh Agent**, and **NavMesh Obstacle** components will be configured separately but, at the end of this section, they are going to work together in our RTS game.

The NavMesh component

As we just saw, **NavMesh** is a data structure that contains all the walkable surfaces of the game world. There are a few different ways to use it: we could let the game generate all these data dynamically at runtime, or we could pre-generate these on our scene, using a process called **baking**. By baking the **NavMesh** data, we are optimizing our scene and calculating the pathfinder much faster, using the data pre-generated by the baking process.

Another optimization that we are going to use is to limit the **NavMesh** to a few layers. By including only the layers that we want the **NavMesh** to use in the calculations, we are also improving the CPU usage and reducing the pressure on both the CPU and the memory. Let us create a new layer, add the **NavMesh** component, and create a Prefab for the **Plane** GameObject:

1. Click on **Edit | Project Settings…** and, in the **Project Settings** view, click on the item named **Tags and Layers** on the left side panel.

2. On the right side panel, expand the **Layers** list and add `Plane` as the value for **User Layer 8**, right below the **Unit** layer.

3. Open **Scenes | Playground** and, in the **Hierarchy** view, click on the **Plane** item to select it.

4. On the **Inspector** view, change the **Layer** value from `Default` to `Plane`.

5. Click on the **Add Component** button and add the **NavMeshSurface** component.

6. In the **NavMeshSurface** component, expand the **Object Collection** item and, in the property called **Include Layers**, select only the **Unit** and **Plane** layers.

7. On the **Hierarchy** view, drag the **Plane** GameObject and drop it into the **Prefabs | Level** folder to create a Prefab.

8. Click on the **Plane** GameObject in the **Hierarchy** view again to select it and, in the **Inspector** view, click on the **Bake** button, located in the **NavMeshSurface** component.

9. On the **Hierarchy** view, click on the **Main Camera** GameObject and, in the **Inspector** view, click on the **Occlusion Culling** property of the **Camera** component, and also select the **Plane** layer.

10. Save the scene.

Once we bake the **NavMesh** data, Unity will create a new folder inside the scene's location, with the same name as the scene in which we baked the data, which is **Playground**. This folder has an asset named **NavMesh-Plane**, which is used by the AI Navigation package to calculate the paths in the walkable area. Do not delete this folder or the asset inside it; it is part of the pathfinder settings. Each scene and GameObject with the **NavMeshSurface** component will have one corresponding file. We can always clear and bake the data again if we change any settings in the **NavMeshSurface** component attached to **Plane**.

The next figure shows **Plane** after changing **Layer** to **Plane** and adding and configuring the **NavMeshSurface** component. The **NavMeshSurface** component is new to the AI Navigation package and only works on Unity version 2022.2 and higher:

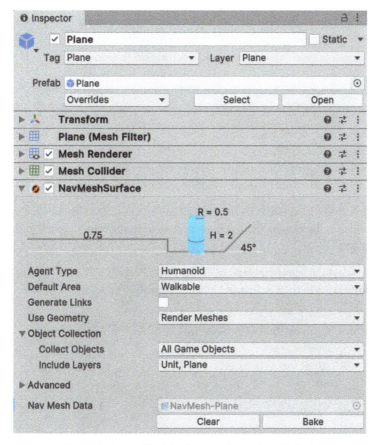

Figure 8.6 – NavMeshSurface configured and baked

Since the **NavMesh** data is unique to each **NavMeshSurface** instance, it is important to bake the data when adding this Prefab to the scene. That is why the **Nav Mesh Data** property has the blue line on the left; it indicates that this value is not the same as the default in the Prefab – and it should not be.

Now that we have a Prefab with the **NavMesh** set up, we need to add it to the level scenes as well:

1. Open **Scenes | Level01** and, in the **Hierarchy** view, delete the **Plane** GameObject.

2. Drag the **Plane** Prefab, located in **Prefabs | Level** in the **Project** view, and drop it in the **Hierarchy** view to add the Prefab to the scene.

3. In the **Hierarchy** view, click on the **Plane** GameObject, and in the **Inspector** view, click on the **Bake** button located in the **NavMeshSurface** component.

4. In the **Hierarchy** view, click on the **Main Camera** GameObject to select it.

5. In the **Inspector** view, click on the **Occlusion Culling** property of the **Camera** component and also select the **Plane** layer.

6. Save the scene.

So far in this book, we have created only two scenes, **Playground** and **Level01**, but if you have created others, do not forget to follow the aforementioned steps on any other scene as well.

There is one last change that we will need to make because of the new **Plane** layer. Besides the collision, we are also using layers to define what each camera will draw from the worldview. For this reason, we had to add the **Plane** layer in the **Occlusion Culling** property of each scene's **Main Camera**, as we did in the preceding steps, so **Main Camera** will render our plane object.

We have one more camera that needs to have the **Plane** layer selected in the **Occlusion Culling** property as well, which is the camera used to render the level in the mini map UI, in the bottom-left corner of the interface.

In the **Project** view, select the **MiniMapCamera** Prefab in the **Prefabs | UI** folder and, in the **Inspector** view, select the **Plane** layer in the **Occlusion Culling** property. The figure shows the selected layers in this camera:

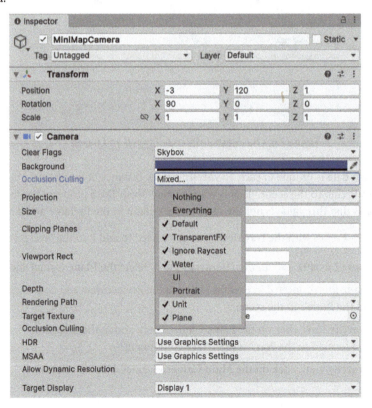

Figure 8.7 – The mini map camera with the updated layers in the Occlusion Culling property

Layers and tags are very powerful features from the Unity Engine that can be used to find or even restrict GameObjects on systems such as the camera's Occlusion Culling or the NavMesh's surface. It is good practice to use them whenever possible, and there are no performance issues should you have many in the Unity Engine. Now that we are all set on the surface, we can move to the agents.

The NavMesh Agent component

The second component that we are going to add to our project is the **NavMesh Agent** component, which is a representation of a character in the game world and is used to navigate the **NavMesh** surface while avoiding other agents and obstacles. It is as simple as adding a new component to a GameObject; however, we will make a few changes in one script that is currently attached to the unit when spawned.

Currently, the **UnitComponent** script updates the `Transform` position to move the selected unit toward the destination. However, the **NavMesh Agent** component is also able to move the agent to the destination. When using the **NavMesh Agent** component, we cannot update the `Transform` position since the path will be calculated by the **NavMesh**.

We could replace the code in the **UnitComponent** script to use **NavMesh**, but that would limit the script usage to **NavMesh** only, and we might want to use it in the future without **NavMesh** since it works quite well in moving a GameObject to a destination without using AI. Instead, we are going to extend the script and override the method used to update the unit position.

First, we need to make a few adjustments in the **UnitComponent** script to override certain methods and add a new one. Open **Scripts** | **Unit** | `UnitComponent.cs` in your IDE and, in the `UpdatePosition` method, move the last two lines to a new method that has the same name but no parameters. Then, where we remove the two lines, add a new line to call the new method, as shown in the following code snippet:

```
private void UpdatePosition(float range, UnitAnimationState state)
{
    ...

    UpdatePosition();
}

protected virtual void UpdatePosition()
{
    Vector3 direction =
        (_movePosition - transform.position).normalized;
    transform.position += direction * Time.deltaTime * WalkSpeed;
}
```

Here we have two important keywords to make this method work as required:

- `protected` means that this method is only visible to classes that extend this one through inheritance
- `virtual` makes it possible to override this method completely when extending the class

Both keywords together mean that we can either call the method or rewrite it when extending the class, indicating a great degree of flexibility. Note that we did not change the way it works; we only extracted the two lines to a new method called `UpdatePosition`, without parameters, and we are using this new method in the same place where those two lines were before.

Next, we need to make a few more changes to this script to have everything we need. Let's change the visibility of the `OnCollistionEnter` method, make it overridable as well, and expose one private variable using a protected method:

```
protected Vector3 GetFinalPosition()
{
    return _movePosition;
}

protected virtual void OnCollisionEnter(Collision collision)
{
    …
}
```

The `GetFinalPosition` method returns the value of the private `movePosition` variable, which is the position set when the player clicks on the map. We will need to access this position for the **NavMesh** to calculate the path. This method is also protected because we only need to access this information when extending the class.

Note that we did not change any code in the `OnCollistionEnter` method; we only added the `virtual` keyword and replaced the `private` keyword with `protected`, with both changes on the method signature.

Now we can create our new class, which will extend the `UnitComponent` class by changing only a few methods to use **NavMesh**, instead of the Transform to move the selected unit. Create a new script in **Scripts | Unit** and name it `UnitComponentNavMesh`. Here is the full code for the class:

```
using UnityEngine;
using UnityEngine.AI;
namespace Dragoncraft
{
    [RequireComponent(typeof(NavMeshAgent))]
    public class UnitComponentNavMesh : UnitComponent
    {
```

```
    private NavMeshAgent _agent;
    private void Start()
    {
      _agent = GetComponent<NavMeshAgent>();
    }

    protected override void UpdatePosition()
    {
      _agent.destination = GetFinalPosition();
    }

    protected override void OnCollisionEnter(Collision collision)
    {
      base.OnCollisionEnter(collision);
      if (!collision.gameObject.CompareTag("Plane"))
      {
        _agent.isStopped = true;
      }
    }
  }
}
```

Although our UnitComponent class has many lines, this new UnitComponentNavMesh class only needs to implement that class by using the two dots in the class declaration. With that line added, this class now behaves in the same way as UnitComponent, but as a new class of the UnitComponentNavMesh type. Our good old friend, the RequireComponent attribute, is also here to make sure that the GameObject with this script will also have the NavMeshAgent component. We need to include the Unity.AI package at the beginning of our script to have access to the NavMeshAgent class and get its component in the Start method.

Now the true power of the virtual keyword we added to the other class becomes evident. Using the same method signature, but replacing virtual with override, we can rewrite the method content. This means that every time the overridden UpdatePosition method is called from the UnitComponentNavMesh class, this method is executed instead of the UpdatePosition virtual method in the UnitComponent class. It does not change the behavior of the UnitComponent class; only the executed method is rewritten. The only line here is to set the destination agent to be the return value of the GetFinalPosition method, so the **NavMesh** knows where the agent wants to go to calculate the path.

The last method here, OnCollistionEnter, also has the override keyword to redefine its content. However, in this case, we want to execute both the original method content and the new code that we just added. To do this, we can use the base keyword and call the OnCollisionEnter method from the UnitComponent base class, with the same collision parameter. The base. OnCollisitionEnter line will execute the method from the UnitComponent class before executing the following lines. In the following lines, we have a conditional validation, using the `if` statement, to check whether the collision is different from a GameObject with the `Plane` tag. If this is the case, we use the `isStopped` property to make the agent stop moving when colliding with an object; otherwise, the **NavMesh** will continue trying to move the agent to the destination.

Extending and exiting the class using the virtual method is good practice in order to keep the base class functionalities and only to change or add more features to a new class. It is very powerful and easy, especially when our code is well structured and sufficiently modular so that we can extend classes to add even more flexibility to the code base.

The only thing left is to make changes to both the Warriors and Mages so that they can use the **NavMesh** version of the script when they are spawned. Open both scripts, `BasicWarriorSpawner.cs` and `BasicMageSpawner.cs`, located in the **Scripts | Spawner** folder, and replace `UnitComponent` with `UnitComponentNavMesh` in the callback methods:

```
UnitComponentNavMesh unit =
    warrior.GetComponent<UnitComponentNavMesh>();
if (unit == null)
{
    unit = warrior.AddComponent<UnitComponentNavMesh>();
}
```

The preceding code snippet is from the `BasicWarriorSpawner` class, but the `BasicMageSpawner` class has the same code, with the variable named `mage`, instead of `warrior`. The important part here is to replace the `UnitComponent` type with the `UnitComponentNavMesh` type.

Once those two scripts are modified to use the **NavMesh** version, we can go ahead and test everything we have done so far. Open either the **Playground** or **Level01** scene and play it in the Editor. Spawn a few Warriors and Mages using the debug menu and move them around. Since we only have the **NavMesh** surface and agents, we can move the selected units anywhere on the map, but all the obstacles are ignored because we still need to add that last component, and we are going to do that next.

The NavMesh Obstacle component

The last component we are going to add is the **NavMesh Obstacle**, which is added to a moving or stationary obstacle that the agents should avoid while navigating the game world. When we add the **NavMesh Obstacle** to a GameObject such as a rock, for example, it will create a hole in the GameObject that has the **NavMesh** surface, the plane in our game, which is then used by the **NavMesh Agent** to find the best path, avoiding the obstacles. This is how the three components are connected in AI Navigation.

Now that we have two different scripts for the units, the **UnitComponent** instance that uses Transform for the movement and the **UnitComponentNavMesh** instance that uses the **NavMesh Agent**, we also need different components in the elements placed on the map. We need the flexibility to either use a Rigidbody for the collision or a **NavMeshObstacle** to find the path around it.

The Rigidbody component could be used on things such as mountains or specific trees that delimit the map area, so the units stop moving when colliding with them. On the other hand, objects placed within the map, such as rocks or houses, could have the **NavMesh Obstacle** component so the agents can find a path around the obstacles to reach their destination.

As we can see, there are two options we need to accommodate in our code, so we can configure our objects to use either Rigidbody or **NavMeshObstacle** components. The option to choose which component to use will be added as a new property to the level configuration scripts, so we can have the option of selecting the collision type on each level item added to the list in the **LevelConfiguration** ScriptableObject. Let's create a new script in the **Scripts | Configuration** folder and name it `LevelItemCollisionType`. In the following code, we define an enum of each supported collision type:

```
using System;
namespace Dragoncraft
{
    [Serializable]
    public enum LevelItemCollistionType
    {
        None,
        Rigidbody,
        NavMesh
    }
}
```

It is always good practice to have the first item of the enum as `None`, which by default has a value of zero. Then we have our `Rigidybody` and `NavMesh` options for the configuration. We should not forget the `Serializable` attribute, otherwise these options will not be available for the ScriptableObject in the **Inspector** view.

After creating the aforementioned script, we can add it to the class responsible for each level item configuration, and it is used later by the ScriptableObject that has the configuration for the level. Open the script located at **Scripts | Configuration | LevelItem.cs** and add the following property as the last line of the class:

```
public class LevelItem
{
    ...

    public LevelItemCollistionType CollistionType;
}
```

The `LevelItem` class contains `LevelItemType`, which tells you the configuration, for example, if the item is a rock and a reference to the Prefab of that GameObject is given. We also added the collision type that should be used on the same GameObject. The level items are stored in the list in the **LevelConfiguration** ScriptableObject and then used by the **LevelComponent** script to instantiate the GameObjects in the game world, based on the position defined by the **LevelData** ScriptableObject.

In this last piece of code, we are going to use `CollisionType` from the previous script to determine which component will be added to our GameObject in the level. To do so, we will update the existing `Initialize` method inside the `LevelComponent.cs` script, located at **Scripts | Level**, by adding the following code:

```
private void Initialize(Vector3 start, float offsetX, float offsetZ)
{
    foreach (LevelSlot slot in _levelData.Slots)
    {
        ...

        GameObject item = Instantiate(levelItem.Prefab,
            position, Quaternion.identity, transform);

        switch (levelItem.CollistionType)
        {
            case LevelItemCollistionType.Rigidbody:
                item.AddComponent<BoxCollider>();
                break;
            case LevelItemCollistionType.NavMesh:
                item.AddComponent<NavMeshObstacle>();
                break;
            case LevelItemCollistionType.None:
                default:
                break;
        }
    }
}
```

In the `Initialize` method, we are changing the current `Instantiate` line by adding the return of the method into a variable named `item`. We are using the `item` local variable to store the reference to the GameObject that we just instantiated, and, in the `switch` statement, we are checking the `CollisionType` property to add the corresponding component to the `item` variable. If we want to have the Rigidbody, to stop the unit from moving when a collision is detected, we need to add the `BoxCollider` component. Or, if we want the **NavMesh** to use the item as an obstacle to be avoided by the agents when calculating the path, we need to add the `NavMeshObstacle` component.

After we have finished editing that last script, we are all set to update the level configuration file and set the `CollisionType` property on each item:

1. In the **Project** view, open **Data | Configuration** and left-click on the **DefaultConfiguration** asset to select it.

2. In the **Inspector** view, expand the **Level Items** list and change **Collision Type** to **Nav Mesh** on each element of the list.

We are setting all level items to use **Nav Mesh** for **Collision Type** because we want to use the level items we have created so far; **Tree**, **House**, and **Rock** are the obstacles that we want the selected units to avoid rather than having to stop moving when colliding with them.

The next figure shows our current **Level Configuration** after setting **Collision Type** to each element on the list:

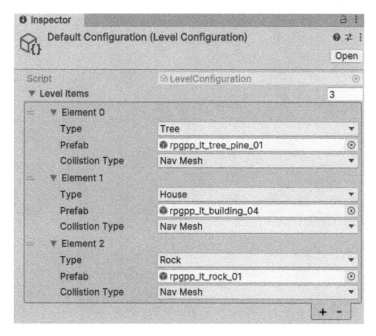

Figure 8.8 – The default configuration with the collision type defined on each item as NavMesh

The ScriptableObject here is used by another ScriptableObject of the **Level Data** type, which we already have in the project, named **Level01** inside the **Data | Level** folder. The **Level01** ScriptableObject will use the **DefaultConfiguration** ScriptableObject to find the **Prefab** and **Collision Type** values to instantiate and configure each level item properly based on the **Type** property.

Let us update **Level01** to have more columns and rows:

1. In the **Project** view, select the asset located at **Data | Level | Level01**.

2. In the **Inspector** view, change both **Columns** and **Rows** to **10**.

3. Click on the **Update** button to apply the changes.

Now that we have more columns and rows in **Level01** and we can update each level item. The following figure shows a configuration in **Level Item per position** of rows and columns that we can use to add a few trees and rocks to the top-left corner and a house to the bottom-right corner of the map:

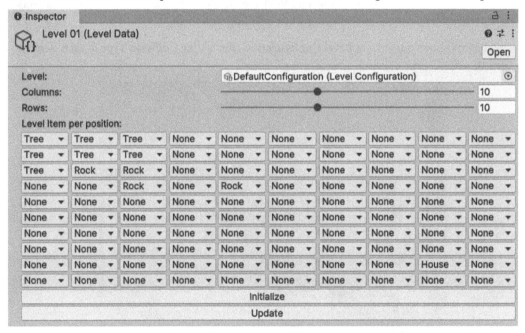

Figure 8.9 – Updated ScriptableObject for the Level01 data

We just edited **Level01** but remember that we can add as many **LevelData** files as we want by right-clicking on the **Data | Level** folder and selecting the **Create | Dragoncraft | New Level** option. This file is currently loaded by the **LevelComponent** script, which is currently only used in the **Playground** scene. The other scene we have, **Level01**, does not use the **LevelComponent** script to define and instantiate each element; instead, all items were placed manually. This is not an issue because we want to have scenes that use different ways to build the map, and we could even have a third variant that would use both at the same time.

It is finally time to test our pathfinder in the game. Open the **Playground** scene and hit the play button in the Unity Editor. Spawn a few Warriors or Mages, select all of them, and start moving toward the top-left corner of the map. Once you reach the rocks and trees, start clicking around them to see the selected units finding the best path around them, as we can see here:

Figure 8.10 – The selected units avoiding the rock to find the best path

The more Warriors and Mages you spawn in the level, the more fun you will have when testing the pathfinder we just created using the **NavMesh**. You can edit **Level01** or create more configurations to test in the **Playground** scene; just remember to update **LevelData** at the GameObject level, located in the **Hierarchy** view, if you want to use a different one. You can also duplicate an existing level configuration by copying and pasting the asset in the same place, so you do not lose a configuration that you liked, and you could also create more levels based on that.

All of this applies to the **Playground** scene that uses the **LevelComponent** script to instantiate all the level items, but we can also do the same manually. Let's open the **Level01** scene and, in the **Hierarchy** view, expand the **Houses** GameObject and select all the children's GameObjects. Then, in the **Inspector** view, we will click on **Add Component** and add the **NavMeshObstacle** component, as shown here:

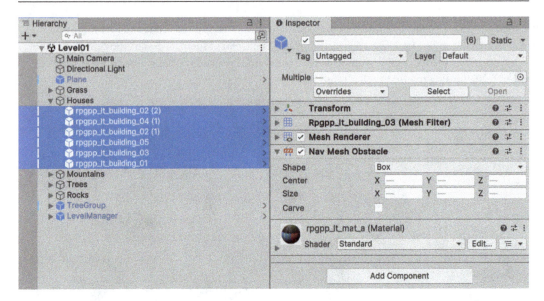

Figure 8.11 – Manually adding the NavMeshObstacle component to the items in scene Level01

Do the same for the children of the **Trees** and **Rocks** GameObjects and remember to save the scene afterward. You can run the **Level01** scene and check out the **NavMesh** on a scene with more elements.

Now that the pathfinder is working using **NavMesh**, let's look at the visual debugging tool that also comes with the AI Navigation package to help us see how **NavMesh** works in real time.

Debugging the NavMesh

The AI Navigation package has a very good visual debugging tool that shows on the Editor the calculations that the **NavMesh** is doing to find the best path around one obstacle in real time.

However, before starting to play with the visual debugging tool, we first need to change our Unity layout so we can see both the **Scene** view and the **Game** view at the same time. In the top-right corner of the Unity Editor, click on **Layout** and select the **2 by 3** option. Your layout will change to the one shown in the following figure:

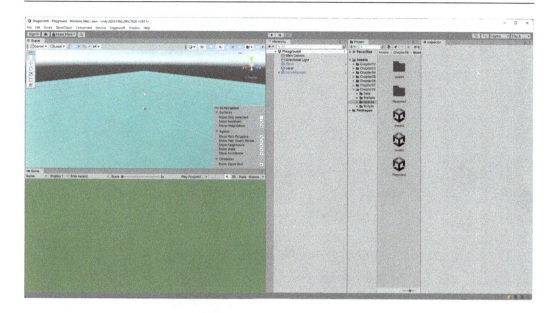

Figure 8.12 – The Unity Editor with the 2 by 3 layout

You can always change back to the **Default** layout but, for now, the **2 by 3** layout will work better because we need to play the game in the **Game** view while looking at the visual debugging tool in the **Scene** view.

Now go ahead and hit the **Play** button in the Unity Editor, spawn a few Warriors, select them, and click somewhere in the map to make the selected units start moving to that destination. Then, click on the **Pause** button in the Unity Editor and, in the **Hierarchy** view, expand the **LevelManager | ObjectPools | BasicWarriorObjectPool** path, before clicking on the first **FootmanHP(Clone)** GameObject to select it.

The GameObject selected in the **Hierarchy** view is shown as follows, as well as the **AI Navigation** overlay on the left side, which is in the **Scene** view:

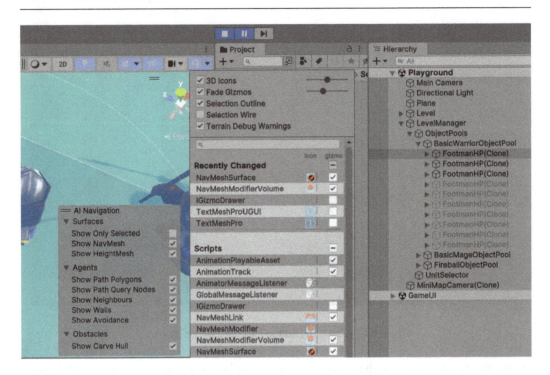

Figure 8.13 – Clicking on the NavMesh items is a workaround to show the visual debug

Click on the checkboxes to enable the same options as shown in the preceding figure for the **AI Navigation** overlay and in the **Gizmos** menu, which are visible when clicking on the sphere icon in the top-right corner of the **Scene** view.

With all those checkboxes enabled, we can see the visual debugging. As we can see in the following figure, the AI Navigation draws the agent cylinder; the blue arrow, which represents the direction; the red arrow and red circle, which represent the NavMesh's search for a path; and the big white circle, which is the area surrounding the agent. If you are not able to see the visual debugging, try to uncheck and check the **NavMeshSurface** option in the **Gizmos** menu, as shown in the following figure:

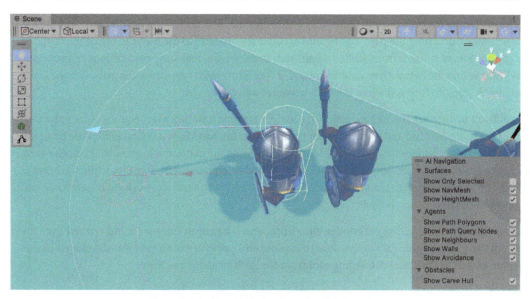

Figure 8.14 – The AI Navigation visual debugging

Without an obstacle, the visual debugging will only show the direction but, if we move the selected units toward a rock or three on the map, we are going to see more information in the **Scene** view. When we add the **NavMeshObstacle** component, the default shape is a box, represented in the following figure around the rock with orange circles and rectangles drawn on the ground; that is the edge of the obstacle used by the **NavMesh** algorithm to make the agent walk around it while trying to find the best path toward the destination:

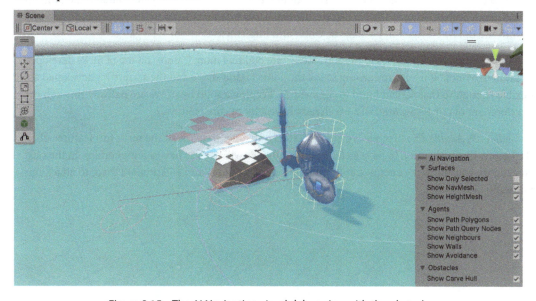

Figure 8.15 – The AI Navigation visual debugging with the obstacle

In the preceding figure, we see the blue arrow showing the direction to the destination, and the red arrow is now calculating the path, taking into account the obstacle in the way. On top of the obstacle, we can see lots of squares from white to light red; these are the real-time calculations made by the **NavMesh** to find the best path around the obstacle. The white color represents the path with the lower cost, and it starts to turn red when the path becomes more expensive.

It is difficult to visualize the AI Navigation visual debugging in a static image; you should try and take some time to understand and see what is happening when the **NavMesh** is trying to find the best path to the destination, with or without obstacles. When running the game and looking at the **Scene** view, without pausing the editor, you will be able to see the white squares changing a lot while the agent is moving. You will be able to see the struggle to find the best path until the obstacle is overcome and the path to the destination is clear.

The **NavMesh** is an amazing pathfinder algorithm, and it has been improved and refined over the years in the Unity Editor. Take some time to explore and use everything that was developed in this chapter, as well as the visual debugging tool from the AI Navigation package.

Summary

Well done for reaching the end of this chapter. The pathfinder is one of the core elements present in an RTS game, one thing that we must have in our game, and now it is working fully. The units can move freely on the map, avoiding obstacles, and even large groups can move through rocks and trees without being blocked by them, always trying to find the best path to the destination.

In this chapter, we learned what the pathfinder algorithm is, and we saw the theory and base concepts behind the most common implementations, from the relatively simple Greedy Best-First to the complex A*. We ended up looking at the **NavMesh**, the data structure most used in the games industry, and how to use the AI Navigation package in Unity and implement it in our RTS game.

The *Dragoncraft* game continues to expand, and we are building more features on top of those developed so far in this book. It is very rewarding to see the game taking shape and the scripts we developed are flexible enough so that we can build more features, keeping a solid base. We are halfway through the development of our RTS game, and we already have a lot of core features.

In *Chapter 9, Adding Enemies*, we are going to start implementing one of the most fun features of an RTS game: the battle. And, to have a challenging fight, we will need to have a few enemies. In the next chapter, we will add some enemies, fully configured as our units, and will prepare them to attack our units and vice versa.

Further reading

Although we saw a brief explanation of the concepts and theory behind different pathfinder algorithms, I recommend checking out the following books and links to learn more about them:

- Madhav, Sanjay. *Game Programming Algorithms and Techniques*. Addison-Wesley (2014)
- Buckland, Mat. *Programming Game AI by Example*. Wordware Publishing Inc. (2005)
- *Navigation System in Unity*: `https://docs.unity3d.com/Packages/com.unity.ai.navigation@1.1/manual/NavigationSystem.html`
- *Inner Workings of the Navigation System*: `https://docs.unity3d.com/Packages/com.unity.ai.navigation@1.1/manual/NavInnerWorkings.html`
- *Virtuals (C# reference)*: `https://learn.microsoft.com/en-us/dotnet/csharp/language-reference/keywords/virtual`

Part 3:
The Battlefield

In this third part of the book, you will learn how to configure different types of enemies and spawn them on the map, as well as how to create a basic AI to patrol and chase the player's units. You will also learn how to implement a battle between the units and enemies, using collision and damage calculation based on the configuration of each character, and creating their life cycle with animations and feedback.

After having finished the battle implementation, you will see how we can configure groups of enemies and add them to the map using a spawn point. Finally, you will learn how to cover the map with fog and unveil the areas that are explored by the player, using units.

This part includes the following chapters:

- *Chapter 9, Adding Enemies*
- *Chapter 10, Creating an AI to Attack the Player*
- *Chapter 11, Adding Enemies to the Map*

9

Adding Enemies

An RTS game has maps, units, and many other things, but there is one game mechanic that cannot be left out, which is combat. However, to have a good and challenging battle system in place, we need more than units; we need to have enemies.

In this chapter, we are going to implement different enemy types, such as Orcs and Golems, against whose attacks it will be a real challenge for the player to defend themself, and the mighty Red Dragon, which we are going use later as the boss of our level.

By the end of this chapter, you will have learned how to create a flexible system for adding different enemies, spawning the enemies in the existing maps, and reusing most of the unit components to expand and create unique features for the enemies. You will also learn how to create damage feedback in the UI that will display the damage done to each unit and enemy, as well as how to calculate the damage considering the configuration.

In this chapter, we will cover the following topics:

- Configuring the enemies
- Spawning the enemies
- Damage feedback in the UI

Technical requirements

The project setup in this chapter with the imported assets can be found on GitHub at https://github.com/PacktPublishing/Creating-an-RTS-game-in-Unity-2023 in the Chapter09 folder inside the project.

Configuring the enemies

So far, we have already built a solid foundation for our RTS game, with a flexible UI and controllable units. Now, we are going to start adding all the enemies the player will face in the game and prepare another layer of a solid strategy game. The enemies will be an Orc, fast but a bit weak on its own; a powerful Golem, which is slower than the Orc but much stronger and resistant; and our mighty boss, the Red Dragon, which will be a challenge for the player.

Figure 9.1 – The Orc, Golem, and Red Dragon enemies

All the enemies in the preceding figure have their Prefabs located inside packages that were already imported into the project in *Chapter 2, Setting Up Unity and the Project*, so we have all the Prefabs that we need. However, we will need to tweak their animation states, as we did for both Warrior and Mage. That is because the animation states are set up as a loop through all states, and we want to control each animation state using our own code.

So, let us remove the animation state connections:

1. Open the **Animator** view by clicking on **Window | Animation | Animator** and keep the view open. You can drag and drop the window to dock it in the Editor if you prefer.

2. Next, in the **Project** view, search for the Prefab name GruntHP, and left-click to select it. We are going to see the animator details in the **Animator** view.

3. With the Prefab selected, back in the **Animator** view, left-click on any white line connecting two animation states to select it and press the *Delete* button on your keyboard to remove the connection. Repeat the process for every white line connecting two animation states and save the changes.

4. Then, in the **Project** view, search for the Prefab name `HP_Golem`, and left-click to select it. Repeat *step 3* to remove the animation state connections.

5. Still in the **Project** view, search for the Prefab name `DragonSoulEaterRedHP`, and left-click to select it. Repeat *step 3* to remove the animation state connections.

Once all animation state connections are removed, we can play any animation state using the Unity API and the animation will keep playing that state without moving to any other. The following figure shows the Red Dragon animation states after all the connections have been removed:

Figure 9.2 – The Red Dragon animation states after removing all the while line connections

The Red Dragon has more animation states than the Orc and the Golem, but the important thing is there are no connections between the animation states. That was the only change needed in the existing Prefabs before starting to add the scripts to represent the enemy configuration.

The enemy configuration shares a lot of things in common with the unit configuration, and most of the scripts we are going to create next are quite similar both in name and content. However, enemies are a bit simpler than units, especially because they are not controlled by the player.

Before adding the script for the enemy, we need to create a script to define each enemy type we are going to have in the game. Let us add a new folder named **Scripts | Enemy** and create a new script called EnemyType. Open the script in your IDE and replace the content with the following code snippet:

```
using System;
namespace Dragoncraft
{
    [Serializable]
    public enum EnemyType
    {
        Orc,
        Golem,
        Dragon
    }
}
```

The script is quite simple – just a list of the types for the enemies. The Serializable attribute is required in this enum so we can use it as a property type in a ScriptableObject, which is what we are going to do in the next script.

In the same **Scripts | Enemy** folder, add a new script, name it EnemyData, and change the content according to the following code snippet:

```
using UnityEngine;
namespace Dragoncraft
{
    [CreateAssetMenu(menuName = "Dragoncraft/New Enemy")]
    public class EnemyData : ScriptableObject
    {
        public EnemyType Type;
        public float Health;
        public float Attack;
        public float Defense;
        public float WalkSpeed;
        public float AttackSpeed;
        public Color SelectedColor;
        public string AnimationStateAttack01;
        public string AnimationStateAttack02;
        public string AnimationStateDefense;
        public string AnimationStateMove;
        public string AnimationStateIdle;
        public string AnimationStateDeath;
    }
}
```

The EnemyData class is a ScriptableObject with the CreateAssetMenu attribute to make it easier to create a new asset in our project using the shortcut in the menu.

Besides the EnemyType enum, all the other properties were also present in the UnitData class, but EnemyData has fewer properties for stats and animation states. One of the reasons is that the unit can be upgraded by the player to become stronger in the game, while the enemies have their fixed values.

Another similarity in the newly created EnemyData class is the GetAnimationState method to retrieve the name of the animation state based on UnitAnimationState. Add the following method to the same class:

```
public string GetAnimationState(UnitAnimationState animationState)
{
  switch (animationState)
  {
    case UnitAnimationState.Attack01:
      return AnimationStateAttack01;
    case UnitAnimationState.Attack02:
      return AnimationStateAttack02;
    case UnitAnimationState.Defense:
      return AnimationStateDefense;
    case UnitAnimationState.Move:
      return AnimationStateMove;
    case UnitAnimationState.Idle:
      return AnimationStateIdle;
    case UnitAnimationState.Death:
      return AnimationStateDeath;
    case UnitAnimationState.Collect:
    default:
      return AnimationStateIdle;
  }
}
```

Even though this is a class responsible for holding the enemy data, we are going to reuse the UnitAnimationState enum because they share the same animation states as the units. One difference is that the enemies will not be able to collect items in the game; that is why the Collect animation state has the default same value, which is the Idle animation state.

After saving the script, we can go to the Unity Editor and start creating our enemies by creating one ScriptableObject configuration for each enemy type. Let us start with the Orc:

1. In the **Project** view, create a new folder inside **Data** called **Enemy**. Then, right-click on it and select **Create | Dragoncraft | New Enemy**, and name it as BasicOrc.

2. Left-click on the newly created BasicOrc file and, in the **Inspector** view, set **Type** to **Orc**; **Health** to **10**; **Attack** to **2**; **Defense** to **1**; **Walk Speed** to **5**; and **Attack Speed** to **2**.

3. In the **Selected Color** property, click on the color field and set the **Hexadecimal** value to **FF6961**.

4. Set **Animation State Attack 01** to **Attack01**; **Animation State Attack 02** to **Attack02**; **Animation State Defense** to **Victory**; **Animation State Move** to **Run**; **Animation State Idle** to **Idle**; and **Animation State Death** to **Die**.

5. Save the project.

The Orc has characteristically low values for most stats, besides **Walk Speed** which has a higher value. **BasicOrc** is a default configuration for the Orc, and we can always tweak or create new variations from it. The following figure shows the **BasicOrc** file configured:

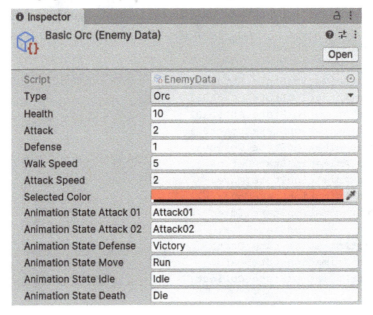

Figure 9.3 – The BasicOrc asset configured

The next enemy on our list is the Golem, which is bigger than the Orc, making it both powerful and slow. This is reflected in the basic configuration that we are going to create for it:

1. In the **Project** view, open the folder inside **Data** called **Enemy**. Then, right-click on it to select the **Create | Dragoncraft | New Enemy** option, and name it `BasicGolem`.

2. Left-click on the newly created `BasicGolem` file and, in the **Inspector** view, set **Type** to **Golem**; **Health** to **20**; **Attack** to **5**; **Defense** to **2**; **Walk Speed** to **2**; and **Attack Speed** to **1**.

3. In the **Selected Color** property, click on the color field and set the **Hexadecimal** value to **FF6961**.

4. Set **Animation State Attack 01** to **Attack01**; **Animation State Attack 02** to **Attack02**; **Animation State Defense** to **Victory**; **Animation State Move** to **Walk**; **Animation State Idle** to **Idle**; and **Animation State Death** to **Die**.

Compared with the Orc, the Golem has much higher health and attack, making it a bit more difficult to defeat. However, it is also slower than the Orc, which will force the player to use different strategies when combating a Golem, or even both enemies together.

The following figure shows **BasicGolem** configured, which is also a base line for future adjustments or variations:

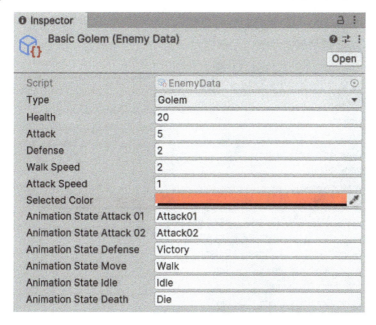

Figure 9.4 – The BasicGolem asset configured

The last enemy is our mighty Red Dragon, the boss that the player must defeat on the map. The configuration for this one is quite overpowered, but we need to keep in mind that this is the most challenging of the three enemy types and we will only have one in the level, where there will be plenty of Orcs and Golems for the player to fight the way to the boss. Let us create the configuration for the Red Dragon:

1. In the **Project** view, open the folder inside **Data** called **Enemy**. Then, right-click on it to select the **Create | Dragoncraft | New Enemy** option, and name it RedDragon.

2. Left-click on the newly created **RedDragon** file and, in the **Inspector** view, set **Type** to **Dragon**; **Health** to **100**; **Attack** to **10**; **Defense** to **3**; **Walk Speed** to **3**; and **Attack Speed** to **2**.

3. In the **Selected Color** property, click on the color field and set the **Hexadecimal** value to **FF6961**.

4. Set **Animation State Attack 01** to **Basic Attack**; **Animation State Attack 02** to **Tail Attack**; **Animation State Defense** to **Defend**; **Animation State Move** to **Run**; **Animation State Idle** to **Idle**; and **Animation State Death** to **Die**.

Our Red Dragon has the highest overall stats, only losing in **Walk Speed** to the Orc, which is lighter and faster than the Red Dragon. The boss will be quite tough to beat, but that is the point of having a mighty Red Dragon as the main enemy of the level.

As always, the values in the following figure are a baseline, and we can change them or use them to create more dragons of different colors, for example:

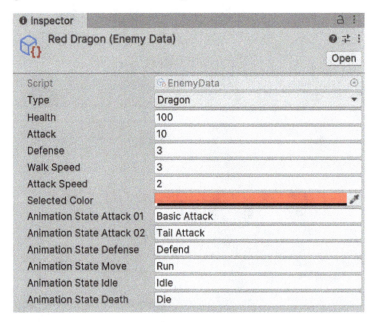

Figure 9.5 – The RedDragon asset configured

The player will need to upgrade the units and surround the Red Dragon to defeat it while holding the attacks from Orcs and Golems. The combat between units and enemies will be quite fun to implement, test, and play in the final game, but for now, let us focus on creating a solid base for the enemies.

Now that we have our enemies configured and with the proper animation state set up, we can start adding the scripts to spawn enemies in the level.

Spawning the enemies

We built a very solid and flexible code base to spawn more objects in the game, using our Object Pool to manage the instances triggered by our message queue. It all started with the Warrior unit, and then we reused everything we had to configure and add the Mage unit as well. Now, we are also going to reuse both systems, but with a few changes since the enemies are a bit different from the units.

The first scripts that we are going to create are the messages to spawn the enemies, and then, in the next sections, the other scripts for the enemy component and spawner. Each enemy will have its own message; however, they will all share the same parameter to determine the position on the map at which the enemy will be spawned. Since it is a common property there is no need to add it to all enemy messages, instead, we are going to create a base enemy message and then implement it on all enemy messages.

Creating the enemy messages

Create a new folder inside **Scripts | MessageQueue | Messages** and name it Enemy. In the newly created folder, add a new script called `BaseEnemySpawnMessage` and add the following content:

```
using UnityEngine;
namespace Dragoncraft
{
  public class BaseEnemySpawnMessage : Imessage
  {
    public Vector3 SpawnPoint;
  }
}
```

The `BaseEnemySpawnMessage` class implements the `Imessage` interface, so we can use it as a message, and have the `SpawnPoint` parameter, which, as mentioned, will hold the position at which the enemy must be spawned on the map.

Next, in the same folder, add a new script named `BasicOrcSpawnMessage` and add this code:

```
namespace Dragoncraft
{
  public class BasicOrcSpawnMessage : BaseEnemySpawnMessage
  {
  }
}
```

Note that the `BasicOrcSpawnMessage` class does not have any parameter or method declared; however, since it implements the `BaseEnemySpawnMessage` class, the class has the `public SpawnPoint` property available.

Now, we need to add the messages for the other enemy types as well. Add a new script in the same folder, name it `BasicGolemSpawnMessage`, and replace the content with the following code snippet:

```
namespace Dragoncraft
{
  public class BasicGolemSpawnMessage : BaseEnemySpawnMessage
  {
```

```
    }
}
```

As seen in the previous class, the only the `BasicGolemSpawnMessage` class is that it implements the `BaseEnemySpawnMessage` class.

For the last message, add a script called `RedDragonSpawnMessage` and add the following code:

```
namespace Dragoncraft
{
    public class RedDragonSpawnMessage : BaseEnemySpawnMessage
    {
    }
}
```

After creating the `RedDragonSpawnMessage` class, we now have one message to spawn each enemy type, without duplicated code for each class. We could use a different approach by adding a property with the enemy type in the `BaseEnemySpawnMessage` class; however, having a class represent each enemy type gives us more flexibility to send and listen to one specific message type when we want to spawn an enemy.

Now that we have created all the messages, we can move the enemy component that will be responsible for managing the enemy behavior onto the map.

Creating the enemy component

The component that we are going to create to manage the enemy is a bit similar to the **UnitComponent** script. However, the unit and the enemy do not share most of the logic already in that script. While the unit is reactive to the player input, the enemy will have its own logic to move and attack the units, which we are going to implement in *Chapter 10, Creating an AI to Attack the Player*.

For that reason, if we were to inherit and override the methods, we would end up doing that for all methods, without any real benefit from it. Instead, it makes more sense for us to create a new script for the enemy that will give us more flexibility to change the code without being limited by the **UnitComponent** script.

In the **Scripts | Enemy** folder, add a new script named `EnemyComponent` and add this code snippet:

```
using System;
using UnityEngine;
namespace Dragoncraft
{
    [RequireComponent(typeof(BoxCollider), typeof(Animator))]
    public class EnemyComponent : MonoBehaviour
    {
```

```
    public string ID;
    public EnemyType Type;
    public float Health;
    public float Attack;
    public float Defense;
    public float WalkSpeed;
    public float AttackSpeed;
    public Color SelectedColor;

    private Animator _animator;
    private Renderer _renderer;
    private Color _originalColor;
    private EnemyData _enemyData;
  }
}
```

The EnemyComponent class has a couple of required components, BoxCollider and Animator, both used in the class to detect collisions and change the animation state, respectively. Compared with the UnitComponent class, this EnemyComponent class has fewer properties than the unit, which is one good reason to have a script tailored to the enemy, besides it being good practice to have different classes for different use cases.

All the public properties are the same as those available in the **EnemyData** script, except for the addition of ID, which we are going to generate a random value to make sure the instance is unique, even when reused from the Object Pool. private properties are there to help make the next methods easier to handle, as we are going to see in the next methods.

The first method we are going to define in this class is Awake, as we can see in the next code snippet, which is executed once the GameObject is activated by the Unity Engine:

```
private void Awake()
{
  _animator = GetComponent<Animator>();
  _renderer = GetComponentInChildren<Renderer>();
  _originalColor = _renderer.material.color;
}
```

In the Awake method, we are going to get the components for the Animator component in the current GameObject, and the Renderer component in the child GameObject. This is not a general rule of any sort; instead, we are getting the Renderer from the children because it is how the Prefabs for all enemies (and units as well) were set up. We are also storing the originalColor value of the Renderer material, so can freely manipulate the material color and always return to the original value when needed.

It is a good practice to initialize variables or get components from the GameObject in this method. However, since it is the first thing executed when the class is instantiated by the engine, it is better to keep it short and lightweight, without any method executions.

The enemy instance will be managed by the Object Pool, which means that we are likely to reuse instances in the runtime. This is why we have the CopyData method, defined next, which works as a setup method to reset all the enemy properties so we can have a reused object but with updated values that will act as a new instance:

```
public void CopyData(EnemyData enemyData, Vector3 spawnPoint)
{
    ID = System.Guid.NewGuid().ToString();
    Type = enemyData.Type;
    Health = enemyData.Health;
    Attack = enemyData.Attack;
    Defense = enemyData.Defense;
    WalkSpeed = enemyData.WalkSpeed;
    AttackSpeed = enemyData.AttackSpeed;
    SelectedColor = enemyData.SelectedColor;
    _enemyData = enemyData;
    transform.position = spawnPoint;
    _animator.Play(
        _enemyData.GetAnimationState(UnitAnimationState.Idle));
}
```

The CopyData method primarily sets the information from the EnemyData parameter in the public properties in this class. As we did in the UnitComponent class, we are assigning a Guid value to the ID property, and we generate a new Guid every time the CopyData method is called. Another important thing here is the spawnPoint parameter, which is set to the GameObject's position. Later, in *Chapter 11*, *Adding Enemies to the Map*, we are going to set up spawn points for the enemies, and that information will be used in this method.

Besides setting the value for the public properties, we also do a few more things here. The enemyData parameter is stored in the private property so we can use it to call the GetAnimationState method when we need to play a different animation state, as we are doing here changing the animation to be in the Idle state. Since the Animator is a required component, and we got the component reference in the Awake method, we can safely call the Play method here.

The next couple of methods are going to be executed when the mouse pointer is over the GameObject, which in this case is the enemy, and we can use them to change the color to indicate to the player that the mouse pointer is hovering over an enemy:

```
private void OnMouseEnter()
{
    _renderer.material.color = SelectedColor;
```

```
}

private void OnMouseExit()
{
  _renderer.material.color = _originalColor;
}
```

These methods, `OnMouseEnter` and `OnMouseExit`, basically change the color of the material attached to the GameObject, respectively, to `SelectedColor`, defined in the `EnemyData` ScriptableObject, and back to `originalColor`, stored in the `Awake` method. The enemy will have a nice tinted color to show to the player that the mouse is hovering over it and can be set as a target for the selected units to attack, for example.

For now, we are only changing the material color, and the functionality to select the enemy as a target to be attacked will be added in *Chapter 10, Creating an AI to Attack the Player*. We are done with the **EnemyComponent** script for now, and we can move to the spawner class to start using it.

Creating the enemy spawner

We already created a few spawner classes and we are familiar with them. All spawner classes must implement the `BaseSpawner` class, which has some code to avoid duplication and make our lives easier when setting up a new spawner.

The new spawner class we are going to create for the enemy is quite like the one we have created for the unit. However, we have a few changes here because instead of creating a spawner class for each enemy type, as we did for each unit type, we are going to create a single spawner class that will handle all the enemy types.

Add a new script in the **Scripts | Spawner** folder and name it `EnemySpawner`. Then, the following is the first code snippet for this new class:

```
using UnityEngine;
namespace Dragoncraft
{
  public class EnemySpawner : BaseSpawner
  {
    [SerializeField] private EnemyData _enemyData;
  }
}
```

The `BaseSpawner` class will give the new `EnemySpawner` class access to the Object Pool, which we are going to set up in a moment with a reference to the enemy Prefab. Although we are not going to create a spawner class for each enemy type, we will still need a spawner object for each enemy Prefab. We are going to use the `enemyData` property to determine what the enemy type is and listen to the correct message, which is the trigger to spawn the enemy in the map.

We can start with the OnEnable method, which is going to add the listener to the proper message based on the EnemyType enum that is defined in the enemyData ScriptableObject:

```
private void OnEnable()
{
  switch(_enemyData.Type)
  {
    case EnemyType.Orc:
      MessageQueueManager.Instance.
        AddListener<BasicOrcSpawnMessage>(OnEnemySpawned);
      break;
    case EnemyType.Golem:
      MessageQueueManager.Instance.
        AddListener<BasicGolemSpawnMessage>(
          OnEnemySpawned);
      break;
    case EnemyType.Dragon:
      MessageQueueManager.Instance.
        AddListener<RedDragonSpawnMessage>(OnEnemySpawned);
      break;
    default:
      break;
  }
}
```

In the previous code snippet, we are basically using switch to add the listener based on Type. Regardless of the message type, the callback will be the same, the OnEnemySpawner method. We can do that because all the enemy messages implement the same class, BaseEnemySpawnMessage.

The OnDisable method is defined next, which is the counterpart of the OnEnable method and has almost the same code but, instead of adding it, we are going to remove the message listener based on Type:

```
private void OnDisable()
{
  switch (_enemyData.Type)
  {
    case EnemyType.Orc:
      MessageQueueManager.Instance.
        RemoveListener<BasicOrcSpawnMessage>(
          OnEnemySpawned);
      break;
    case EnemyType.Golem:
      MessageQueueManager.Instance.
```

```
        RemoveListener<BasicGolemSpawnMessage>(
            OnEnemySpawned);
        break;
    case EnemyType.Dragon:
        MessageQueueManager.Instance.
            RemoveListener<RedDragonSpawnMessage>(
                OnEnemySpawned);
        break;
    default:
        break;
    }
}
```

Now that we have both methods to add and remove the message listeners, we can implement the main method of this class, which is the OnEnemySpawned callback method. The following code snippet defines the method, which is also similar to the callback for the unit spawner classes, but with a few differences:

```
private void OnEnemySpawned(BaseEnemySpawnMessage message)
{
    GameObject enemyObject = SpawnObject();
    enemyObject.SetLayerMaskToAllChildren("Unit");

    EnemyComponent enemyComponent =
        enemyObject.GetComponent<EnemyComponent>();
    if (enemyComponent == null)
    {
        enemyComponent =
            enemyObject.AddComponent<EnemyComponent>();
    }

    enemyComponent.CopyData(_enemyData, message.SpawnPoint);
}
```

The parameter of this method is BaseEnemySpawnMessage, which works perfectly for any enemy message. The OnEnemySpawned method does not need to know what the enemy that is going to be spawned is since the SpawnObject method, defined in the BaseSpawner class, will take care of that.

The SpawnObject method will get an instance of the enemy based on the configuration attached to the GameObject, which is the Prefab and the **EnemyData** ScriptableObject. This code can be considered consistent with the **Prototype** pattern, where we have a generic method to instantiate copies of one other object without really knowing the type – everything is configured so that the Prefab is used as a prototype for the spawner, and no specific code is required for that.

We are going to set the layer mask in `enemyObject` to `Unit` because we set up the collision detection for the units to use that layer mask and, since we need to detect collisions between enemies and units as well, we can safely use the same value.

The `EnemyComponent` component is going to be added to this GameObject if it does not exist already. If the instance returned by the Object Pool is an existing one, the component is there; otherwise, we add a new one. `EnemyComponent` is used in the last line of the method, which calls the `CopyData` method with the `enemyData` parameters, which is a property in this class, and `SpawnPoint` from the message we received in the callback.

That will be all we need to add to this script before starting to set up the Prefab with this new spawner for each enemy type that we are going to do next.

Configuring the enemy spawner

Now that we have everything required to set up the spawner, we can modify the Prefab that already has the spawner for the units to also include the spawners for the enemies since they will all be used in the map. Let us start by adding one for the Orc enemy:

1. In the **Project** view, search for the `ObjectPools` Prefab and double-click to open it in Edit Mode.

2. In the **Hierarchy** view, left-click on the **ObjectPools** GameObject, select the **Create Empty** menu option, and name it `EnemyOrcObjectPool`.

3. Select the new **EnemyOrcObjectPool** GameObject and, in the **Inspector** view, click on the **Add Component** button, search for `EnemySpawner`, and double-click to add it.

4. In **Object Pool Component**, click on the circle button on the right side of the **Prefab** property. In the new window, select the **Assets** tab, search for the Prefab named `GruntHP`, and double-click to add it.

5. Change **Pool Size** to **10** and enable the checkbox in the **Allow Creation** property.

6. In the **Enemy Spawner** component, click on the circle button on the right side of the **Enemy Data** property. In the new window, select the **Assets** tab, search for the ScriptableObject named `BasicOrc`, and double-click to add it.

7. Save the Prefab changes.

Once the changes are saved, the setup should look like the following figure. The Orc is represented by the **GruntHP** Prefab and will use the settings that we configured in the corresponding ScriptableObject.

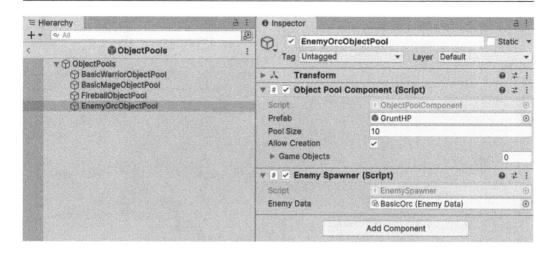

Figure 9.6 – The Object Pool component for the Orc spawner

Next, still in the same **ObjectPools** Prefab in Edit Mode, we will add the second enemy, the Golem:

1. In the **Hierarchy** view, left-click on the **ObjectPools** GameObject, select the **Create Empty** menu option, and name it EnemyGolemObjectPool.

2. Select the new **EnemyGolemObjectPool** GameObject and, in the **Inspector** view, click on the **Add Component** button, search for EnemySpawner, and double-click to add it.

3. In **Object Pool Component**, click on the circle button on the right side of the **Prefab** property. In the new window, select the **Assets** tab, search for the Prefab named HP_Golem, and double-click to add it.

4. Change **Pool Size** to **5** and enable the checkbox in the **Allow Creation** property.

5. In the **Enemy Spawner** component, click on the circle button on the right side of the **Enemy Data** property. In the new window, select the **Assets** tab, search for the ScriptableObject named BasicGolem, and double-click to add it.

6. Save the Prefab changes.

The Golem is very similar to the Orc and, besides the Prefab and **EnemyData**, the only difference is the number for **Pool Size**, which pre-instantiates the Prefabs when the level is loaded. We are going to have more Orcs than Golems in the map, and that is reflected in **Pool Size** for both enemies. The following figure shows the Golem after the Object Pool is set up:

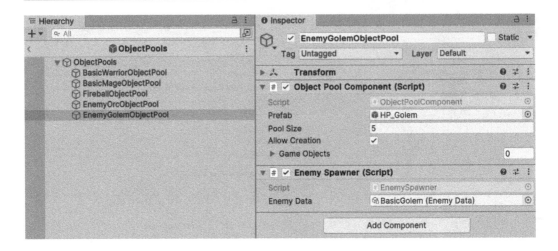

Figure 9.7 – The Object Pool component for the Golem spawner

Finally, we are going to add the Red Dragon to the **ObjectPools** Prefab. Since the Red Dragon is the boss enemy, we will have only one in the map, but we are still going to use the Object Pool to pre-instantiate it when the level is loaded:

1. In the **Hierarchy** view, left-click on the **ObjectPools** GameObject, select the **Create Empty** menu option, and name it EnemyRedDragonObjectPool.

2. Select the new **EnemyRedDragonObjectPool** GameObject and, in the **Inspector** view, click on the **Add Component** button, search for EnemySpawner, and double-click to add it.

3. In **Object Pool Component**, click on the circle button on the right side of the **Prefab** property. In the new window, select the **Assets** tab, search for the Prefab named **DragonSoulEaterRedHP**, and double-click to add it.

4. Change **Pool Size** to **1** and do not enable the checkbox in the **Allow Creation** property.

5. In the **Enemy Spawner** component, click on the circle button on the right side of the **Enemy Data** property. In the new window, select the **Assets** tab, search for the ScriptableObject named RedDragon, and double-click to add it.

6. Save the Prefab changes.

Besides the Prefab and **EnemyData**, the Red Dragon has a couple of differences when compared with both the Orc and Golem setups. We are going to have only one instance of the Red Dragon; to ensure that, **Pool Size** has the value **1** and the Object Pool is not allowed to create a new Red Dragon if there is already one on the map. We can see in the following figure the Object Pool configured for the Red Dragon:

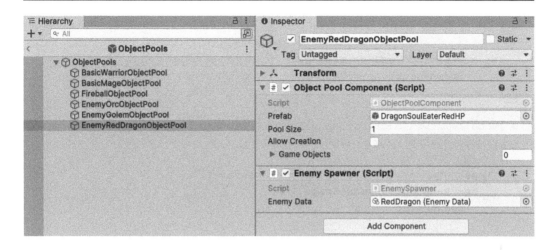

Figure 9.8 – The Object Pool component for the Red Dragon spawner

We now have the Object Pool set up for all three enemy types and ready to spawn them on the map. Next, we are going to create a new debug script to spawn the enemies, like the one we created for units as well.

Testing the enemy spawner

The easiest way to test our scripts is to create a debug menu option. Even when the game is all done and has all the features implemented, the debug menus are still great tools for speeding up the development and testing of changes or new features. In the **Scripts | Editor | Debug** folder, add a new script named EnemyDebugger, and then replace the content with this code:

```
using UnityEditor;
using UnityEngine;
namespace Dragoncraft
{
  public static class EnemyDebugger
  {
    [MenuItem("Dragoncraft/Debug/Enemy/Spawn Orc")]
    private static void SpawnOrc()
    {
      MessageQueueManager.Instance.SendMessage(
        new BasicOrcSpawnMessage() {
        SpawnPoint = new Vector3(-6, 0, 0) });
    }

    [MenuItem("Dragoncraft/Debug/Enemy/Spawn Golem")]
    private static void SpawnGolem()
```

```
    {
        MessageQueueManager.Instance.SendMessage(
            new BasicGolemSpawnMessage() {
                SpawnPoint = new Vector3(6, 0, 0) });
    }


    [MenuItem("Dragoncraft/Debug/Enemy/Spawn Red Dragon")]
    private static void SpawnRedDragon()
    {
        MessageQueueManager.Instance.SendMessage(
            new RedDragonSpawnMessage() {
                SpawnPoint = new Vector3(0, 0, 6) });
    }
  }
}
```

In the new EnemyDebugger class, we added the SpawnOrc, SpawnGolem, and SpawnRedDragon methods to send different spawning messages for each enemy type. In contrast with the units, the enemy message has a parameter that is required to be set, SpawnPoint; otherwise the default value of (0, 0, 0) will spawn all the enemies in the same spot at the center of the screen.

Here, we are setting some hardcoded values for our debug tool to spawn the enemies at different locations, all of them within the visible area. SpawnPoint values will spawn the Orc on the left side of the screen, the Golem on the right side of the screen, and the Red Dragon at the top of the screen.

Once we save the script, we can go back to the Unity Editor, run the **Level01** or **Playground** scene, and use the new **Debug** menu, as shown in the following figure, to test each of the enemy spawns:

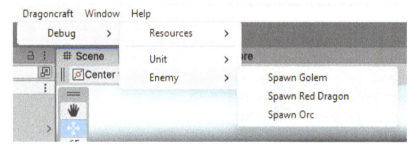

Figure 9.9 – The enemy debug menu

After spawning one of each enemy on the map, we are going to see them as shown in the following figure. Notice that, since this is just a **Debug** menu for testing the enemy spawn, if we click on any option more than once, a new instance will be added but in the same place, so we might not be able to see it on the map.

Figure 9.10 – All the enemies spawned on the map

We just finished configuring the messages, components, spawner, object pool, and debug menu for the three enemies that we added to our RTS game. They are all ready to be used and expanded, but before doing that, we will first add feedback in the UI to show the damage that the enemy has received when attacked, as well as play the **Die** animation state once the enemy life reaches zero.

Damage feedback in the UI

In an RTS game, we will have many units and enemies at the same time on the battlefield, and we need good feedback to show the player what is happening, giving the player enough information to think and react fast with a different strategy or update the current one. When selecting the units, the player will be able to see the current **Health Points (HP)** of each unit, but the enemies are not going to be selectable by the player, which means that we need visual feedback that they are being damaged the units, as well as the damage enacted by the enemies on the units, even when not selected.

We are going to create some simple UI feedback that will be displayed on top of the enemy or unit when they take damage. This UI feedback, which will be text about the amount of damage taken, will be visible for a couple of seconds and fade away eventually. We are going to position the UI feedback on top of the 3D model, but with a small random range so that when multiple attacks happen, we can see multiple damage values, without them overlapping with each other.

Finally, we are going to use our trusty Message Queue system to tell the UI to add the feedback, and our flexible Object Pool system to pre-instantiate and reuse the feedback object.

Managing the damage feedback

Let us start by adding a new script to the **Scripts | MessageQueue | Messages | UI** folder, naming it `DamageFeedbackMessage` and replacing the content with this code snippet:

```
using UnityEngine;
namespace Dragoncraft
{
  public class DamageFeedbackMessage : IMessage
  {
    public float Damage;
    public Vector3 Position;
  }
}
```

The `DamageFeedbackMessage` class has two important properties that we will use to configure the UI feedback – the first one is `Damage`, which is the value of the attack that is being inflicted on the enemy or unit, and the second is `Position`, which will tell the UI where the GameObject is in the world so the UI can add the text on top of it. It is worth noting that while the enemies and units are on the level scene, the feedback will be handled by the UI scene because we are going to need the **Canvas** component to draw the UI text.

Now that we have the message defined, we can create the script that will be used by the Object Pool in **SceneUI** to instantiate the feedback text with the damage taken by the enemies or units. In the **Scripts | UI** folder, create a new script named **DamageFeedbackUI** and add the following code snippet:

```
using TMPro;
using UnityEngine;
namespace Dragoncraft
{
  public class DamageFeedbackUI : MonoBehaviour
  {
    [SerializeField] private ObjectPoolComponent _objectPool;
  }
}
```

The `DamageFeedbackUI` class is like a spawner but not quite the same. As we are already familiar with, the spawner is attached to the Object Pool and listens to a specific message to spawn a new GameObject. However, `DamageFeedbackUI` is similar to the `DetailsUpdater` class, which uses an Object Pool but the script responsible for managing the GameObjects is in **Canvas**, and not in the Object Pool.

That is why we are implementing the MonoBehaviour class instead of the BaseSpawner class, and ObjectPoolComponent is the property that we are going to set up in the GameObject as soon as we finish the **DamageFeedbackUI** script.

ObjectPoolComponent will be used when this script receives one specific type of message, and to receive the message, we need to add the code for adding and removing listeners for the DamageFeedbackMessage message type to this class, in the OnEnable and OnDisabled methods:

```
private void OnEnable()
{
  MessageQueueManager.Instance.
    AddListener<DamageFeedbackMessage>(OnDamageFeedback);
}

private void OnDisable()
{
  MessageQueueManager.Instance.
    RemoveListener<DamageFeedbackMessage>(
      OnDamageFeedback);
}
```

Both methods will add and remove the listener, respectively, for the DamageFeedbackMessage message and execute the callback method.

The next and last method in this script is OnDamageFeedback, which does a few things to position the text in the correct position but uses a random range so as not to overlap with other GameObjects:

```
private void OnDamageFeedback(DamageFeedbackMessage message)
{
  GameObject damageFeedback = _objectPool.GetObject();
  damageFeedback.transform.SetParent(transform, false);

  Vector3 position =
    Camera.main.WorldToScreenPoint(message.Position);
  RectTransform rectTransform =
    damageFeedback.GetComponent<RectTransform>();
  rectTransform.anchoredPosition = position +
    new Vector3(Random.value * 100, Random.value * 10, 0);

  TMP_Text damageText =
    damageFeedback.GetComponentInChildren<TMP_Text>();
  damageText.text = $"-{message.Damage}";
}
```

The first couple of lines in this method call the GetObject method from the Object Pool to retrieve a new instance or reuse an existing one, and then use the SetParent method to make sure that damageFeedback is positioned under the GameObject that will have the DamageFeedbackUI script attached to it.

Once we have damageFeedback, we need to find the corresponding position to which to move the text UI. The 3D world and the UI **Canvas** component have different coordinates, so a Vector3 position from the level will not work in a UI element unless we convert the coordinates for the system used by the UI. That is why we are using the WorldToScreenPoint method from the Camera API to convert the position from the message, that is, the enemy or the unit in the 3D world, into Vector3, which we can use in the **Canvas** to position a UI element.

Another difference between a GameObject in the 3D world and a UI element in the **Canvas** is that the component used changes the position, rotation, and scale. While the 3D world uses the Transform component, the **Canvas** uses the RectTransform component, which has a slightly different API but works in the same way. Once we get the RectTransform component, we can change anchoredPosition to the converted position. We are also adding a Vector3 variable with a random range to avoid overlapping with the text in the UI when adding multiple instances of damage feedback.

The Random.value API will return a random floating value between 0 and 1, and we are going to multiply it by 100 for position X and by 10 for position Y. The Z position is not used, so we can simply set it to 0. The resulting Vector3 added to the converted position will be the new anchoredPosition of the GameObject.

Finally, once we have the GameObject in the correct place, we get the TMP_Text component from the child GameObject and change text to be the Damage value received in the message. Note that we are adding the negative sign (-) before the Damage value in the string message, so it will have the visual effect of a decrease in the enemy's or unit's HP.

With the **DamageFeedbackUI** script finalized, we can create the last script of this chapter, the one responsible for fading the damage text over time.

Fading the text over time

We have just finished a nice script that will take care of displaying the damage value in UI text on top of whichever enemy or unit took that damage. However, if we use the scripts we have made so far, the UI text will stay on the **Canvas** pretty much forever. A nice effect that we can add to the UI text is making it fade away after a few seconds, making it more dynamic and professional-looking.

Let us start by adding a new script to the **Scripts | UI** folder, naming it TimeFadeComponent, and replacing the content with the following code snippet:

```
using TMPro;
using UnityEngine;
```

```
namespace Dragoncraft
{
  public class TimeFadeComponent : MonoBehaviour
  {
    [SerializeField] private float _timeToLive = 2;
    private float _counter;
    private Color _originalColor;
    private TMP_Text _text;
  }
}
```

This script is quite simple, but we have a few properties that are required to make it work properly. timeToLive, also commonly known as TTL, is a value in seconds that represents how long the GameObject will be visible. The goal of this script is to fade away the text by changing the transparency over time, making it proportional to the TTL. We are going to reach total transparency when the TTL reaches the zero value, and the GameObject is disabled. The counter property will take care of the remaining time.

Then, the last couple of properties, text and originalColor, will be used to display the text that we want to fade away and the color for which to change the transparency over time for the fade effect, respectively.

The next methods in this script, Awake and OnEnable, have the responsibility of initializing the text and originalColor properties and resetting the counter and text.color properties, respectively:

```
private void Awake()
{
  _text = GetComponentInChildren<TMP_Text>();
  _originalColor = _text.color;
}

private void OnEnable()
{
  _counter = _timeToLive;
  _text.color = _originalColor;
}
```

The Awake method gets the TMP_Text component from the children GameObject and, after getting it from the same object, it stores the original color in the originalColor property. Since the Awake method is executed only once, we can safely use it to get the component and default settings before changing the GameObject. The OnEnable method will reset counter to have the TTL value, and text.color to the value of the originalColor property.

The GameObject is returned to the Object Pool when disabled, and when the **DamageFeedbackUI** script requests a new GameObject, the OnEnable method is executed, and resets the settings required to restart the fade-away effect. We do not need the OnDisable method because there is nothing we want to do when disabling this GameObject.

The next and last method in this script is Update, which is responsible for the fading effect over time, and for disabling the GameObject when the TTL is reached:

```
private void Update()
{
  _counter -= Time.deltaTime;
  if (_counter < 0)
  {
    gameObject.SetActive(false);
    return;
  }

  _text.color = new Color(
    _text.color.r, _text.color.g, _text.color.b,
    _text.color.a - (Time.deltaTime / _timeToLive));
}
```

In the Update method, the most used information is how much time has passed every time an update is called internally by Unity Engine, and we can get that information from the Time.deltaTime API. The first thing we do here is to decrease the value of counter by the current value of deltaTime. We initialized counter with the TTL value in the OnEnable method, so now we decrease its value until it reaches 0 or less, and then we can disable the GameObject by calling the SetActive method with a value of false. Once that line is executed and the GameObject is disabled, the Update method is no longer executed.

If counter still has a valid value above zero, we can change text.color by creating a new Color object with the same values for R, G, and B, but with A decreased using the value of deltaTime divided by the value of timeToLive. This will change the transparency bit by bit over time until it is fully transparent. Now that we have finished the last script, we can move to **GameUI** and start adding the damage feedback components there.

Setting up the Prefabs for the damage feedback

We are going to need to create a new Prefab for the feedback UI that will have the text for the damage value and then add this Prefab to a new entry in the Object Pool, which will be used by our new **DamageFeedbackUI** component to display the feedback when requested. We can start by creating the Prefab in the project:

1. Open **GameUI** and, in the **Hierarchy** view, right-click on the **Canvas** GameObject, select the **Create Empty** option, and name the new GameObject DamageFeedback.

2. Click on the new **DamageFeedback** GameObject and, in the **Inspector** view, change **Anchor Preset** to **bottom** on the left and **left** on the top sides. Make sure to set both **Pos X** and **Pos Y** back to **0** after changing **Anchor Preset**.

3. Click on the **Add Component** button, search for the **TimeFadeComponent** script, and add it to the GameObject.

4. In the **Hierarchy** view, right-click on the **DamageFeedback** GameObject, and select the **UI |Text – TextMeshPro** menu option. Leave the default name **Text (TMP)**.

5. Click on the **Text (TMP)** GameObject and, in the **Inspector** view, change the text to **-10**, **Font Size** to **24**, and **Vertex Color** to the **Hexadecimal** value of **FF0000** (red).

6. Drag the **DamageFeedback** GameObject, from the **Hierarchy** view, and drop it into the **Prefabs |UI** folder, in the **Project** view, to create the Prefab.

7. Right-click on the **DamageFeedback** object, in the **Hierarchy** view, and select the **Delete** menu option to remove it. We no longer need the GameObject there since we created the Prefab.

The **DamageFeedback** Prefab will be used as a visual representation of the feedback in the UI, with the damage value in the text, and will also have the fade-away effect once is enabled on top of the enemy or the unit GameObject. The Prefab will be managed by the Object Pool, so we do not need it in the scene. Now that we have the Prefab, we can go ahead and create the new Object Pool in **GameUI**:

1. In the **Hierarchy** view, right-click on the **ObjectPoolsUI** GameObject, select the **Create Empty** menu option, and name it `DamageFeedbackObjectPool`.

2. Click on the new **DamageFeedbackObjectPool** GameObject and, in the **Inspector** view, click on the **Add Component** button, search for the `ObjectPoolComponent` script, and add it to the GameObject.

3. In **Object Pool Component**, click on the circle button on the right side of the **Prefab** property and, in the new window, click on the **Assets** tab, and search for the `DamageFeedback` Prefab. Double-click to select it.

4. Change the **Pool Size** property to **10** and click on the checkbox on the right-hand side of the **Allow Creation** property.

5. Save the changes in **GameUI**.

After creating the new Object Pool for the damage feedback, we will now have two object pools in **GameUI**, as we can see in the following figure. Note that this is not the same Object Pool used by the units and enemies, which is a Prefab added to each level scene.

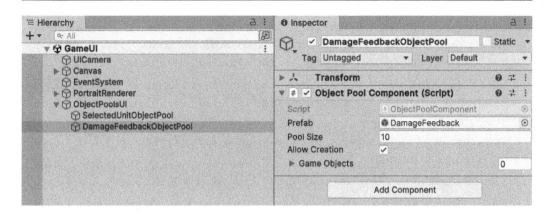

Figure 9.11 – DamageFeedbackObjectPool set up

The only thing left now is to create the GameObject inside the **Canvas** that will be responsible for displaying the actual damage text in the UI:

1. In the **Hierarchy** view, right-click on the **Canvas** GameObject, select the **Create Empty** menu option, and name it DamageFeedback.

2. Click on the new **DamageFeedback** GameObject and, in the **Inspector** view, change **Anchor Preset** to **stretch** on the left and **stretch** on the top sides. After changing **Anchor Preset**, change the **Left**, **Top**, **Right**, and **Bottom** values to **0**.

3. Click on the **Add Component** button, search for the **DamageFeedbackUI** script, and add it to the GameObject.

4. Drag the **DamageFeedbackObjectPool** GameObject from the **Hierarchy** view and drop it into the **Object Pool** property of the **Damage Feedback UI** component, in the **Inspector** view.

5. Save the changes in **GameUI**.

The following figure shows the result of these steps, and that will be the last change we are going to make in **GameUI**. We are all set up now to start using the new damage feedback system.

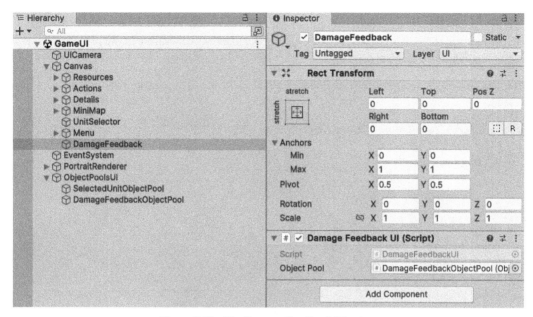

Figure 9.12 – The DamageFeedback UI set up

We just finished adding all components and UI changes to accommodate the damage feedback, and now it is time to wrap up this chapter by testing what we just finished implementing in our RTS game.

Testing the damage feedback

We do not have a combat system in place just yet, that will be done in the next chapter, *Chapter 10, Creating an AI to Attack the Player*. However, we can still test the damage feedback system by adding a test code to one of our scripts. Let us edit the **EnemyComponent** script and add a couple of methods that will help us to see the damage feedback in action. Open the script and add the following `Selected` method, which has temporary content:

```
public void Selected()
{
  // Test code for the damage
  TakeDamage(Attack);
}
```

The first line of the `Selected` method is a comment to indicate that the following line is just temporary code that we are using to test our damage feature. Here, we are calling the `TakeDamage` method, which will be defined next, using the own enemy `Attack` value as the damage, so we can have different values for each enemy type when testing it.

It is always a good practice to add comments to our code, especially to indicate something that may be tricky to understand at first sight, or just a quick note like the one we have in the preceding case, mentioning that this is just test code.

In the next code block, we have the TakeDamage method, which has the damage logic and calculation applied to the enemy:

```
public void TakeDamage(float attack)
{
    if (Health <= 0)
    {
        return;
    }

    float damage = attack - Defense;
    if (damage > 0)
    {
        Health -= damage;
        // The offset is 1/4 of the model size
        Vector3 position = transform.position +
          (_renderer.bounds.size * 0.25f);

        MessageQueueManager.Instance.SendMessage(
            new DamageFeedbackMessage() {
                Damage = damage, Position = position });
    }

    if (Health <= 0)
    {
        _animator.Play(
          _enemyData.GetAnimationState(
            UnitAnimationState.Death));
    }
}
```

The TakeDamage method will perform a few things when called. We are going to update the Health property, so the first validation we will do is to check whether Health, or HP, is 0 or less, and then avoid continuing with the method execution by calling the return keyword.

Next, we do the damage calculation, which is basically the value of the attack parameter minus the Defense property, and then check whether the resulting damage is greater than 0. Weak units will not be able to cause a lot of damage (if any) to the Red Dragon, for example, since its defense is quite high compared to a weak unit attack. The same applies the other way around, where a powerful attack will not be diminished by a low defense value.

If any damage is caused, we can send a DamageFeedbackMessage message with the current GameObject's position and the damage value that we have just calculated. Note that we are adding 1/4 of the model size as an offset, so the damage feedback will be placed roughly on top of the GameObject.

Finally, if Health is 0 or less, that means that the last attack killed this enemy, and we can change the animation state to Death. The next time the TakeDamage method is called, nothing will be executed since Health being 0 or less will skip everything.

Now, the last thing we need to do is to call the Selected method that we just created. We are going to do that in the **UnitSelectorComponent** script, which is also used to select the enemy. Open the **UnitSelectorComponent** script and, in the SelectUnits method, add the following code snippet to the foreach loop as the last thing to be done in the loop:

```
EnemyComponent enemy = collider.GetComponent<EnemyComponent>();

if (enemy != null)
{
    enemy.Selected();
}
```

If any collider found has the EnemyComponent component, we are going to execute the Select method. This will not affect any units since they do not have this component attached to their GameObject.

And that is it! Save all the scripts, return to the Unity Editor, and play either the **Playground** or **Level01** scene. Using the **Dragoncraft | Enemy** debug menu, spawn one of each enemy, Orc, Golem, and Red Dragon. Once we have the three enemies in the mix, left-click on the map and drag to select the enemies. This will execute our test code, which is going to call the TakeDamage method and trigger the damage feedback system.

Figure 9.13 – The damage feedback on the enemies

In the previous figure, we can see the damage feedback system in action, causing **-1** damage to the Orc, **-3** damage to the Golem, and **-7** damage to the Red Dragon. The more you select the enemies, the more damage they will take from our test method, and, after a few instances, they will end up without any HP left and the animation state will change to Death, resulting in the following figure.

Figure 9.14 – The enemies in the Death animation state

Go ahead and test the new enemies and the damage feedback system we just added to our RTS game. We just finished the foundation of our enemy components, and now we are going to continue from this solid foundation and add more features to the enemy variations we have.

Summary

Great job reaching the end of this chapter! We now have three different enemies added to and configured in our RTS game. The trio of Orc, Golem, and Red Dragon has different configurations and will present a real challenge to the player.

In this chapter, we used most of the systems that were developed in the book so far, showing that they were not made only for the units but could also be used for enemies as well. We learned how to modify, expand, and adapt existing code to add enemies, and we can even spawn them using the debug menu, which now has more options to help us develop our RTS game.

We also created a very useful and flexible damage feedback system to show the player how much damage the enemies and the units are taking and learned how to calculate the damage based on a few properties from the enemy configuration. The Dragoncraft game is reaching an important milestone now with enemies added into the project, with different configurations that will make that game challenging and interesting for the player.

In *Chapter 10, Creating an AI to Attack the Player*, we will start implementing one of the most important features of an RTS game: the battle. Our units will be able to attack the enemies, and the enemies will have different behaviors and states for patrolling, chasing, and attacking units. Now that we are past halfway through the book, we are starting to connect the systems and features to turn our project into a real RTS game.

Further reading

In this chapter, we did not introduce many new concepts or Unity features since we mostly reused things we already covered in this book. However, it is worth mentioning at least a couple of links to complement what we just saw in this chapter:

- *Prototype*: `https://gameprogrammingpatterns.com/prototype.html`
- *Time.deltaTime*: `https://docs.unity3d.com/2023.1/Documentation/ScriptReference/Time-deltaTime.html`

10

Creating an AI to Attack the Player

In an RTS game, the battle is one of the most important and challenging features of the game, and it could be the one thing that will make players love or hate the game. Combat must be a fun challenge for the player, and all the actions and decisions must be made by the player during the battle.

In this chapter, we are going to expand the components we have created so far for both units and enemies, adding more physics settings and a new collision component that will work with the NavMesh system to create a solid combat experience for the player. We will also refactor some duplicated code to make it easier for units and enemies to interact with all the new and existing systems.

By the end of this chapter, you will have learned how to set up a solid but flexible physics and collision system, and how it works with a NavMesh system at the same time. You will see how we calculate damage and make the enemy chase the units that are trying to flee or attack from a distance. You will also learn how to identify which code can be refactored and how to do so without breaking compatibility with the whole project, as well as how to ensure that enemies and units have a proper life cycle.

In this chapter, we will cover the following topics:

- Updating the physics settings
- Calculating the damage
- Managing the collision and chase behavior
- Implementing the character life cycle

Technical requirements

The project setup in this chapter with the imported assets can be found on GitHub at https://github.com/PacktPublishing/Creating-an-RTS-game-in-Unity-2023 in the Chapter10 folder inside the project.

Updating the physics settings

In the previous chapter, we included all the variations of enemies in the project and created a couple of scripts to spawn and update them. Now, we are going to expand enemy support by adding a custom layer for them and setting up the layer collision matrix and camera's Occlusion Culling to use the new layer.

Let's start by creating the new layer:

1. Click **Edit | Project Settings...**.

2. In the **Project Settings** window, select the **Tags and Layers** option from the left panel.

3. Add the **Enemy** value to **User Layer 9**.

4. Save the project.

The following figure shows the current custom layers that we have added so far to our project:

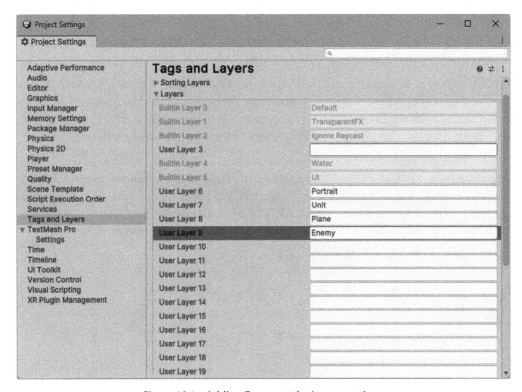

Figure 10.1 – Adding Enemy to the Layers settings

It is important that the **Portrait**, **Unit**, **Plane**, and **Enemy** custom layers are added to the corresponding **User Layer**. When setting a layer to a Occlusion Culling, for example, the Unity Engine uses **User Layer** as an index, and for that reason, it is not recommended to change or move layers once they have been created and used, to avoid unexpected behaviors and issues.

The next setting we are going to update is physics. Besides updating the Layer Collision Matrix, we are also going to change one very special option here that is required to make the collision detection work better with the NavMesh system. Let us set up the new layer in the physics configuration:

1. Open the **Project Settings** window and, in the left panel, click on **Physics**.

2. Change the **Contact Pairs Mode** property to **Enable All Contact Pairs**.

3. Expand the **Layer Collision Matrix** property and uncheck the intersections where the **Enemy** column meets the **Enemy**, **Portrait**, and **Plane** rows.

4. Save the project.

Each new custom layer that we add will create a new intersection in the **Layer Collision Matrix** section, making it a good practice to always check the settings in the **Physics** configuration. The current layer configuration in the following figure is what we have so far, and the goal is to disable the collision between GameObjects within the same **Unit** layer or in the same **Enemy** layer:

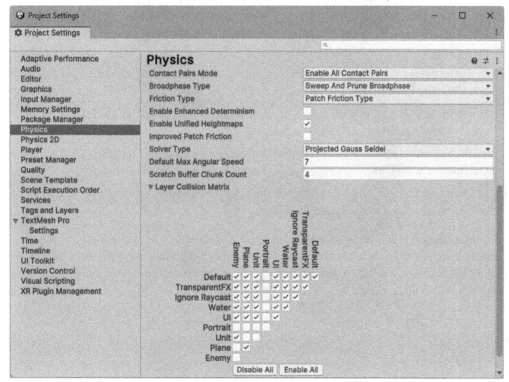

Figure 10.2 – Updating the Physics settings

Note that we have previously unchecked all the intersections between the **Portrait** layer and any other layer in the project because the **Portrait** layer is only applied to the selected unit shown on the UI and no physics is required there since it is only used for the portrait camera rendering.

Besides the **Layer Collision Matrix**, there is another **Physics** setting that we need to change as well, **Contact Pairs Mode**, which is how the engine manages the collision and trigger events between two GameObjects with `Rigidbody` components attached. The **Default Contact Pairs** option, which was the previous default value, will ignore the collision between two kinematic `Rigidbody` components. However, as we are going to see later, in the *Managing the collision and chase behavior* section, we are going to enable the kinematic property for our units and enemies; otherwise, the default collision will conflict with the NavMesh by pushing and pulling the GameObjects at the same time as the pathfinder is trying to do so.

To avoid collision detection issues, we changed **Contact Pairs Mode** to **Enable All Contact Pairs** – this will make collisions between any combinations of kinematic and non-kinematic GameObjects be detected but without the physics force being applied. The intersection between the **Enemy** and **Unit** layers with the **Plane** layer can also be unchecked because, without physics at play, we do not need collision detection with the ground.

Now that we have created the new **Enemy** layer, we also need to add it to the Occlusion Culling property of a few cameras to make the GameObjects visible once we set the layer to them:

1. In the **Project** view, locate the `MiniMapCamera` file in the **Prefabs | UI** path and double-click on it to open the Prefab in Edit Mode.

2. In the **Inspector** view, on the **Camera** component, click on the **Occlusion Culling** property and select the **Enemy** layer.

3. In the **Project** view, double-click to open the **Playground** scene and, in the **Hierarchy** view, click on the GameObject named **Main Camera**.

4. With the camera selected, in the **Inspector** view, on the **Camera** component, click on the **Occlusion Culling** property and select the **Enemy** layer. Save the scene changes.

5. In the **Project** view, double-click to open the **Level01** scene and repeat the previous step to add the **Enemy** layer to **Main Camera** as well.

The **MiniMapCamera** Prefab is responsible for rendering the map from the top, and, without selecting the **Enemy** layer, we would not be able to see the enemies in the mini-map. The same applies to **Main Camera** in each scene – they all must have the **Enemy** layer in **Occlusion Culling** to make the GameObject visible to the player.

The following shows all the layers selected in **Occlusion Culling** for **MiniMapCamera**.

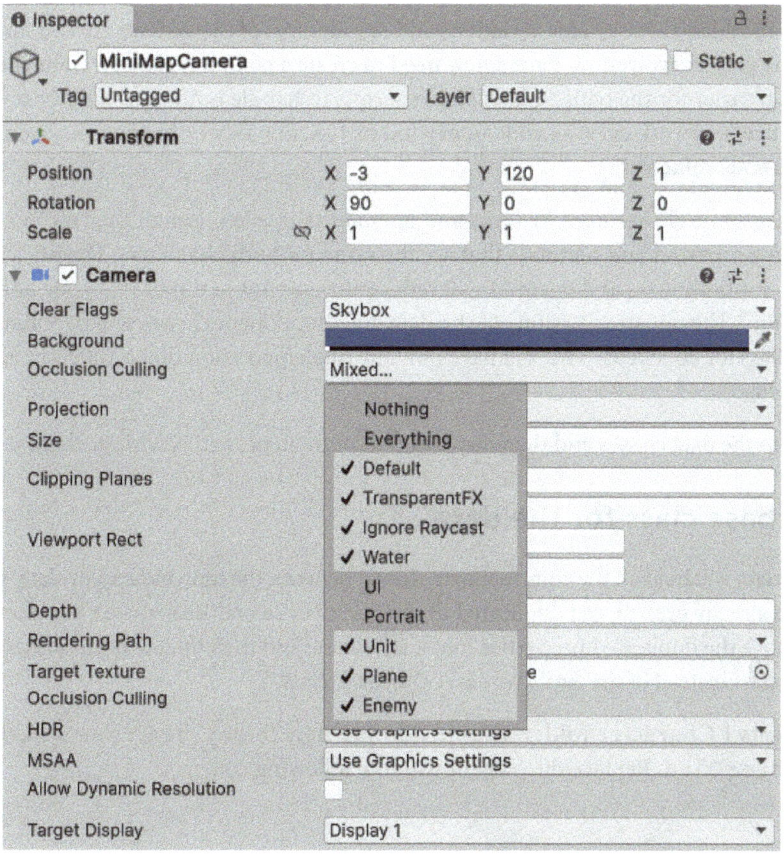

Figure 10.3 – Adding the Enemy tag to Occlusion Culling

The only scene that is not required to have the **Enemy** layer selected in the camera is **GameUI** because that specific scene has only the **UI** layer selected since it is the only thing that should be rendered in this scene. If you have created any other level scenes, remember to select the **Enemy** layer for **Occlusion Culling** as well.

Now that we have added the new layer and updated the physics settings, we can move to the battle code, starting with the damage calculation and a few changes to our existing code for both unit and enemy components.

Calculating the damage

Previously, we included all the enemy variations and reused most of the systems created for the units. However, as the project evolves, we have a new need to create a common class for both enemies and units to make it easier for the battle and collision scripts to handle both types of characters. It is a common practice to rework existing code to expand its functionality or improve its overall quality, which is called code **refactoring**.

We already have some classes used by units and enemies that, even though they are very specific, still share some properties and methods that are the same for both characters. Having a class with common code for both units and enemies will reduce the amount of duplicated code and improve the code usability. The updated versions of the data and component classes will be inherited from the common class for code reuse and will have only the implementation that is different and unique for each character.

Let us start with the data classes and then move to the components and NavMesh classes.

Adding a base class for the data

The new base class will hold all the common data shared between the unit and enemy data, so most of its content was already present and duplicated in the `UnitData` and `EnemyData` classes. Now we are going to move the duplicated properties into a new class, and then update both classes to inherit from the new class instead of the `ScriptableObject` class.

Create a **Scripts | Character** folder and, inside the new folder, create a new script called `BaseCharacterData`. Replace the content with the following code:

```
using UnityEngine;
namespace Dragoncraft
{
    public class BaseCharacterData : ScriptableObject
    {
        public float Health;
        public float Attack;
        public float Defense;
        public float WalkSpeed;
        public float AttackSpeed;
        public Color SelectedColor;
        public string AnimationStateAttack01;
        public string AnimationStateAttack02;
        public string AnimationStateDefense;
        public string AnimationStateMove;
        public string AnimationStateIdle;
        public string AnimationStateCollect;
```

```
        public string AnimationStateDeath;
        public float AttackRange;
        public float ColliderSize;
    }
}
```

The new `BaseCharacterData` class is still a `ScriptableObject` class, and here we can see all the properties that were duplicated for both the `UnitData` and `EnemyData` classes. We are also adding a new property called `ColliderSize`, which we are going to use later in *Managing the collision and chase behavior* to configure the collider size.

There is also one method that has the same code for both unit and enemy data classes that we are going to include in the `BaseCharacterData` class as well. Add the `GetAnimationState` method below the previous code in the newly created script:

```
public string GetAnimationState(
    UnitAnimationState animationState)
{
    switch (animationState)
    {
        case UnitAnimationState.Attack01:
            return AnimationStateAttack01;
        case UnitAnimationState.Attack02:
            return AnimationStateAttack02;
        case UnitAnimationState.Defense:
            return AnimationStateDefense;
        case UnitAnimationState.Move:
            return AnimationStateMove;
        case UnitAnimationState.Idle:
            return AnimationStateIdle;
        case UnitAnimationState.Collect:
            return AnimationStateCollect;
        case UnitAnimationState.Death:
            return AnimationStateDeath;
        default:
            return AnimationStateIdle;
    }
}
```

This method is the same one that we duplicated for both the `EnemyData` and `UnitData` classes, without any changes.

That is the last part of the new `BaseCharacterData` class. Now, we have removed the duplicated code from two different classes, `EnemyData` and `UnitData`, and have it all in the new `BaseCharacterData` class, which can be reused; it is always a good practice to reduce code duplication and use resources such as inheritance to expand classes.

Now, we need to change the `EnemyData` script to inherit from the new `BaseCharacaterData` class. Open the script located at **Scripts | Enemy |** `EnemyData.cs` and replace all content with the following script, which is smaller now since we moved all the duplicated code:

```
using UnityEngine;
namespace Dragoncraft
{
    [CreateAssetMenu(menuName = "Dragoncraft/New Enemy")]
    public class EnemyData : BaseCharacterData
    {
        public EnemyType Type;
    }
}
```

Note that the updated `EnemyData` is now inherited from the `BaseCharacterData` class, making it still a ScriptableObject class with the same properties as before. There is only one property left in this updated script, `Type`, since it is unique to the enemy and cannot be moved into a common class. We also kept the attribute to create the menu option for the same reason.

Once we update and save the changes, we will notice another cool thing: even though we moved all but one property into the `BaseCharacterData` class, the `EnemyData` class still has all these properties due to inheritance. That means that no change is required on any script or ScriptableObject that uses this class, so we have improved our code base by removing duplicated content, and everything is still working flawlessly.

This is one great example of how powerful **Object-Oriented Programming (OOP)** is. **Inheritance**, which is one feature of the OOP, makes it easy to expand and refactor our code without needing to update assets and scripts that are using the existing class.

Now that the `EnemyData` class is updated, let us do the same in the corresponding unit class. Open the `UnitData.cs` script located at **Scripts | Unit** and replace all its content with the following code:

```
using UnityEngine;
namespace Dragoncraft
{
    [CreateAssetMenu(menuName = "Dragoncraft/New Unit")]
    public class UnitData : BaseCharacterData
    {
        public UnitType Type;
        public int Level;
        public float LevelMultiplier;
        public ActionType Actions;
```

```
    }
  }
```

The updated UnitData class is now inherited from the BaseCharacterData class and has most of the properties removed from here. We still have the Type, Level, LevelMultiplier, and Actions properties because they are exclusive to a unit, but it makes no sense to include them in the common class for the enemy since they are not required there.

After saving the changes to the updated UnitData class, the code and ScriptableObject that use this class are still working and showing the correct data.

Now that we are done with the data classes, we can do a similar refactoring for both the UnitComponent and EnemyComponent classes, which also share some duplicated code.

Adding a base class for the characters

We are going to use the same concept of inheritance in the new class, which will have common properties and methods for both enemies and units. The following class is slightly more complex than the one just created, so we are going to address it in parts. First, let us create a new script inside the **Scripts | Character** folder, name it BaseCharacter, and then replace the content with the following code:

```
using System;
using UnityEngine;
namespace Dragoncraft
{
  [RequireComponent(typeof(Animator))]
  public class BaseCharacter : MonoBehaviour
  {
    protected Animator _animator;
    protected Renderer _renderer;
    protected Color _originalColor;
    protected Color _emissionColor;
    protected ActionType _action;

    public string ID;
    public float Health;
    public float Attack;
    public float Defense;
    public float WalkSpeed;
    public float AttackSpeed;
    public Color SelectedColor;
    public float AttackRange;
    public float ColliderSize;
```

```
      public bool IsDead { get; private set; }
   }
}
```

All the properties in the code were already part of both the UnitComponent and EnemyComponent classes. One important change here is that the private properties are now protected, so they can be accessed by the classes that are inheriting from it. We also added the ColliderSize property, which is new and common to both units and enemies, and the IsDead property, which will help to check whether the character is still alive or not. The RequiredComponent attribute of the Animator type was also moved into this base class since the animator property is also here.

Duplicated code might not be an exact copy; it could have a different implementation for the same feature. The following methods were presented for both the UnitComponent and EnemyComponent classes; however, they had slightly different implementations. We are going to keep the version from EnemyComponent because it is cleaner and simpler than the UnitComponent version.

Add the following private methods to our new BaseCharacter class:

```
private void Awake()
{
  _renderer = GetComponentInChildren<Renderer>();
  _animator = GetComponent<Animator>();
  _originalColor = _renderer.material.color;
  _emissionColor =
    _renderer.material.GetColor("_EmissionColor");
}

private void OnMouseEnter()
{
  _renderer.material.color = SelectedColor;
}

private void OnMouseExit()
{
  _renderer.material.color = _originalColor;
}
```

This code is going to change the color of either a unit or enemy when the mouse cursor is hovering over it and return to the original color when the mouse cursor is not over the GameObject. This already applied to enemies, but the one in the UnitComponent class was a bit different and kept all the selected units as SelectedColor when they were selected. We are also initializing the emissionColor property here by getting the EmissionColor property from the material in the GameObject.

Moving both units and enemies to have the same implementation is a simple and consistent way to make sure the same feature is written and used in the same way, without confusing the player with two different types of feedback for basically the same action.

The next protected method is one of the reasons we created the `BaseCharacterData` class previously – so that we can use it as a generic parameter to copy the data from the ScriptableObject into the properties in the class:

```
protected void CopyBaseData(BaseCharacterData data)
{
  ID = System.Guid.NewGuid().ToString();
  Health = data.Health;
  Attack = data.Attack;
  Defense = data.Defense;
  WalkSpeed = data.WalkSpeed;
  AttackSpeed = data.AttackSpeed;
  SelectedColor = data.SelectedColor;
  AttackRange = data.AttackRange;
  ColliderSize = data.ColliderSize;
}
```

In the `CopyBaseData` protected method, we are setting the values of each `public` property in this class, copying the value from the ScriptableObject represented by the `BaseCharacterData` class. All the properties here are common and shared between both unit and enemy components, and this method will be called from the classes that are inheriting from the `BaseCharacter` class.

The next couple of protected methods are quite unique because they are `virtual`, which means they can be executed or overridden by the class inheriting from this one, but there is a catch in one of them. Add the following methods to the class:

```
protected virtual void UpdateState(ActionType action)
{
  if (IsDead || _action == action)
  {
    return;
  }
  _action = action;
}

protected virtual void PlayAnimation(UnitAnimationState state)
{
  throw new NotImplementedException();
}
```

The two protected methods are new, and they will be overridden by both the UnitComponent and EnemyComponent classes, but each one differently. The UpdateState method does exactly that – updates the action property that reflects the current state – and will have that piece of code executed alongside some unique code in each class. However, the PlayAnimation method must be overridden; otherwise, it will throw a NotImplementedException error, which will cause an error stating that this method was not implemented. It is a good way to ensure that this method is overridden when executed.

The next public method is the main one responsible for calculating the damage taken by a unit or enemy, and it will be called by the class responsible for collision detection and attacking later, in *Managing the collision and chase behavior*. Add the following method to the class:

```
public bool TakeDamage(float damage)
{
    if (IsDead)
    {
        return true;
    }

    Health -= damage;
    if (Health > 0)
    {
        return false;
    }

    gameObject.AddComponent<DeadComponent>();
    PlayAnimation(UnitAnimationState.Death);
    IsDead = true;
    return true;
}
```

The public TakeDamage method defined here will receive the total amount of damage that the enemy or unit has suffered and will return a bool value that represents whether the unit or enemy is still alive after receiving this damage. The first part of the method checks whether the unit or enemy is dead or not and returns true, exiting the method if the character is already dead. The second part subtracts damage from Health and, if the value is still greater than 0, the method returns false indicating that the unit or enemy is still alive and we can exit the method.

However, if Health becomes equal to or less than zero, the last four lines of the method are executed, where we add new a component, DeadComponent, change the animation state, set the value of the IsDead property, and finally exit from the method. The new DeadComponent component is added to the GameObject, but keep in mind that this class does not exist just yet. We are going to create it later, in *Implementing the character life cycle*.

The next line executes the `PlayAnimation` method to change the animation to the `Death` state, which will cause an error if it is not implemented in the class inheriting from the `BaseCharacter` class. Finally, the last couple of lines will change the value of the `IsDead` property to `true` and return `true` as well, indicating that the character is no longer alive.

The last method that we are going to add to this class is used to find the best place to add the damage feedback on top of the character, considering the 3D model sizer and position:

```
public Vector3 GetDamageFeedbackPosition()
{
    return new Vector3(
        transform.position.x + _renderer.bounds.size.x / 2,
        transform.position.y + _renderer.bounds.size.y / 2,
        transform.position.z);
}
```

The `public GetDamageFeedbackPosition` method is new and will be used to find the best position at which to render the text showing the amount of damage taken by the character, represented as a `Vector3` object. This method uses the `position` and `size` properties of the 3D model to place the feedback test just above the character's head.

After adding that last method, we can save the changes to this new base class and start to use it in the updated versions of the `EnemyComponent` and `UnitComponent` classes.

Updating EnemyComponent

In the previous chapter, *Chapter 9, Adding Enemies*, we created the `EnemyComponent` script and added an initial implementation with test code to simulate the damage taken by the enemy so we could test the damage feedback UI that was also created in that chapter. Now, we are going to replace the entire content of the `EnemyComponent` class with the new version, which is going to have everything we need for the battle against the units.

The new version of this script is a bit smaller than the previous version, with most of the old content moved into the `BaseCharacter` class and new methods added here. Open the script located at **Scripts | Enemy** | `EnemyComponent.cs` and replace all the content with this code:

```
using UnityEngine;
namespace Dragoncraft
{
    public class EnemyComponent : BaseCharacter
    {
        public EnemyType Type;
        private EnemyData _enemyData;
    }
}
```

We can already see how much cleaner the class is because all properties are now in the `BaseCharacter` class, and `EnemyComponent` now only has a couple of properties that are specific to it, `Type` and `enemyData`.

All the methods were also moved into the `BaseCharacter` class, but we are keeping the following method with one change. Add this one to the class as well:

```
public void CopyData(EnemyData enemyData, Vector3 spawnPoint)
{
    CopyBaseData(enemyData);
    Type = enemyData.Type;

    _enemyData = enemyData;
    _action = ActionType.None;

    transform.position = spawnPoint;
    PlayAnimation(UnitAnimationState.Idle);
}
```

The `public CopyData` method has the same parameters. However, we are going to reduce the amount of content by setting the initial value of the class properties using the `CopyBaseData` method from the `BaseCharacter` class. After that, we are setting the only `public` property that is unique in this class, which is `Type`. We are also initializing `action` with a default value of `None` and we are using the `PlayAnimation` method instead of directly accessing the `animator` variable, which is now in the `BaseCharacter` class.

The next couple of methods are new to this class, and are also `virtual`, which means we are going to override them in this class. The following method must be included so we can change the animation state based on the action that was executed; this is an example of the **State** design pattern, where we change the behavior based on an internal change – the action makes the animation state change in this case:

```
protected override void UpdateState(ActionType action)
{
    base.UpdateState(action);

    switch (action)
    {
        case ActionType.Attack:
            UnitAnimationState attackState =
                (UnityEngine.Random.value < 0.5f) ?
                    UnitAnimationState.Attack01 :
                        UnitAnimationState.Attack02;
```

```
        PlayAnimation(attackState);
        break;
    case ActionType.Move:
        PlayAnimation(UnitAnimationState.Move);
        break;
    case ActionType.None:
        PlayAnimation(UnitAnimationState.Idle);
        break;
    default:
        break;
    }
}
```

The protected UpdateState method overrides the virtual method with the same name and parameters from the base class. However, we are not ignoring the code that is already inside the virtual method defined in the BaseCharacter base class because we are executing the code in the virtual method by calling the base.UpdateState method with the action parameter. This will make sure we execute that code before moving on to the code in this class, which is the switch statement and basically plays a corresponding animation for each of the three supported actions here: Attack, Move, and None.

In the UpdateState method, we are using the virtual keyword and the override keyword to make it possible to execute the code on both methods, from both the BaseCharacter and EnemyComponent classes.

For the next and last method, we are doing it slightly differently because we do not want to execute the code from the base class since this will throw an exception that the method is not implemented:

```
protected override void PlayAnimation(UnitAnimationState state)
{
    _animator.Play(_enemyData.GetAnimationState(state));
}
```

The protected PlayAnimation method overrides the virtual method and we are not calling the method from the base class because we do not want to execute the code there. This time, we are only executing the line that will find the corresponding animation state based on the state we want to play. We are still using UnitAnimationState because the enemy still shares the same animations with the unit.

Now that we have finished adding the methods to the EnemyComponent class, we can see that it is cleaner and more organized than before since the duplicated code is now in a base class, and we added only what is specific to the enemy. After these changes, the class is still working as it was before, but we do need to make one small change in another class to remove a test code that we added in the previous chapter. So, open the script located at **Scripts | Unit** | `UnitSelectorComponent.cs` and remove the highlighted lines from the `SelectUnits` method, as shown here:

```
EnemyComponent enemy = collider.GetComponent<EnemyComponent>();
if (enemy != null)
{
    enemy.Selected();
}
```

This code must be removed from the last lines of the `SelectUnits` method since we no longer have the `Selected` method in the `EnemyComponent` class and that was only temporary code to allow us to test the damage feedback UI.

Next, we are going to do the same with the `UnitComponent` class; however, this time we are not replacing the whole content but adding and removing the differences since this class already has a lot of specific code for the unit.

Updating UnitComponent

The refactoring in the `UnitComponent` class is a bit different from the previous class because we want to keep most of the code that is already there. In this updated version, we are going to remove the properties and methods that were moved into the `BaseCharacter` class and add a few new methods.

Open the script located at **Scripts | Unit** | `UnitComponent.cs` and remove the `RequireComponent` attribute on top of the class definition since it was moved to the new base class and is not required here anymore. Also, remove the `public ID, Health, Attack, Defense, WalkSpeed, AttackSpeed, SelectedColor`, and `AttackRange` properties, and the `private animator, renderer, originalColor`, and `action` properties.

The following code snippet shows the updated version of the class declaration, inheriting from `BaseCharacter`, and the `public` and `private` properties that we will keep in the class because they are specific to the unit:

```
using UnityEngine;
namespace Dragoncraft
{
    public class UnitComponent : BaseCharacter
    {
        public UnitType Type;
        public int Level;
        public float LevelMultiplier;
        public ActionType Actions;

        private Vector3 _movePosition;
```

```
        private bool _shouldMove;
        private bool _shouldAttack;
        private float _attackCooldown;
        private UnitData _unitData;
        private float _minDistance = 0.5f;
    }
}
```

Now that the properties are cleaned up, we can move forward and remove the methods that were also moved into the BaseCharater class. Remove the Awake, OnMouseEnter, OnMouseExit, and OnCollisionEnter methods. The last one is not present in the new class, but we are going to create a new component for it later, in *Managing the collision and chase behavior*.

There is only one method that we are going to keep but update its content, which is CopyData. As we did in the EnemyComponent class, we are going to remove the code that was setting the properties that are now in the base class and, instead, we are going to call the CopyBaseData method. Update the method to reflect the following code:

```
public void CopyData(UnitData unitData)
{
    CopyBaseData(unitData);
    Type = unitData.Type;
    Level = unitData.Level;
    LevelMultiplier = unitData.LevelMultiplier;
    Actions = unitData.Actions;

    _unitData = unitData;
    _action = ActionType.Move;

    EnableMovement(false);
}
```

Besides removing the properties and calling the CopyBaseData method, we also added a new line here to initialize the action variable with the value Move. The Type, Level, LevelMultiplier, and Actions properties are unique to the unit and not present in the base class, so we are still initializing them in this method.

Since we added a new property to the BaseCharacter class for emissionColor, we need to update the Selected method to use the new property. In the existing Selected method, update the highlighted line with the usage of emissionColor in the following code block:

```
public void Selected(bool selected)
{
    ...
    foreach (Material material in materials)
```

```
{
  if (selected)
  {
    material.SetColor("_EmissionColor",
      SelectedColor * 0.5f);
  }
  else
  {
    material.SetColor("_EmissionColor", _emissionColor);
  }
}
}
```

This change will make sure to revert emissionColor back to the original value before we apply SelectedColor to emissionColor in the GameObject's material property.

The next method is new and is going to be overridden by the NavMesh class, which is why we are defining it as protected and virtual in this class:

```
protected virtual void StopMovingAndAttack()
{
  _shouldMove = false;
  _shouldAttack = true;
}
```

The StopMovingAndAttack method updates both properties that will reflect that the unit should stop moving and start attacking. The NavMesh class will also prevent the NavMesh agent from moving, which is why we are defining it here as a virtual method.

We are going to add the last couple of methods to the updated version of this class. Both were also added to the EnemyComponent class as overridden methods; although the content of the methods might look similar to the enemy version, they have a few differences that justify adding them here as well:

```
protected override void UpdateState(ActionType action)
{
  base.UpdateState(action);

  switch (action)
  {
    case ActionType.Attack:
      EnableMovement(false);
      UnitAnimationState attackState =
        (UnityEngine.Random.value < 0.5f) ?
          UnitAnimationState.Attack01 :
```

```
          UnitAnimationState.Attack02;
      PlayAnimation(attackState);
      break;
  case ActionType.Move:
      EnableMovement(true);
      break;
  case ActionType.None:
      _movePosition = transform.position;
      break;
  default:
      break;
  }
}
```

In this class, the protected `UpdateState` method is also overriding the `virtual` method with the same name and parameters from the base class, as well as executing the code in the `virtual` method by calling the same method with the `base` keyword. The `switch` statement in this class does a few different things other than just playing an animation state.

The `Attack` and `Move` actions will make the unit stop moving and change the animation to `Idle` or `Move`, respectively. However, when the action is `Attack`, the `Attack01` or `Attack02` state will be chosen randomly and played. When the action is `None`, we update `movePosition` to be the current position, so the unit stays there.

The last method to be overridden is `PlayAnimation`, which uses the `unitData` property to find the corresponding animation state and uses the `animator` property to play the animation:

```
protected override void PlayAnimation(UnitAnimationState state)
{
  _animator.Play(_unitData.GetAnimationState(state));
}
```

Once we have added that last method, the updated version of the `UnitComponent` class is done. Now we have classes with less duplicated code that are easy to expand. We can now easily add new features to `UnitComponent` or `EnemyComponent` or make them available to both classes if we decide to include the new feature in the `BaseCharacter` class that is common to them.

We have finished the code refactoring of these two classes, and they are still working as they were before. Their content has been changed but not how they work or interact with the project, which shows once again how powerful OOP can be when we use it wisely. Before we move on to collision, there is one more change we need to make the ranged attack also cause damage to enemies.

Adding damage to ranged attack

When the projectile fireball hits an enemy, we want to cause damage, taking away some health points the same way we expect a melee attack would do. To do that, we will need to include the damage value in the fireball so that when it collides with the enemy, we know how much damage it will cause. The damage value will be calculated soon, in *Managing the collision and chase behavior*, but we need a few changes to make it work.

Open the script located at **Scripts | MessageQueue | Messages | Battle | Fireball** SpawnMessage.cs and add the new Damage property as highlighted in this code:

```
public class FireballSpawnMessage : IMessage
{
  public Vector3 Position;
  public Quaternion Rotation;
  public float Damage;
}
```

While Position and Rotation are values used to determine the spawn position, the Damage value will be used when the projectile collides with something only.

Now, open the script located at **Scripts | Battle | **ProjectileComponent.cs to make a few changes as well. In the Setup method, we are going to add a new damage parameter, which will be stored in the new property named Damage, as we can see in the highlighted changes here:

```
public float Damage;

public void Setup(Vector3 pos, Quaternion rotation, float damage)
{
  transform.position = pos;
  transform.rotation = rotation;
  Damage = damage;

  …
}
```

That is the only change in the ProjectileComponent class, just adding a new parameter to the Setup method so we can store it in a public property to be used when we need it. Now that we have updated the method signature by adding a new parameter, we need to update the script that is using it as well to accommodate the change. Open the script located at **Scripts | Spawner |** FireballSpawner.cs and add the highlighted change:

```
private void OnFireballSpawned(FireballSpawnMessage message)
```

```
{
  ...
  projectile.Setup(
    message.Position, message.Rotation, message.Damage);
}
```

In the `OnFireballSpawned` method, we are adding the `Damage` property from `Fireball SpawnMessage` to the updated `Setup` method from the `ProjectileComponent` class. In the preceding code, we are setting up the projectile with the damage value received in the message, and the last change we are going to make is to add the damage value to the message.

Open the script located at **Scripts | Unit |** `UnitComponent.cs`, which we just finished updating, and in the `UpdateAttack` method, add the highlighted code to `FireballSpawnMessage`:

```
...
  MessageQueueManager.Instance.SendMessage(
    new FireballSpawnMessage
  {
    Position = transform.position,
    Rotation = transform.rotation,
    Damage = Attack
  });
...
```

We are defining the value of `Damage` as the value of the `Attack` property configured for the unit. This means that the projectile fireball will cause the same amount of damage that a melee attack would cause. However, keep in mind that so far, only the Mage can use a ranged attack, especially because their defense is quite low in physical combat when compared with the Warrior, who cannot use ranged attack but has a good defense value for melee attacks.

Now we have everything that we need to move on to the enemy's collision and chase behavior.

Managing the collision and chase behavior

Over the few last chapters, we learned how to build a great foundation for an RTS game. In *Chapter 7, Attacking and Defending Units*, we started to develop the **UnitComponent** script, which has now been updated to have more features. After that, in *Chapter 8, Implementing the Pathfinder*, we used the Unity built-in NavMesh system to move units on the map. Then, in *Chapter 9, Adding Enemies*, we created the **EnemyComponent** script, which we have also updated now, and added enemy variations.

In this chapter, we updated and expanded the scripts from the chapters mentioned, and now we are going to put everything together to set up combat between units and enemies. First, we will need a new script to handle the collision between units and enemies, and vice versa. Once we have the collision detection ready to use, we are going to update and create the NavMesh scripts so that both systems can work together without causing issues.

Creating the collision component

We are going to use the physics system from Unity to detect collisions between two objects. We already set up the layer collision matrix so that enemies don't collide with each other, and did the same thing to avoid collision between two units. The script we will create now will detect a collision, make the unit or the enemy attack one target at a time, and take damage from the opponent from both melee or range attacks.

Create a new script inside the **Scripts | Battle** folder and name it `CollisionComponent.cs`. The following code snippet is the class declaration and the `public` and `private` properties that we are going to use in this class:

```
using System;
using System.Collections;
using UnityEngine;
namespace Dragoncraft
{
    [RequireComponent(typeof(SphereCollider), typeof(Rigidbody))]
    public class CollisionComponent : MonoBehaviour
    {
        private BaseCharacter _character;
        private SphereCollider _sphereCollider;
        private Rigidbody _rigidbody;
        private Coroutine _dealDamageCoroutine;
        private string _targetId;

        public Action<Transform> OnStartAttacking;
        public Action<Transform, bool> OnStopAttacking;
    }
}
```

The `RequireComponent` attribute in this class makes sure to include two dependencies of this script: `SphereCollider` and `Rigidbody`. Both are essential to define the collision boundaries for detection and how physics should affect the GameObject, respectively.

Besides the `private` properties for holding a reference to both `SphereCollider` and `Rigidbody`, we also have one `BaseCharacter` property, which means this script will work with both enemy and unit objects. The most interesting property is `Coroutine`, which can be used to run asynchronous functions, as we are going to see in later methods of this class. The last `private` property is a `string targetId` property that will contain the ID of the character being attacked so that we can avoid attacking multiple characters at the same time, even if more than one has a collision detected.

This class is a bit generic. None of the code here is made for a unit or an enemy. Units and enemies each require some custom code to be executed, and we are adding a couple of `Action` properties that will work as a callback to methods that can be assigned to them when an attack starts and when an attack stops.

The first method we are going to define in this class is `Initialize`, which is responsible for storing the `BaseCharacter` object, as well as getting the components for `SphereCollider` and `Rigidbody`:

```
public void Initialize(BaseCharacter character)
{
  _character = character;

  _sphereCollider = GetComponent<SphereCollider>();
  _sphereCollider.radius = character.ColliderSize;

  _rigidbody = GetComponent<Rigidbody>();
  _rigidbody.isKinematic = true;
}
```

Since both `SphereCollider` and `Rigidbody` are required components for this class, they will never be `null` or invalid, so we can use the return of the `GetComponent` method without checking whether they are valid. `radius` is the size of the sphere to detect the collider around the GameObject, and the value comes from the `ScriptableObject` class that we updated earlier in this chapter. As we learned at the beginning of the chapter, in *Updating the physics settings*, we need to make our `Rigidbody` component kinematic, so physics is not applied when two GameObjects collide but the collision detection still tells us that the objects touched each other's colliders. The `kinematic` property is `false` by default, so we need to change the value to `true` in the initialization.

The next method we are adding to this class is `FixedUpdate`, which comes from the `MonoBehaviour` class. This method is like the `Update` method, which is called once per internal tick in the Unity Engine; however, `FixedUpdate` is used for physics calculations due to its optimizations. In the code here, we are using it to change the `radius` property of the collider when the character is no longer alive:

```
private void FixedUpdate()
{
  if (_character != null && _character.IsDead)
  {
    _sphereCollider.radius = 0;
    return;
  }
}
```

We first check whether the `character` property is not `null` before accessing the value of the `IsDead` property because `FixedUpdate` is likely to run before we call the `Initialize` method that sets the value of the `character` property. If the character is dead, we change the collider's `radius` property to 0, which will make Unity call the OnCollisionExit method once. This is a clever way to ensure OnCollisionExit is called after OnCollisionEnter because removing the collider component or destroying the GameObject would not execute OnCollisionExit, while reducing the radius triggers the exit event.

The next method, OnCollisionEnter, is one of the most important in this class because it is the event triggered by Unity when a collision happens. It is triggered only once when the collision happens, and it is only called again if the two objects stop colliding with each other and then collide again. It is important to note that OnCollisionEnter (and OnCollisionExit as well) is only called if one of the GameObjects has a Rigidbody component attached to it. In our case, we have the component, but because of the kinematic property, we needed to change **Contact Pairs Mode** in **Project Settings** to allow two kinematic GameObjects to detect the collision.

Let us add the following OnCollisionEnter method to our CollisionComponent class:

```
private void OnCollisionEnter(Collision collision)
{
  if (collision.gameObject.TryGetComponent<BaseCharacter>(
    out var opponent))
  {
    if (string.IsNullOrEmpty(_targetId))
    {
      _targetId = opponent.ID;
    }

    if (_targetId.Equals(opponent.ID))
    {
      OnStartAttacking(collision.transform);
      _dealDamageCoroutine =
        StartCoroutine(TakeDamageOverTime(opponent));
    }
  }

  if (collision.gameObject.
    TryGetComponent<ProjectileComponent>(
      out var projectile))
  {
    collision.transform.gameObject.SetActive(false);
    TakeDamageFromProjectile(
      projectile.Damage, collision.transform);
  }
}
```

Let us break down what is happening inside the OnCollisionEnter method. There are two different objects for which we are interested in checking the collision, and both of them have a distinctive component attached:

- The first case is a collision with any GameObject that has the BaseCharacter component attached, which translates into having either UnitComponent or EnemyComponent scripts. If that is the case, the TryGetComponent method will return true and get the BaseCharacter instance in the opponent variable. Then, we have two different validations:

  - If the targetId property is null, that means that we haven't collided with any object that can be attacked yet, and then we set the value to be the ID value of the opponent

  - If the current targetId is the same as the ID value from the GameObject that entered the collision, and if that is true, then we can execute the attack

  The attack executes the OnStartAttack callback and then calls the StartCoroutine method to asynchronously execute the TakeDamageOverTime method. The TakeDamageOverTime method uses the opponent variable as a parameter so we know that the opponent is attacking the current GameObject, which could be a unit or an enemy.

  However, as we can see in the code, it is not required to know who the character being attacked is because we are attacking the unit that has this script attached. When the attack happens, we need to store the return of the StartCoroutine method so we can stop the coroutine when it is not needed anymore.

- The second case is a collision with any GameObject that has the ProjectileComponent attached. In our project so far, only the fireball thrown from the Mage would be valid here, which means that this GameObject took a hit from a fireball. If the TryGetComponent method returns true, we disable the GameObject that was hit by the fireball using the SetActive method so we can return it to the Object Pool to be reused. Then, we call the TakeDamageFromProjectile method with the Damage value and the transform property with the position at which the collision happened.

Even when a collision with a different GameObjects is detected, those are the only two components attached that will be relevant to this method. The next method, which is the opposite event, is triggered when the collision stops – in this situation, we do not care about any projectile; only the BaseCharacter component is relevant here:

```
private void OnCollisionExit(Collision collision)
{
   if (collision.gameObject.TryGetComponent<BaseCharacter>(
      out var opponent))
   {
      if (!string.IsNullOrEmpty(_targetId) &&
         _targetId.Equals(opponent.ID))
```

```
        {
            StopAttacking(collision.transform, opponent.IsDead);
        }
    }
}
```

The `OnCollisionExit` method is executed internally by Unity when two GameObjects stop colliding with each other. It is only triggered if the `OnCollisionEnter` event happened before; otherwise, it is ignored by Unity.

As we did in `OnCollisionEnter`, we need to check whether the GameObject has the `BaseCharacter` component as well, so we know the object that we collided with is a unit or enemy. Due to the **Physics** settings in the Layer Collision Matrix, a unit will never collide with another unit, and the same applies to enemies. When a collision happens, it is likely to be between a unit and an enemy, or it will be ignored.

When the collision stops happening and the `BaseCharacter` component is found in the GameObject, we check whether the GameObject that stopped the collision is the same as the one that attacked this GameObject. If that is the case, we call the `StopAttacking` method, which will receive as parameters the `transform` property with the collision position and the value of the `IsDead` property, so we know whether the collision stopped because the other GameObject moved or whether it is dead.

The next couple of methods are responsible for dealing damage to a GameObject in a collision or the GameObject that a projectile hit. We will start with a method that causes damage over time, which means that while a collision is happening, after the start event and before the exit event, this method will cause damage using the character's attack speed as the delay between each instance of damage:

```
private IEnumerator TakeDamageOverTime(BaseCharacter opponent)
{
    while (!opponent.IsDead && !_character.IsDead)
    {
        float damage = _character.Attack - opponent.Defense;

        MessageQueueManager.Instance.SendMessage(
            new DamageFeedbackMessage()
        {
            Damage = damage,
            Position = opponent.GetDamageFeedbackPosition()
        });

        if (damage <= 0 || opponent.TakeDamage(damage))
        {
            yield break;
        }
```

```
      yield return new
        WaitForSeconds(_character.AttackSpeed);
   }
}
```

A method that returns an `IEnumerator` variable as `TakeDamageOverTime` does, can use the `yield` keyword to return an iterator or break the coroutine. It is important to note that although it is an asynchronous operation, it still runs on Unity's main thread, and it might cause performance issues when used for heavy operations.

This coroutine will run while both the opponent and the character are still alive. We can also stop the coroutine using the variable that stores it after it has started, and we are going to use this in the last method of this class to stop the attack.

To calculate `damage`, we are considering the value of the character's `attack` minus the opponent's `defense`. This means that if the attack is 2 and the defense is 1, the total damage will be 1. Once the damage has been calculated, we can send a `DamageFeedbackMessage` message to display the damage taken on top of the opponent's GameObject using `GetDamageFeedbackPosition` to find the best position for the UI feedback.

After sending the message, we check whether the damage is 0 or less, or if the opponent is dead after taking damage, which is indicated by `true` being returned from the `TakeDamage` method. If one of the two validations is `true`, we stop the coroutine using the `yield break` keywords.

The last line is simple but one of the cool features of using coroutines. We can use the `WaitForSeconds` object to basically wait in the middle of the loop for the time that we want, which in this case is defined by the character's `AttackSpeed`. It might look confusing, but `yield return` does not exit the method and just moves to the next iterator. `yield` is holding the loop for a certain number of seconds, which in this case is the value of the `AttackSpeed` variable, and then iterating again as soon as the `WaitForSeconds` method is done waiting.

The next method also causes damage, but this time it is just a regular method that takes the opponent's `attack` value and the `transform` property with the position at which the collision happened when the projectile hit the GameObject:

```
private void TakeDamageFromProjectile(float opponentAttack,
    Transform target)
{
    float damage = opponentAttack - _character.Defense;
    MessageQueueManager.Instance.SendMessage(
      new DamageFeedbackMessage()
      {
        Damage = damage,
        Position = _character.GetDamageFeedbackPosition()
```

```
  });

  _character.TakeDamage(damage);
  StopAttacking(target, false);
}
```

In the `TakeDamageFromProjectile` method, we calculate the damage using the same formula, but this time we consider the opponent's `Attack` value and the character's `Defense` value. Once calculated, we send the `DamageFeedbackMessage` message as well to display the damage feedback on the UI.

The last couple of lines cause damage to the character, using the `TakeDamage` method, and make the `target` character stop attacking (if it was attacking any GameObject) and move toward the GameObject that fired the projectile, using the `StopAttacking` method. Note that we do not check whether any of the two parts are still alive since that does not influence a moving projectile even if the GameObject that fired it died afterward.

In the current game context, this means that when an enemy is hit by a fireball, even if a Warrior is attacking the enemy, the enemy will stop attacking the Warrior to move toward the Mage that is attacking from the distance. Considering that the Mage has a weak defense but a powerful attack, it will make the player think carefully about when and how to attack the enemy. Sometimes, luring the enemy could be a good strategy, as well as attacking a slow enemy with an army of Mages to deal as much damage as possible before running aways.

The last method in this class will stop the attack, by stopping the coroutine and executing the callback to let the unit or the enemy know that the attack was stopped:

```
private void StopAttacking(Transform target, bool opponentIsDead)
{
  _targetId = null;
  if (_dealDamageCoroutine != null)
  {
    StopCoroutine(_dealDamageCoroutine);
    _dealDamageCoroutine = null;
  }
  OnStopAttacking(target, opponentIsDead);
}
```

The `StopAttacking` method is called when the `OnCollisionExit` event is received, and we receive as parameters the `transform` property from the opponent GameObject that is no longer colliding and the information on whether the opponent is dead. These are the same parameters sent in the `OnStopAttacking` callback; however, we do not execute the callback directly because we need to clean up the `targetId` variable and stop the coroutine that is executing the `TakeDamageOverTime` method. If `dealDamageCoroutine` is not `null`, we call the `StopCoroutine` method to dismiss the coroutine and set it to `null`. This will stop the coroutine immediately.

This was quite a long class with lots of new things, but we finally reached the end of it. Now, we are going to see, in the next two scripts, how the enemy will use `OnStopAttacking` to chase the unit that is trying to flee or the unit that is attacking from a distance.

Creating the Enemy NavMesh

We just finished the **CollisionComponent** script and now we can use it with the NavMesh so that they work together. We are going to create a new script that will be responsible for the enemy NavMesh, which is slightly different from the unit NavMesh logic, which we will update after finishing the enemy script. So, create a new script in the **Scripts | Enemy** folder, name it `EnemyComponentNavMesh`, and add the content from the following code block:

```
using UnityEngine;
using UnityEngine.AI;
namespace Dragoncraft
{
    [RequireComponent(typeof(NavMeshAgent),typeof(CollisionComponent))]
    public class EnemyComponentNavMesh : EnemyComponent
    {
        private NavMeshAgent _agent;
        private CollisionComponent _collisionComponent;
        private Transform _targetToFollow;
    }
}
```

`RequireComponent` in this class will include the two most important components to make this class work: `NavMeshAgent` from the Unity Engine and the `CollisionComponent` component that we just created.

The `EnemyComponentNavMesh` class is inherited from `EnemyComponent`, which is a `BaseCharacter` class. Technically, it would be nice to also inherit from `CollistionComponent`; however, only one `MonoBehaviour` class can be inherited and both `CollisionComponent` and `BaseCharacter` are `MonoBehaviour`. This is not a problem since we can still use the `CollisionComponent` component attached to the GameObject.

The first couple of `private` properties in this class are used to store the `agent` and `collision Component` references to `NavMeshAgent` and `CollisionComponent`, respectively, and the `private targetToFollow` property is used to store the information we need to make the enemy chase the unit.

Now, let us add the `Start` method, where we are going to get the components for `Collistion Component` and `NavMeshAgent`, as we can see in the following code block:

```
private void Start()
{
    _collisionComponent = GetComponent<CollisionComponent>();
    _collisionComponent.Initialize(this);
    _collisionComponent.OnStartAttacking += OnStartAttacking;
    _collisionComponent.OnStopAttacking += OnStopAttacking;

    _agent = GetComponent<NavMeshAgent>();
    _agent.speed = WalkSpeed;
}
```

In the `Start` method, we get both required components:

- The `CollisionComponent` is initialized using the current enemy class as the character and we assign two methods to the two callback events that are defined in this class, `OnStartAttacking` and `OnStopAttacking`

- In the last couple of lines, we get the `NavMeshAgent` component and set `speed` to be the `WalkSpeed` property that we define in `ScriptableObject`

For every event that we assign a callback, we must always unassign it when is not needed anymore. If the callback will be used while the script is active, we can assign it at the very first opportunity, which is the `Start` method in our case. When the script is not active anymore, we can unassign the callback, as we are doing in the `OnDestroy` method next:

```
private void OnDestroy()
{
    if (_collisionComponent != null)
    {
        _collisionComponent.OnStartAttacking -=
            OnStartAttacking;
        _collisionComponent.OnStopAttacking -=
            OnStopAttacking;
    }
}
```

It is important to control when we assign and unassign events and callbacks because any missing part could lead to multiple duplicated callbacks being received at the same time, or even none.

To assign a method to an action, we use the plus equal sign (+=), and to unassign, we use the minus equal sign (-=). It is also a good practice to check whether the property is not null before using it in a method such as `OnDestroy`, which is called when the GameObject is about to be destroyed.

The next method we are going to add to this class is our good old friend `Update` – for the enemy, this will help chase a unit that is fleeing or attacking from a distance:

```
private void Update()
{
    if (!IsDead && _targetToFollow != null)
    {
        if (Vector3.Distance(transform.position,
            _targetToFollow.position) < WalkSpeed)
        {
            return;
        }
        _agent.destination = _targetToFollow.position;
    }
}
```

If the enemy is not dead and there is a target to follow, when it is not `null`, we check the distance between the enemy and the unit using the `Vector3.Distance` method. If the distance is less than `WalkSpeed`, we just ignore it using the `return` keyword. Otherwise, we set `NavMeshAgent`'s `destination` to be the `position` property of the target unit. Since this is an `Update` method, even if the unit changes position, the enemy will keep chasing it using NavMesh to find the best path to reach the unit and attack it. This is a very aggressive chase behavior because the enemy will not stop until defeating the unit or being defeated.

Now we have reached the last couple of methods in this class, which are the callbacks that `CollisionComponent` will execute when the attack starts or stops. Let us first define the `OnStartAttacking` method here:

```
private void OnStartAttacking(Transform target)
{
    transform.LookAt(target.position);
    _agent.isStopped = true;
    _targetToFollow = null;
    UpdateState(ActionType.Attack);
}
```

The `OnStartAttacking` method is very straightforward: first, the enemy rotates toward the unit using the `LookAt` method; next it stops the NavMesh (otherwise the enemy will continue to move until reaches the destination position); then `targetToFollow` is set to `null` so we know there is no unit to follow; and finally, the state is updated using the `Attack` action. All this means that the enemy is preparing to attack the unit.

When the enemy is no longer colliding with the unit, the following method, `OnStopAttacking`, is called with a `target` parameter, which is the unit's `transform` property, and information on whether the unit is dead. Using both parameters, we can decide to chase the unit or just stay idle:

```
private void OnStopAttacking(Transform target, bool opponentIsDead)
{
  if (IsDead)
  {
    return;
  }

  if (!opponentIsDead)
  {
    transform.LookAt(target.position);
    _agent.isStopped = false;
    _targetToFollow = target;
    UpdateState(ActionType.Move);
  }
  else
  {
    UpdateState(ActionType.None);
  }
}
```

The first validation we are doing in this method is to check whether the enemy is still alive; otherwise, we can just return and ignore the rest of the method. Then, we check whether the unit is still alive to decide whether we are going to chase it or just change the state to the `None` action, which will play the `Idle` animation state.

If both the enemy and the unit are alive, we will make the enemy chase the unit. First, we rotate the enemy to look at the current unit's position, then we resume the `NavNeshAgent` movement, set the unit's `transform` property to `targetToFollow`, and change the state to the `Move` action. The combination of these lines means that the enemy is now hunting the unit until either is dead.

Now that we have finished the `EnemyComponentNavMesh` class, we need to update one script that spawns the enemy to use it instead of the `EnemyComponent` class. Open the script located at **Scripts | Spawner** | `EnemySpawner.cs`, then in the `OnEnemySpawned` method, replace `EnemyComponent` with `EnemyComponentNavMesh` in the following highlighted code:

```
private void OnEnemySpawned(BaseEnemySpawnMessage message)
{
  GameObject enemyObject = SpawnObject();
  enemyObject.SetLayerMaskToAllChildren("Enemy");
```

```
EnemyComponentNavMesh enemyComponent =
    enemyObject.GetComponent<EnemyComponentNavMesh>();
if (enemyComponent == null)
{
    enemyComponent =
        enemyObject.AddComponent<EnemyComponentNavMesh>();
}

enemyComponent.CopyData(_enemyData, message.SpawnPoint);
}
```

In the OnEnemySpawned method, we are changing the three places that were using EnemyComponent to use the new EnemyComponentNavMesh and updating the layer to be Enemy instead of the previous value, Unit.

Now we can update the existing NavMesh component for the unit to use the new CollisionComponent class as well.

Updating Unit's NavMesh

We are going to update one existing script and replace all its content to use CollisionComponent together with the NavMesh system. This script is a bit simpler than EnemyComponentNavMesh and has a couple of new methods. Open the script located at **Scripts | Unit |** UnitComponentNavMesh. cs and replace the content with the following code:

```
using UnityEngine;
using UnityEngine.AI;
namespace Dragoncraft
{
    [RequireComponent(typeof(NavMeshAgent), typeof(CollisionComponent))]
    public class UnitComponentNavMesh : UnitComponent
    {
        private NavMeshAgent _agent;
        private CollisionComponent _collisionComponent;
    }
}
```

The UnitComponentNavMesh class continues to be inherited from UnitComponent, and now that means that this class is also a BaseCharacter class. Besides NavMeshAgent, we are also adding CollisionComponent as a required component for this class, and we have a couple of private properties to hold a reference to both components. Note that different from the enemy, we do not have a target to follow here because the unit is very passive and only obeys commands from the player.

Two private properties, agent and collisionComponent, are set in the following Start method, which has similar content to the Start method that we just added to the EnemyComponentNavMesh class, initializing CollisionComponent and changing a property of NavMeshAgent:

```
private void Start()
{
    _collisionComponent = GetComponent<CollisionComponent>();
    _collisionComponent.Initialize(this);
    _collisionComponent.OnStartAttacking += OnStartAttacking;
    _collisionComponent.OnStopAttacking += OnStopAttacking;

    _agent = GetComponent<NavMeshAgent>();
    _agent.speed = WalkSpeed;
}
```

CollisionComponent is initialized using the current unit class as the character and we assign two methods to the two callback events that are defined in this class, OnStartAttacking and OnStopAttacking. In the last couple of lines, we get the NavMeshAgent component and set speed to WalkSpeed.

In the next method, OnDestroy, we unassign both OnStartAttacking and OnStopAttacking methods from the CollisionComponent callbacks:

```
private void OnDestroy()
{
    if (_collisionComponent != null)
    {
        _collisionComponent.OnStartAttacking -=
            OnStartAttacking;
        _collisionComponent.OnStopAttacking -=
            OnStopAttacking;
    }
}
```

In the OnDestroy method, before unassigning the OnStartAttacking and OnStopAttacking methods, we need to validate whether the collisionComponent property is not null to avoid null reference errors when trying to access a variable from the collisionComponent object.

The next couple of methods are both virtual in the UnitComponent class and we are going to override them in this class; they are quite simple and the only reason to override them is to be able to control NavMeshAgent as well when updating the unit position or when stopping moving and starting to attack an enemy:

```
protected override void UpdatePosition()
{
```

```
    _agent.isStopped = false;
    _agent.destination = GetFinalPosition();
}


protected override void StopMovingAndAttack()
{
    base.StopMovingAndAttack();
    _agent.isStopped = true;
}
```

In the UpdatePosition method, we are resuming the agent and setting the updated destination using the GetFinalPosition method. We are overriding the method but ignoring the base content of it because we want to execute only the code here. This will allow the unit to move even while attacking to flee from the enemy or band together with more units and form a new attack.

The StopMovingAndAttack method does exactly that – we stop moving and start attacking the enemy. Even with the override keyword, this time, we want to execute the content from the base method before changing the agent property to stop it from moving to the destination that was previously set.

The last couple of methods in this class are quite simple, and both are executed by the respective callbacks in the CollisionComponent class:

```
private void OnStartAttacking(Transform target)
{
    transform.LookAt(target.position);
    _agent.isStopped = true;
    UpdateState(ActionType.Attack);
}
```

Here, OnStartAttacking prepares the unit to attack the enemy by rotating the GameObject to face the target, using the LookAt method, preventing the NavMeshAgent from moving, and updating the state, changing it to the Attack action. These changes are all we need to do when an attack starts, but we also need to do a few things when the attack stops, in the OnStopAttacking method.

The OnStopAttacking method is so modest in the unit class that we do not even use the two parameters received here because they are not required. We basically check whether the unit is still alive and, if that is the case, we update the state to the None action, which represents the Idle animation state:

```
private void OnStopAttacking(Transform target, bool opponentIsDead)
{
    if (IsDead)
    {
```

```
        return;
    }
    UpdateState(ActionType.None);
}
```

When a unit is attacking an enemy, we just look at the enemy and start the attack, but when the attack stops and the unit is still alive, this means that we have defeated the enemy, and the unit will become idle until the player gives it a new command.

Now we have both enemy and unit classes that are fully prepared to move using the pathfinder, attack each other using the collision system, and even chase units that are trying to flee or attack from a distance. The only thing left now is to clean up both enemy and unit GameObjects when they are dead and prepare them to be reused when a new unit or enemy is spawned.

Implementing the character life cycle

We have everything we need to create an epic battle between units and enemies; however, we need to add one last component to make it possible to reuse them once they are dead in combat. We need to return the GameObjects to the Object Pool, but also clean up all the scripts that we attached to it so that the next time we spawn a new GameObject, there is nothing left from the previous data – it is still the same GameObject instance reused, but with fresh scripts.

We are going to add a new component that will be attached to the GameObject when it is dead. The new component will take care of waiting a few seconds before adding the GameObject back to the Object Pool and removing all the scripts attached since it was spawned. So, create a new script in the **Scripts | Battle** folder and name it DeadComponent. Then, replace the content with the following code:

```
using UnityEngine;
using UnityEngine.AI;
namespace Dragoncraft
{
    public class DeadComponent : MonoBehaviour
    {
        private float _timeToLive = 5;
        private float _counter;
    }
}
```

The DeadComponent class has two private properties that will manage how long this script should wait before destroying itself. The default value for the timeToLive property is 5 seconds, which is enough to see the Death animation and wait a bit before removing the unit or enemy from the battlefield. The other property, counter, will be used as a supporting variable to count the time in seconds until it reaches the value of the timeToLive property.

The counting process is done in the Update method, defined next, which will wait until counter is greater than timeToLive and then execute a sequence of methods to remove all the components, reset the transform property, deactivate the GameObject, and destroy itself:

```
private void Update()
{
  _counter += Time.deltaTime;
  if (_counter > _timeToLive)
  {
    RemoveComponents();
    ResetComponents();
    gameObject.SetActive(false);
    Destroy(this);
    return;
  }
}
```

Destroying itself means that this component will be removed once the time is up, but the GameObject that represents the unit or enemy will still exist. SetActive will deactivate the GameObject and return it to the Object Pool, while the other methods do the cleanup.

The ResetComponents method, defined as follows, resets the position and rotation of the GameObject to be the default values:

```
private void ResetComponents()
{
  transform.position = Vector3.zero;
  transform.rotation = Quaternion.identity;
}
```

The last method in this class is RemoveComponents, which will literally remove the components that were added to the GameObject after it was spawned:

```
private void RemoveComponents()
{
  BaseCharacter character = GetComponent<BaseCharacter>();
  Destroy(character);

  CollisionComponent collision =
    GetComponent<CollisionComponent>();
  Destroy(collision);

  NavMeshAgent navMeshAgent = GetComponent<NavMeshAgent>();
  Destroy(navMeshAgent);
```

```
    Rigidbody rigidbody = GetComponent<Rigidbody>();
    Destroy(rigidbody);

    SphereCollider sphereCollider =
      GetComponent<SphereCollider>();
    Destroy(sphereCollider);
  }
```

Removing `BaseCharater` will also remove any component that is inherited from it, like the `UnitComponent` and `EnemyComponent` scripts.

The other components, `CollisionComponent`, `NavMeshAgent`, `Rigidbody`, and `SphereCollider`, were all added using the `RequiredComponent` attributes, and for that reason, we must remove the `BaseCharacter` component first; otherwise, we will get errors saying it is not possible to remove because they are required by `BaseCharacter`.

And that is all we need to clean up and reset a unit or enemy after they were dead in battle. There is one last change we need to make, though, because we also have a similar cleanup method to remove all components from the 3D model that are instantiated in the portrait of the details panel once we select any unit in the map. So, open the script located at **Scripts | UI |** `DetailsUpdated.cs` and replace the content of the `RemoveUnitComponent` method with the following code:

```
    private void RemoveUnitComponent(GameObject portrait)
    {
      BaseCharacter character =
        portrait.GetComponent<BaseCharacter>();
      Destroy(character);

      CollisionComponent collision =
        portrait.GetComponent<CollisionComponent>();
      Destroy(collision);

      NavMeshAgent navMeshAgent =
        portrait.GetComponent<NavMeshAgent>();
      Destroy(navMeshAgent);

      Rigidbody rigidbody = portrait.GetComponent<Rigidbody>();
      Destroy(rigidbody);

      SphereCollider sphereCollider =
        portrait.GetComponent<SphereCollider>();
      Destroy(sphereCollider);
    }
```

This code is basically the same vode that we added to `DeadComponent` to remove the components, which makes sense since we are doing the same cleanup process for both use cases.

Now, before testing everything we created and updated in this chapter, we need to tweak a few properties of our units' and enemies' ScriptableObjects:

1. In the **Project** view, select the ScriptableObject located at **Data | Unit | BasicWarrior** and, in the **Inspector** view, change the **ColliderSize** property to **1.5**, **WalkSpeed** to **3**, and **AttackSpeed** to **1.5**.

2. In the **Project** view, select the ScriptableObject located at **Data | Unit | BasicMage** and, in the **Inspector** view, change the **ColliderSize** property to **1.5**, **WalkSpeed** to **2**, and **AttackSpeed** to **1.8**.

3. In the **Project** view, select the ScriptableObject located at **Data | Enemy | BasicOrc** and, in the **Inspector** view, change the **ColliderSize** property to **1.5**, **WalkSpeed** to **2.5**, and **AttackSpeed** to **2**.

4. In the **Project** view, select the ScriptableObject located at **Data | Enemy | BasicGolem** and, in the **Inspector** view, change the **ColliderSize** property to **2.5**, **WalkSpeed** to **1.5**, and **AttackSpeed** to **1**.

5. In the **Project** view, select the ScriptableObject located at **Data | Enemy | RedDragon** and, in the **Inspector** view, change the **ColliderSize** property to **3.5**, **WalkSpeed** to **1**, and **AttackSpeed** to **2**.

Now we can play the game using the **Playground** or **Level01** scene, spawn a few enemies and units, and select the units to attack the enemies:

Figure 10.4 – Units and enemies attacking each other

Feel free to spawn different enemies and units and try to beat all the enemies with different units. You can also make the enemies follow you to an ambush with a bigger army waiting for them. Do not forget to tweak the values of each ScriptableObject and test different combinations for the enemies and units.

This was a long chapter full of new features and concepts, and a lot of code! Our project is taking shape and looking even more like an RTS game full of features and with a solid foundation to expand upon.

Summary

Well done on reaching the end of this chapter – it was one of the most challenging so far with many new features and concepts introduced, and a lot of code written. Our RTS game is getting bigger and full of great solid features that we are starting to put together to interact with each other. We are getting so close to having challenging and fun gameplay with many mechanics implemented.

In this chapter, we not only used systems developed in the previous chapters but also refactored and expanded the main scripts used by enemies and units, and, finally, we have a first version of the battle between them. Units can attack enemies with a melee or ranged attack, and, if the player chooses to flee, the enemies will chase down the units until one of the sides is defeated.

We also learned how to set up the physics settings in Unity and how to create a collision script that works nicely with the NavMesh system. Many new concepts and resources were introduced, such as the powerful coroutines, the action callback, inheritance, and virtual methods. We are not only creating a solid Unity game but also learning about the best practices to get the most out of what the engine and the C# programming language can offer us.

In *Chapter 11, Adding Enemies to the Map,* we are going to learn how to create and configure spawn points across the map so we can dynamically add different enemies and groups of enemies to the map, expanding our level manager tool. We will also learn how to add patrol behavior to enemies so that they will attack the units when they get close, as well as add a fog effect to the map so the player can only see in the mini-map the areas explored by the units.

Further reading

In this chapter, we covered a lot of new concepts and features in C# and Unity. The following links will help you to dig a bit deeper into the main topics we have learned:

- *Physics Manager*: https://docs.unity3d.com/2023.1/Documentation/Manual/class-PhysicsManager.html
- *C# OPP*: https://www.w3schools.com/cs/cs_oop.php
- *State*: https://gameprogrammingpatterns.com/state.html

- *Coroutines*: `https://docs.unity3d.com/2023.1/Documentation/Manual/Coroutines.html`

- *Action Delegate*: `https://learn.microsoft.com/en-us/dotnet/api/system.action`

- *Using NavMesh Agent with Other Components*: `https://docs.unity3d.com/2023.1/Documentation/Manual/nav-MixingComponents.html`

11
Adding Enemies to the Map

Most RTS games have groups of enemies that work together to attack the player units while they are exploring the map covered by the Fog of War, making it difficult to predict where the enemies could be hidden, patrolling an area, and waiting for the right moment to attack from the shadows.

In this chapter, we are going to learn how to create enemy groups, with different configurations, and how to add them to the map using predefined spawn points. We will also learn how to implement a fog to cover all of the map but leaving the initial area clear, as well as how to clear the fog above the units while they are moving to covered areas of the map. To make the enemies even more dynamic, we are going to add a new behavior that will allow them to patrol an area between two points, creating more challenges for the player when exploring the map.

By the end of this chapter, you will learn how to create a flexible solution to add one or multiple enemies in a group, and how to reuse it across the map using spawn points defined in the level configuration. You will learn how to create a Fog of War by adding a plane on top of the map that will be clear using Raycast detection between the units on the map and the camera position, changing the alpha channel of a built-in shader to make the vertex transparent when there is a unit exploring the map. You will also learn how to use the NavMesh system to implement patrol behavior, expanding the current enemy component that already has the attack and chase behaviors.

In this chapter, we will cover the following topics:

- Creating spawn points
- Adding fog to the map
- Patrolling the area

Technical requirements

The project setup in this chapter, with the imported assets, can be found on GitHub at `https://github.com/PacktPublishing/Creating-an-RTS-game-in-Unity-2023` in the `Chapter11` folder inside the project.

Creating spawn points

In *Chapter 9, Adding Enemies*, we learned how to configure different types of enemies and how to spawn them in a fixed position using the **Debug** menu. Later, in *Chapter 10, Creating an AI to Attack the Player*, we learned how to set up the physics and NavMesh systems, creating attack and chase behaviors against the units. Now, we are going to use everything we have developed for the enemies so far to set up spawn points, which are configurable locations on the map where we can spawn one or more enemies when the player starts to play the level.

We are going to create a new configuration file using `ScriptableObject` to define a group of enemies that are going to be spawned in one specific position. Create a new script in the **Scripts | Enemy** folder, name it `EnemyGroupData`, and replace the content with the following:

```
using System.Collections.Generic;
using UnityEngine;
namespace Dragoncraft
{
  [CreateAssetMenu(
    menuName = "Dragoncraft/New Enemy Group")]
  public class EnemyGroupData : ScriptableObject
  {
    public string Name;
    public List<EnemyData> Enemies = new List<EnemyData>();
  }
}
```

Since this script is a ScriptableObject, we need to include the `CreateAssetMenu` attribute data before the class definition so Unity will display the option to create a new asset of the type `EnemyGroupData` when clicking with the right button on any folder in the **Project** view. The **Create | Dragoncraft | New Enemy Group** menu will be displayed to create a new file.

The `EnemyGroupData` class is quite simple, with only two properties: the `Name` of the group, so we can identify it easily; and the list of `Enemies` that will be in the group. Each enemy is defined by an `EnemyData` object, which we created in *Chapter 9, Adding Enemies*, for each enemy type: `Orc`, `Golem`, and `Dragon`.

The previous script defines the group of enemies with any combination and quantity of enemies per group. Now, we need to create a new script that will place each enemy group on each spawn point. This new script is required because it will make it possible to reuse enemy groups in different locations on the map, so we do not need to duplicate an enemy group to add it more than once to the map. In the **Scripts | Enemy** folder, create a new script, name it `EnemyGroupConfiguration`, and replace the content with the following code snippet:

```
using System;
using UnityEngine;
```

```
namespace Dragoncraft
{
  [Serializable]
  public class EnemyGroupConfiguration
  {
    public EnemyGroupData Data;
    public Vector3 Position;
  }
}
```

Although this script is used to configure the enemy group with position, it is not required to be a ScriptableObject. However, it is required to have the Serializable attribute because we are going to add this configuration as a property of the existing LevelData script.

Now that we have both scripts required to set up an enemy group and define the position of each enemy group, we need to add a new property to the existing ScriptableObject that holds the data used to configure the level. In the **Project** view, open the script located at **Scripts | Level |** LevelData.cs and add the following property as the last one in this class, before the class-closing curly bracket:

```
public class LevelData : ScriptableObject
{
  ...
  public List<EnemyGroupConfiguration> EnemyGroups =
    new List<EnemyGroupConfiguration>();
}
```

The EnemyGroups property is a list of EnemyGroupConfiguration objects that we are going to use for spawning the enemy groups at each desired position on the map. Since the EnemyGroupConfiguration class has the Serializable attribute, we will be able to set up the EnemyGroupData asset and position in the **Inspector** view.

However, we are using a different class to render all the properties from the LevelData class in the **Inspector** view, so just adding the EnemyGroups property will not make it available for us yet. To make the EnemyGroups property visible in the **Inspector** view, we need to add a few lines in the existing LevelDataEditor script. In the **Project** view, open the script located at **Scripts | Editor |** LevelDataEditor.cs and add the following code lines as the last ones in the AddLevelDetails method before the method-closing curly bracket:

```
private void AddLevelDetails(LevelData levelData)
{
  ...
  SerializedProperty enemyGroups =
    serializedObject.FindProperty("EnemyGroups");

  EditorGUILayout.PropertyField(enemyGroups,
```

```
    new GUIContent("Enemy Groups"), true);

  serializedObject.ApplyModifiedProperties();
}
```

Since the type of the EnemyGroups property is a class created by us, with the Serializable attribute, we are going to find it in the LevelData ScriptableObject using the FindProperty method, which uses the name of the property as the search parameter. This method is already defined for us in the serializedObject class variable, which is available because our LevelDataEditor class inherits from the Editor class. The return of the FindProperty method is an object of the type SerializedProperty that we are going to use to render the information in the **Inspector** view and manipulate the property value in the same view.

After finding the EnemyGroups property, we will use the EditorGUILayout.PropertyField method to render it in the **Inspector** view. The first parameter is SerializedProperty enemyGroups; the second is a new GUIContent object with the text "Enemy Groups" that will be displayed next to the property; and the last one is a Boolean with the value true, which means the EditorGUILayout object must include all the children of this property, which makes sense since it is a list of objects.

The last new line is to call the ApplyModifiedProperties method to make sure that any changes we make in the ScriptableObject, through the **Inspector** view, will be reflected and stored in the asset file. Without that line, we would be able to see the property, but no changes would be applied to the asset. It is also important to have this line as the last one in the method, specifically if we have more than one SerializedProperty attribute.

We have just finished creating the scripts to hold the enemy group definition and position, as well as updating the class responsible for rendering the LevelData properties to include the new EnemyGroupConfiguration object. Before we start creating new ScriptableObjects for the enemy groups, and add them to the LevelData object, we need to implement the code that will be responsible for reading the enemy groups and positions and spawn each enemy on the respective spawn point.

We are going to update another existing script to spawn the enemy groups based on the LevelData ScriptableObjects. In the **Project** view, open the script located at **Scripts** | **Level** | LevelComponent. cs and add the following line as the last one in the Start method, before the method-closing curly bracket:

```
private void Start()
{
    ...
    SpawnEnemyGroups();
}
```

The line we just added executes the new SpawnEnemyGroups method, which we are going to define next, as the last line of the Start method. That is the only change to an existing method in this class. Now we can add the new methods at the end of the LevelComponent class. Let us start with the SpawnEnemyGroups method. Add the following code snippet to the end of the class:

```
private void SpawnEnemyGroups()
{
    foreach (EnemyGroupConfiguration group in _levelData.EnemyGroups)
    {
        SpawnEnemyGroup(group);
    }
}
```

In the SpawnEnemyGroups method, we are accessing the EnemyGroups property from the levelData class variable and, for each EnemyGroupConfiguration we find in the list, we are calling another SpawnEnemyGroup method with the group configuration as the parameter. Since it is likely that we are going to have multiple enemy groups on each level, this method is responsible for spawning each enemy group from the list.

The next SpawnEnemyGroup method, which is the one we just called in the previous method, will take care of spawning each individual enemy from the enemy group in a proper position. Add the following private class variable and method at the end of the class:

```
private float _distanceBetweenEnemies = 3.0f;

private void SpawnEnemyGroup(EnemyGroupConfiguration enemyGroup)
{
    int rows =
        Mathf.RoundToInt(Mathf.Sqrt(
            enemyGroup.Data.Enemies.Count));
    int counter = 0;

    for (int i = 0; i < enemyGroup.Data.Enemies.Count; i++)
    {
        if (i > 0 && (i % rows) == 0)
        {
            counter++;
        }

        float offsetX = (i % rows) * _distanceBetweenEnemies;
        float offsetZ = counter * _distanceBetweenEnemies;

        Vector3 offset = new Vector3(offsetX, 0, offsetZ);
```

```
        Vector3 spawnPoint = enemyGroup.Position + offset;

        SpawnEnemy(enemyGroup.Data.Enemies[i].Type,
            spawnPoint);
    }
}
```

This method is a bit long, but it is also familiar to us. We used a similar approach in *Chapter 6*, *Commanding an Army of Units*, in the MoveSelectedUnits method defined in the UnitSelectorComponent class, to position the selected units side by side in a grid formation, without overlapping each other after selecting a few units to move to a location on the map. Although the approach is similar, here we are going to adapt it to our current needs.

First, we will calculate the number of rows we will have to organize the enemies. This is done by calculating the square root of the total of Enemies, using the Sqrt method, and then converting this float point value to an integer by using the RoundToInt method.

Next, inside the for loop, we will check whether a new row is required by using the % mod operator, which will return the remainder after dividing the i variable by the rows variable. If the remainder is zero, this means a new row is required, then the counter variable is increased. There is also a validation to avoid increasing the counter variable on the very first row if the value of i is 0.

The offsetX value is the distance between the enemies multiplied by the remainder of the division of i by rows, which will change offsetX based on the column of each enemy in our imaginary grid. The offsetZ variable only changes if a new row is required, when the counter value is increased, and multiplied by the distance between the enemies.

Once we calculate both offsetX and offsetZ, we can use them to create a new Vector3 offset, which will be added to the enemyGroup.Position property so we can spawn each one of the enemies to a different position near each other. The result is a Vector3 spawnPoint variable that we are going to use in the last line of the method, calling another SpawnEnemy method, using the enemy Type and spawnPoint as parameters.

We can now add the last method required in the LevelComponent class, which is the SpawnEnemy method that we used in the previous code snippet. So, add the following method at the end of this class:

```
private void SpawnEnemy(EnemyType enemyType, Vector3 spawnPoint)
{
    switch (enemyType)
    {
    case EnemyType.Orc:
        MessageQueueManager.Instance.SendMessage(
            new BasicOrcSpawnMessage {
                SpawnPoint = spawnPoint });
        break;
```

```
    case EnemyType.Golem:
      MessageQueueManager.Instance.SendMessage(
        new BasicGolemSpawnMessage {
          SpawnPoint = spawnPoint });
      break;
    case EnemyType.Dragon:
      MessageQueueManager.Instance.SendMessage(
        new RedDragonSpawnMessage {
          SpawnPoint = spawnPoint });
      break;
    default:
      break;
  }
}
```

This method is similar to what we created in *Chapter 9, Adding Enemies*, in the `EnemyDebugger` class, in which we added methods to test the spawn of each enemy type into a fixed position hardcoded there. This works well as a debugging tool to test enemy spawning, but now we need a more flexible solution to spawn each enemy in a proper position.

The `SpawnEnemy` method has a couple of parameters. The first one is `EnemyType`, used in a conditional `switch` keyword to send the correct spawn message for each enemy type. Using the `SendMessage` method from **Message Queue**, we create a new message of the corresponding enemy type. Each enemy spawn message has a `SpawnPoint` parameter that is set by the value of the second parameter received in the `SpawnEnemy` method. Using `switch`, we can send the proper message based on the enemy type to spawn an enemy at the spawn point received.

That was the last script we had to add or modify to set up the enemy groups. Now we can move on to creating the ScriptableObject for the enemy groups we want. Let us start with one simple enemy group that will have only two Orcs:

1. In the **Project** view, right-click on the `Data` folder, select **Create | Folder**, and name the new folder `EnemyGroup`.

2. Right-click on the new `EnemyGroup` folder, select **Create | Dragoncraft | New Enemy Group**, and name the new asset `OrcDuo`.

3. Left-click on the new **OrcDuo** asset and, in the **Inspector** view, set the **Name** value as **OrcDuo**.

4. On the **Enemies** property, change the **0** value to **2**, which will create two elements in the list.

5. On **Element 0**, click on the small circle button on the right side and, in the new window, select the **BasicOrc** asset by double-clicking on it.

6. Repeat the last step for **Element 1** as well, selecting the same **BasicOrc** asset.

7. Save the changes using the **File | Save Project** menu.

We just created our first enemy group containing two Orcs, as we can see in *Figure 11.1*. This is a very simple enemy group but very useful to add just a couple of Orcs on the map. Sometimes it is better to have two groups with two Orcs a bit far away from each other than a single group with four Orcs that are limited to one position. It all depends on the level configuration and the change we want to add for the player.

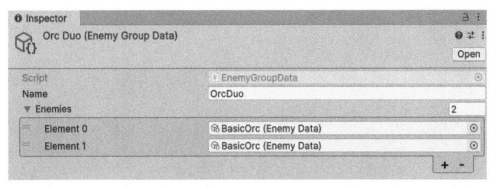

Figure 11.1 – Enemy Group Data with two Orcs

Now let us create a second enemy group, a bigger one with two different types of enemies, Orcs and a Golem:

1. In the **Project** view, right-click on the **EnemyGroup** folder, select **Create | Dragoncraft | New Enemy Group**, and name the new asset GolemAndOrcs.

2. Left-click on the new GolemAndOrcs asset and, in the **Inspector** view, set the **Name** value as **GolemAndOrcs**.

3. On the **Enemies** property, change the **0** value to **9**, which will create two elements in the list.

4. On **Element 4**, click on the small circle button on the right side and, in the new window, select the **BasicGolem** asset by double-clicking on it.

5. Repeat the last step for all other elements, but select the **BasicOrc** asset instead.

6. Save the changes using the **File | Save Project** menu.

The second enemy group we have created is much bigger and more powerful than the first one, and we are taking advantage of the changes we made in the LevelComponent script that will spawn the enemies in a grid formation. This means that the Golem will be spawned at the center and will be surrounded by Orcs to protect it.

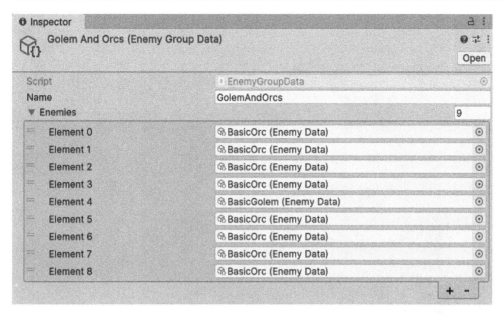

Figure 11.2 – Enemy Group Data with eight Orcs and one Golem

Now that we have two enemy groups created, we can move on to the last setup step, which is adding both enemy groups to the level and defining the respective spawn point for them. We are going to update an existing asset to add the enemy groups:

1. In the **Project** view, left-click to select the asset located at **Data | Level | Level01**.

2. In the **Inspector** view, on the **Enemy Groups** property, change the **0** value to **2**, which will create two elements in the list.

3. On **Element 0**, in the **Data** property, click on the small circle button on the right side and, in the new window, select the **OrcDuo** asset by double-clicking on it. Leave the **Position** property as (**0, 0, 0**).

4. On **Element 1**, in the **Data** property, click on the small circle button on the right side and, in the new window, select the **GolemAndOrcs** asset by double-clicking on it. Change the **Position** property to (**20, 0, 10**).

5. Save the changes using the **File | Save Project** menu.

This is all we need to do to add enemy groups to our level. Note that only **Enemy Groups** is new in the **Level01** asset; all the other properties were set in *Chapter 3, Getting Started with Our Level Design*, when we created all the level assets and scripts. *Figure 11.3* shows the current state of the **Level01** asset with the two enemy groups added.

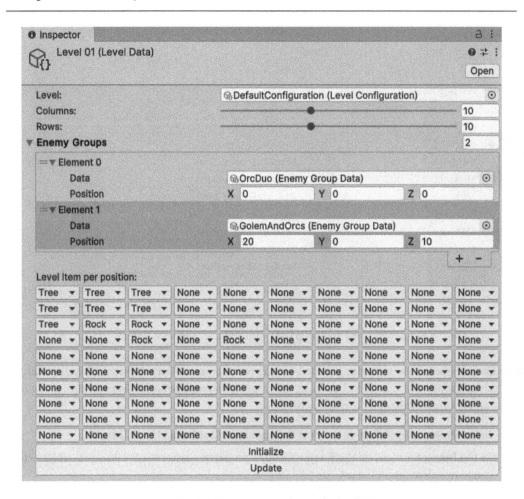

Figure 11.3 – Adding the enemy groups to Level Data

The **Level01** asset is already attached to our **Playground** scene, in the **Level** GameObject. No extra changes are required, so if we open the **Playground** scene and run the game on the Editor, we will be able to see the two enemy groups spawned in the formation that was set on their respective assets, as we can see in *Figure 11.4*:

Figure 11.4 – The two enemy groups and the respective enemies on the map

We just finished creating a very powerful configuration that will enable us to add multiple enemy groups, or even reuse enemy groups at different locations on the map. We could even create enemy groups with only one enemy to position on the map if that is something we decide to do in our level.

Figure 11.4 shows both enemy groups, but we can also spot an issue there: currently, it is possible for the player to see all the enemies on the minimap, located in the bottom-left corner of the UI. We need one gameplay element that is required to make a great RTS game, Fog of War, that will hide the enemies until the player explores the map by sending units in all directions, and that is what we are going to add next.

Adding fog on the map

Fog of War is a term used in strategy games that indicates that part of the map is hidden from the player, and it is only revealed once the player sends units to explore the covered map. There are some variations. For example, the fog could reappear when no unit is in the explored area, but for our RTS game, we are going to use a more classic approach where any area that is explored stays without the fog covering the map.

Adding a new layer and updating cameras

We are going to need a few things to set up the fog, which includes creating a new layer, a new material, a new mesh, and a new script that will manipulate the material color using Raycast and update a couple of existing scripts in the project. Let us start with the new layer that will be used to render the fog:

1. Open the **Project Settings** window using the **Edit | Project Settings…** menu option.

2. On the left panel, click on **Tags and Layers**, and, on the right panel, expand the **Layers** property.

3. Add the new **Fog** layer as the value of **User Layer 10**.

4. Close the **Project Settings** window.

This is the fifth custom layer that we have added to the project so far, as we can see in *Figure 11.5*, which has all of them: **Portrait**, **Unit**, **Plane**, **Enemy**, and **Fog**.

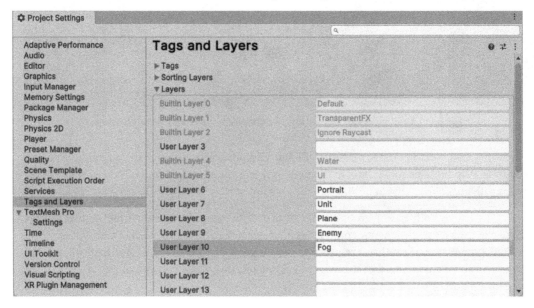

Figure 11.5 – Adding a new Fog layer

Now that we have the new layer for the fog, we need to make sure that the new layer is rendered on the cameras that we want to display the fog, which are the minimap camera and the main camera. The minimap camera is already in a Prefab, which makes it easier to add the new layer:

1. In the **Project** view, left-click to select the asset located at **Prefabs | UI | MiniMapCamera**.

2. With the Prefab selected, in the **Inspector** view, on the **Camera** component, click on the dropdown in the **Occlusion Culling** property and select the **Fog** layer name as well.

3. Save the changes using the **File | Save Project** menu.

The **MiniMapCamera** Prefab is used by the `LevelManager` script to instantiate it in the current scene, which means that we do not need to update any other asset for the minimap camera to work with the new **Fog** layer.

However, this is not quite the case for the main camera because we do not have a Prefab for it yet, but we can easily create a new one and update the scenes:

1. In the **Project** view, double-click to open the scene at **Scenes | Playground** and, in the **Hierarchy** view, select the **Main Camera** GameObject.

2. In the **Inspector** view, on the **Camera** component, click on the dropdown in the **Occlusion Culling** property and select the **Fog** layer name as well.

3. On the **Transform** component, change the **Position X** property to **0**, **Y** to **60**, and **Z** to **0**, and change the **Rotation X** property to **90**, **Y** to **0**, and **Z** to **0**.

4. To create a Prefab, drag the **Main Camera** GameObject from the **Hierarchy** view, and drop it in the **Prefabs | Level** folder in the **Project** view.

5. Save the changes using the **File | Save** menu.

These steps will create a new Prefab for the main camera that will make it easier to add changes to it when needed without having to update all the level scenes. However, we still have one scene to replace the main camera with the new Prefab:

1. In the **Project** view, double-click to open the scene at **Scenes | Level01** and, in the **Hierarchy** view, right-click on the **Main Camera** GameObject and select the **Delete** menu option to remove the GameObject from the scene.

2. Drag the Prefab located at **Prefabs | Level | Main Camera**, from the **Project** view, and drop it into the **Hierarchy** view to add it to the scene.

3. Save the changes using the **File | Save** menu.

That is all we need to have the same main camera for both scenes, using the Prefab that we just created. Now, both cameras – the main camera and the minimap camera – are Prefabs and easier to update if needed because we no longer need to update the scenes to change the camera values, which is handy when we have multiple scenes in the project.

The new `Fog` layer was created to separate the renderer but also to optimize the Raycast detection, and we will see next how to create the new `FogComponent` script.

Creating the fog component

Now we are going to write the code for the main part of the fog logic, which is the component responsible for initializing everything we need, check whether a unit is below the fog to clear the area, and clear the initial area so that the player can see part of the map when the game starts.

Create a new script in the **Scripts** | **Level** folder, name it `FogComponent`, and replace the content with the following code snippet:

```
using System.Collections.Generic;
using UnityEngine;
namespace Dragoncraft
{
  public class FogComponent : MonoBehaviour
  {
    [SerializeField] private float _radius = 7;
    [SerializeField] private float _initialArea = 14;
    [SerializeField] private LayerMask _layerMask;

    private Mesh _mesh;
    private Vector3[] _vertices;
    private Color[] _colors;
  }
}
```

In the class, we have a few properties exposed in the **Inspector** view using the `SerializeField` attribute, which are variables we can tweak when needed. The `radius` property is the size of the circumference around the unit that is used to clear the fog while the selected units are moving into an area with fog. At the beginning of the game, the player will have no units to clear the area yet, so it is important to have an `initialArea` property defined that will be used to clear the fog in the initial position of the map. Since we are using the `Fog` layer to optimize the Raycast detection, we need to set `layerMask` in the **Inspector** view with the `Fog` layer, which we are going to do after finishing this script.

The other private variables are here to help with the calculations we are going to do. The `mesh` property is a `Plane` GameObject that we are going to position on top of the map, with a black material to cover everything. To clear the fog, we are going to use the `vertices` of the mesh and calculate the Raycast between the camera position and the units on the map, and when the hit is detected, we will change the alpha channel of the `colors` material of the corresponding vertices to make it transparent and the map area visible for the player.

Moving to our first method in this class, we are going to override the `Start` method from the `MonoBehaviour` class, as we can see here:

```
private void Start()
{
  Initialize();
  ClearInitialArea();
}
```

In the Start method, which is executed by Unity when the GameObject that has the previous script attached is enabled, we are going to initialize our class variables. Then, using the following Initialize method, we will clear the fog in the initial area using the ClearInitialArea method, as defined in the following code snippet:

```
private void Initialize()
{
  _mesh = GetComponent<MeshFilter>().mesh;
  _vertices = _mesh.vertices;

  _colors = new Color[_vertices.Length];
  for (int i = 0; i < _colors.Length; i++)
  {
    _colors[i] = Color.black;
  }

  _mesh.colors = _colors;
}
```

The script that we are writing, FogComponent, will be attached to a GameObject that has a **Plane** with a **MeshRenderer** component. Using the GetComponent method, we can get the reference for the MeshFilter component, which is part of the renderer component, access the mesh property, and store it in a class variable with the same name. From the mesh property, we can get the array of vertices and assign it to our vertices class variable to be used later.

Each vertex has color information, which is why we can safely initialize the array of colors with the same Length property as the array of vertices. Once initialized, we loop over each item to set the Color.black value to each color. After initializing the array of colors with the default black color, we can set colors in the mesh property to be the same as the array of colors we just created. This is the color information that we are going to manipulate on each vertex to make it transparent.

Now that we have initialized all class variables, we can move on to the next step, which is to clear the initial area so the player can see something without the fog when the game starts. Add the following code snippet to our class:

```
private void ClearInitialArea()
{
  float fogPosY = transform.position.y;
  float distance = _initialArea;

  List<Vector3> area = new List<Vector3>
  {
    new Vector3(0, fogPosY, distance), //top-center
    new Vector3(-distance, fogPosY, distance), //top-left
```

```
        new Vector3(distance, fogPosY, distance), //top-right

        new Vector3(0, fogPosY, 0), //mid-center
        new Vector3(-distance, fogPosY, 0), //mid-left
        new Vector3(distance, fogPosY, 0), //mid-right

        new Vector3(0, fogPosY, -distance), //bot-center
        new Vector3(-distance, fogPosY, -distance), //bot-left
        new Vector3(distance, fogPosY, -distance), //bot-right
    };

    foreach (Vector3 point in area)
    {
        RemoveFog(point);
    }
}
```

We are going to reuse the same RemoveFog method to clear the initial area. This method receives a Vector3 point parameter to change the alpha channel of the color on each vertex that matches the Vector3 point parameter. When the point is from a unit, we will use the information from the Raycast detection. However, in our case, we need to predefine the points for our initial area.

In Vector3 list, which is the area variable, we have nine points. Imagine that each point corresponds to a location in a 3 x 3 grid. The first three points are the top row, the next three points are the middle row, and the last three points are the bottom row. For position X and position Z, we are ranging from a negative distance variable, 0, and a positive distance variable, which is based on the value of the initialArea property. Position Y does not need to be updated, so we can keep the value from the transform position assigned to the fogPosY variable. By changing the value of the initialArea property on the **Inspector** view we can make a bigger or smaller area to clear the fog when the game starts.

The comments on the right side of each Vector3 object in the area list help to identify the position of each point in our 3 x 3 grid. Once we have the list of points, we can loop on each one and execute the RemoveFog method to clear the fog in that area. This method is defined in the following code snippet:

```
private void RemoveFog(Vector3 point)
{
    for (int i = 0; i < _vertices.Length; i++)
    {
        Vector3 position =
          transform.TransformPoint(_vertices[i]);

        float distance = Vector3.SqrMagnitude(
          position - point);
```

```
    float radiusSquare = _radius * _radius;

    if (distance < radiusSquare)
    {
       float alpha = Mathf.Min(
          _colors[i].a, distance / radiusSquare);
       _colors[i].a = alpha;
    }
  }

  _mesh.colors = _colors;
}
```

The RemoveFog method is the most important code in the FogComponent class because it is responsible for calculating what color in the mesh we need to change the alpha channel to, to make it transparent and visible as we are looking through the fog. This method has Vector3 point as a parameter, and we need to compare this point against all the vertices objects in the mesh property, that is why we are doing a loop in the array of vertices.

First, we need to convert the current vertex from local space to the world space using the TransformPoint method from the Transform API. Now that we have the position from the current vertex, we will calculate the distance between this position and the point parameter received in this method. We will do it using the SqrMagnitude method, from the Vector3 API, to calculate the distance by subtracting both Vector3 variables, position and point, and getting the magnitude of the resulting Vector3 variable, which is a faster way to compare distances using vectors.

Now that we have the distance between the current vertex and the point, we consider that the fog should be removed if this distance is less than the square of the radius, which is the property we are using to define the size of the area around the unit to clear the fog. If this statement is true, we can finally change the value of the alpha channel to make it transparent.

To make sure we only change the alpha value to add more transparency instead of making it darker, we consider the minimum value between the alpha channel in the color of the current vertex and the distance divided by the square of the radius. Using the Mathf.Min method, we can assign to the alpha variable the minimum value between those two, and then assign it back to the alpha channel in the color of the current vertex, which is accessible through the a property.

Note that we are manipulating the array of colors and, when we finish looping through all vertices objects we can assign the updated array of colors to the mesh colors properties. All the mathematical operations presented look a bit complex, but once we understand what each line is doing, it all makes sense to achieve the desired outcome.

We have finished writing the method to remove the fog from a specific point, which so far has been used by the ClearInitialArea method. Now, we are going to use the same RemoveFog method to also remove the fog above the units when they are exploring the map. First, let us add the `Update` method, which overrides the method with the same name from the `MonoBehaviour` class:

```
private void Update()
{
  foreach (GameObject unit in LevelManager.Instance.Units)
  {
    UpdateUnit(unit);
  }
}
```

In the `Update` method, we are going to loop through all the units that are currently on the map and update the fog for each unit using the `UpdateUnit` method, defined next. We are accessing the list of units using the Singleton from the `LevelManager` class; however, the `Instance` property and the `Units` list do not exist yet. We are going to modify the `LevelManager` class as soon as we finish the `FogComponent` class. You can comment out this method to avoid compilation errors on Unity until we add it but remember to uncomment it once we have updated the `LevelManager` class.

There is only one method left to be added in the `FogComponent` class, which is the `UpdateUnit` method defined here. Add the following code snippet to the class:

```
private void UpdateUnit(GameObject unit)
{
  Ray ray = new Ray(Camera.main.transform.position,
    unit.transform.position -
      Camera.main.transform.position);

  RaycastHit hit;
  if (Physics.Raycast(ray, out hit, 1000, _layerMask,
    QueryTriggerInteraction.Collide))
  {
    RemoveFog(hit.point);
  }
}
```

The last method in this class has one parameter, `GameObject`, which is the unit that we want to check whether it is under the camera view before we attempt to remove the fog from above the unit. The first thing to do is create a new `Ray` variable to define the line between two points, using the class constructor that has two `Vector3` variables: the position that we are shooting the ray from, which is the camera position; and the difference between the unit position and the camera position.

Here, we are using the Unity API's `Camera.main` property to access the main camera of the scene instead of adding a reference to the main camera in this class, which is a safe way to access the main camera since each scene can have only one main camera that is defined by the **MainCamera** tag in the **Main Camera** GameObject on that scene.

Next, we will create a new `RaycastHit` variable, which will hold the information on whether the Raycast hit something. The `Raycast` method, from the `Physics` API, has five parameters used internally to identify whether the ray hit a GameObject and the point where it was hit. The first parameter is the `ray` variable, which we created a few lines previously. Next, we have the `out` keyword, which indicates that the `hit` variable might have a value once the `Raycast` method is executed. The third parameter is the max distance that the ray should check for collision, and `1000` is a safe and standard value to use. The `layerMask` property is used for optimization purposes because it helps the ray to ignore the collision check on all other layers. As the last parameter, we have `QueryTriggerInteraction`, which specifies whether this query should hit collision triggers, and using the `Collide` value, we make sure that the `Raycast` method will always report any collision triggers that were hit.

If the return of the `Physics.Raycast` method is true, the `hit` variable will have the value of the point that the ray hit, something between the main camera and the unit, which in our case is the plane used as the fog. The last thing to do is to call the `RemoveFog` method, sending as a parameter the point where the hit happened, so we can remove the fog above the unit.

Now that we have finished writing the `FogComponent` class, we can move on to the last couple of things we need to do before we start using the fog in our game, which is creating the material and Prefab for the fog.

Creating the material and Prefab for the fog

We have a new script that is prepared to clear the fog on the map where the units are exploring. Now, we need to create a Prefab that will have this script attached to a GameObject plane and the material that will represent the fog.

Before jumping into the Prefab, we need to create a material that will have a very specific shader from Unity, which will allow us to manipulate the alpha channel:

1. In the **Project** view, left-click on the existing **Materials** folder, select the **Create | Material** option, and name it `BlackFog`.

2. Left-click on the new `BlackFog` material to select it and, in the **Inspector** view, click on the dropdown on the right side of the **Shader** property and select **Legacy Shaders | Particles | Alpha Blended**.

3. On the right side of the **Tint Color** property, click on the color button and, in the new **Color** window, change the values of **R** to **0**, **G** to **0**, **B** to **0**, and **A** to **255**.

4. Save the changes using the **File | Save Project** menu.

We are using a legacy built-in shader from Unity to change the alpha channel of **Tint Color** using a script. Although this shader is in the **Legacy Shaders** category, they are all still used in many games, and it is not an indication that it is going to be deprecated. One of the differences when compared with the other shader categories is that legacy shaders have some limitations, such as the lack of support for Shader Graph and Universal Render Pipeline, with Alpha Blended being simple and fast enough for our usage.

The following screenshot shows the material configured:

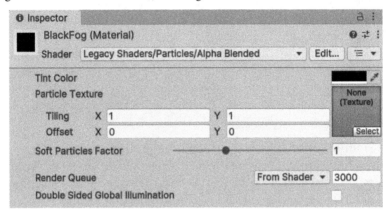

Figure 11.6 – The BlackFog material

Now that we have the material, we can go ahead and create the new Prefab. Since every level scene will have fog on top of the map, it makes sense to add it to the Prefab that is already present on both the **Playground** and **Level01** scenes, which is the **LevelManager** Prefab. Let us create the new Prefab for the fog:

1. In the **Project** view, double-click on the Prefab located at **Prefabs | Level |** LevelManager to edit it.

2. In the **Hierarchy** view, right-click on the **LevelManager** GameObject, select the **3D Object | Plane** option, and name it Fog.

3. Left-click on the new **Fog** GameObject and, in the **Inspector** view, click on the checkbox located on the left side of the GameObject name to disable it. This checkbox is in the top-left corner of the **Inspector** view and will change the Prefab icon from blue to gray to indicate that it was disabled.

4. In the **Layer** configuration, in the top-right corner of the **Inspector** view, click on the dropdown and select the layer named **Fog**.

5. On the **Transform** component, change **Position** to **(0, 20, 0)** and **Scale** to **(12, 12, 12)**.

6. On the **Mesh Renderer** component, expand the **Materials** property, click on the circle button on the right side of **Element 0** and, in the **Select Material** window, double-click on the BlackFog material to select it.

7. On the **Inspector** view, click on the **Add Component** button, and search for and add the FogComponent script. After adding the script, on the **Layer Mask** property, select the option named **Fog**.

8. To create a Prefab, drag the Fog GameObject from the **Hierarchy** view, and drop it in the existing **Prefabs | Level** folder in the **Project** view.

Now we have our Fog Prefab set up and ready to be used.

Note that this Prefab is disabled by default, which means it is not visible unless enabled. We saved the Fog GameObject as a Prefab, and it was also included in the LevelManager Prefab. These are called **Nested Prefabs**, where one Prefab is linked to another Prefab, but we can still update them separately. This is very useful because it allows us to change the Fog Prefab without having to edit the LevelManager Prefab or any scene. *Figure 11.7* shows the Fog Prefab with all the changes we made.

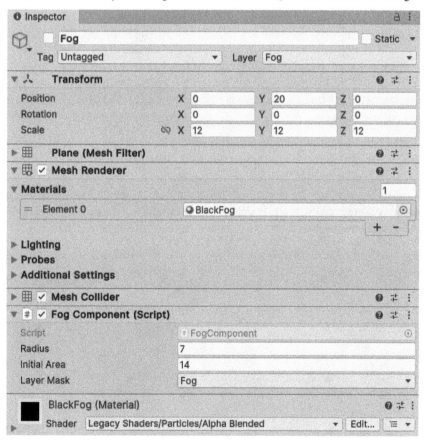

Figure 11.7 – The Fog Prefab configured

We have finished creating the script, material, and Prefab to add the fog to the map. They are all ready to be used, so let us update the scripts and Prefab that will be responsible for managing the fog in our levels.

Updating the Level Manager

In *Chapter 4, Creating the User Interface and HUD*, we added a new script called `LevelManager` to load the **SceneUI** scene additively and to instantiate the minimap Prefab. Now, we are going to expand this script to also take care of the fog in our levels.

Open the existing script located at **Scripts | Level |** `LevelManager.cs` and add the following lines inside the class, before the existing `Start` method:

```
[SerializeField] private GameObject _fog;
public static LevelManager Instance { private set; get; }
public List<GameObject> Units { private set; get; }
```

The changes we are making to this class are to enable the **Fog** GameObject when the game starts and to provide single access to the list of all units in the game. The `fog` property, exposed in the **Inspector** view, using the `SerializeField` attribute, will hold a reference to the **Fog** GameObject so we can enable it when the level is loaded.

We are also making this class a Singleton, and for this reason, we need the static `Instance` variable, which will be a reference to this class once initialized. In the following code snippet, we will see how this Singleton is different from the one we created in *Chapter 5, Spawning an Army of Units*, because this time, this class is a `MonoBehaviour`, which is internally managed by Unity. The last new property is the list of `Units`, which we are going to access in the `FogComponent` script to check the position of all units on the map so we can calculate and clear the fog above them.

Adding a new method in this class, we are going to override the `Awake` method from the `MonoBehaviour` class as we can see in the following code snippet:

```
private void Awake()
{
    if (Instance != null && Instance != this)
    {
        Destroy(this);
        return;
    }
    Instance = this;
    Units = new List<GameObject>();
}
```

The Awake method is an example of how to implement the Singleton pattern in a `MonoBehaviour` class. The first thing we do is check whether the static `Instance` variable is not `null` and if it is different from the current class. In the case where both validations are true, we need to destroy the current class because it means that we already have an instance of the class, and it is not this one.

Since Unity instantiates the `MonoBehaviour` classes internally, instead of creating a new instance, we can assign the value of the `this` keyword to the static `Instance` variable, which represents an instance of the `LevelManager` class that was already created by Unity when the GameObject was enabled. The last thing to do is to initialize the list of `Units`, so we can use it without having to initialize the list first.

The last change we need to make in this class is to add a single line to the existing `Start` method so we can enable the `Fog` GameObject when the level is loaded. Add the following line as the last thing inside the existing `Start` method:

```
private void Start()
{
    ...
    _fog.SetActive(true);
}
```

Now that we have updated the **LevelManager** script, we can also update the **LevelManager** Prefab with the latest changes. We are going to update more things on the **LevelManager** Prefab and remove a few GameObjects from the scenes so the **LevelManager** Prefab can really manage all the things that are common and required in our levels and we will have fewer things to do when setting up a new level scene. Let us first update the existing **LevelManager** Prefab:

1. In the **Project** view, double-click on the Prefab located at **Prefabs | Level |** `LevelManager` to edit it.

2. Drag the existing Prefab located at **Prefabs | Level |** `Plane` from the **Project** view, and drop it inside the **LevelManager** GameObject, in the **Hierarchy** view.

3. In the **Hierarchy** view, left-click on the **LevelManager** GameObject to select it. Drag the **Fog** GameObject from the **Hierarchy** view and drop it into the **Fog** property in the **Level Manager (Script)** component, in the **Inspector** view.

4. In the **Inspector** view, click on the **Add Component** button and search for and add the **LevelComponent** script. After adding the script, click on the circle button on the right side of the **Level Data** property and, in the **Select LevelData** window, double-click on the **Level01** asset to add it.

5. Drag the **Plane** GameObject from the **Hierarchy** view and drop it into the **Plane** property in the **Level Component (Script)** component, in the **Inspector** view.

With these changes, we are going to have a better Prefab that will make our lives easier when we need to create new levels in our game. *Figure 11.8* shows how the **LevelManager** Prefab looks like after the changes are made in the **Hierarchy** view and scripts are updated in the **LevelManager** GameObject that is selected in the **Inspector** view.

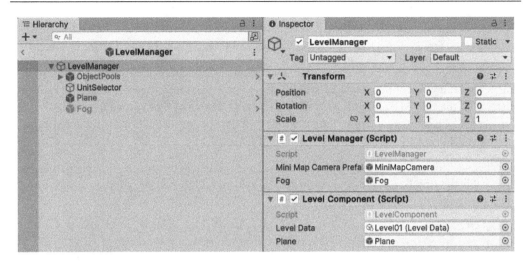

Figure 11.8 – The updated LevelManager Prefab

Now that we have our **LevelManager** Prefab updated, we need to reflect the changes on the existing scenes and remove a couple of GameObjects that are now part of the Prefab:

1. Open the scene located at **Scenes | Playground**.

2. In the **Hierarchy** view, right-click on the **Plane** GameObject and select the **Delete** menu option to remove it. Right-click on the **Level** GameObject and select the **Delete** menu option to remove it as well.

3. Expand the **LevelManager** GameObject and click on the **Plane** GameObject to select it. In the **Inspector** view, on the **NavMeshSurface** component, click on the **Bake** button.

4. Save the changes using the **File | Save** menu.

5. Open the scene located at **Scenes | Level01**.

6. In the **Hierarchy** view, right-click on the **Plane** GameObject and select the **Delete** menu option to remove it.

7. Expand the **LevelManager** GameObject and click on the **Plane** GameObject to select it. In the **Inspector** view, on the **NavMeshSurface** component, click on the **Bake** button.

8. Save the changes using the **File | Save** menu.

These changes will help make our two scenes, **Playground** and **Level01**, more organized and speed up the process of adding new levels to the game. Remember to perform the previous steps on all scenes that you have created so far as well, so they are all configured in the same way.

There is one small change that we need to make in a couple of existing scripts to complete all the changes required to make the fog work in our game. In the FogComponent class, we are accessing the list of units on the map to clear the fog above them; however, we still need to add the units to that list once they are spawned on the map.

Open the existing script located in the **Scripts | Spawner** | `BasicWarriorSpawner.cs` folder
and add the following line at the end of the `OnBasicWarriorSpawned` method:

```
private void OnBasicWarriorSpawned(
   BasicWarriorSpawnMessage message)
{
   ...

   LevelManager.Instance.Units.Add(warrior);
}
```

Let us update the second script now. Open the existing `BasicMageSpawner.cs`
script located in **Script | Spawner** and add the following line at the end of the
`OnBasicMageSpawned` method:

```
private void OnBasicMageSpawned(
   BasicMageSpawnMessage message)
{
   ...

   LevelManager.Instance.Units.Add(mage);
}
```

Those are the last changes we need to make in the project to add the fog to the map. Go ahead and
run the game in the Editor, add a few units to the map, select them, and move to areas where the map
is covered to see the fog being removed, as we can see in *Figure 11.9*:

Figure 11.9 – The fog covering part of the map but not the warriors on the left side

As we can see in the minimap on the bottom-left side of *Figure 11.9*, most of the map is now covered by the black fog. Only the initial area, which is the center of the map, and the left side where the two warriors are, do not have the black fog covering the map.

We can also see an enemy group with a couple of Orcs at the center of the minimap, but we cannot see the larger enemy group that has eight Orcs and one Golem on the top-right side of the map because they are covered by the black fog, which will be quite a surprise for the player when the selected units explore that area.

So far, we have enemy groups on the level and black fog covering the unexplored areas of the map. Now, the only thing left is to add patrol behavior to the enemies so they can move around while waiting for the units to find them. Let us see how in the next section.

Patrolling the area

Having different enemy groups on the map and hidden under the fog is already quite a challenge for the player. However, finding idle enemies while exploring the map is not very exciting. We can make it more fun and dynamic by adding patrol behavior to the enemies, so they are moving around between two points on the map.

Since we are using Unity's NavMesh system, it is very simple to modify the existing Enemy ComponentNavMesh script and add the patrol behavior, so the enemy will be able to walk around the area, attack a unit when a collision is detected, and chase the unit if it tries to escape.

Open the existing script located at **Scripts | Enemy |** EnemyComponentNavMesh.cs and add the following variables inside the class, before the declaration of the existing Start method:

```
private int _currentPoint;
private Vector3 _startPosition;
private Vector3[] _points = new Vector3[2] {
   new Vector3(-10, 0, 0), new Vector3(10, 0, 0) };
```

The three new class variables will assist the patrol behavior. The array of points has two Vector3 items, working as offsets added to the startPosition variable, so we know how much the enemy can move on the map considering the spawn point. Since we only change the value of X on each position, the enemies will be able to patrol the area horizontally, moving between those two points. We are going to use the currentPoint property to find out what is the point that the enemy is moving toward.

The next change in this class is to add a few lines to the existing Start method. Add the following three lines at the end of the Start method:

```
private void Start()
{
   ...
```

```
    _agent.speed = WalkSpeed +
      UnityEngine.Random.Range(0.1f, 0.5f);
    _currentPoint =
      UnityEngine.Random.Range(0, _points.Length);
    _startPosition = transform.position;
}
```

The line that is setting the value of the agent.speed property was already there, but it was only using the value of WalkSpeed. Now, we are also adding a small random value between 0.1 and 0.5 so the enemies have a slight variation in their speed when patrolling to make the group movement look less robotic when they are all walking at the same time.

The currentPoint property is also randomly selected so the enemies do not start moving in the same direction when they are spawned. Since it is random, the enemies may move in the same direction sometimes, but it is acceptable since it is how random things work. The last new line is just getting the current position and adding it to the startPosition variable so we can use it later.

The next Update method is also an existing one, but it has changed quite a bit so we can replace all its content with the following code snippet:

```
private void Update()
{
    if (IsDead)
    {
        return;
    }

    if (_targetToFollow != null)
    {
        // Chase the unit
        if (Vector3.Distance(transform.position,
          _targetToFollow.position) < WalkSpeed)
        {
            return;
        }
        _agent.destination = _targetToFollow.position;
    }
    else
    {
        // Patrol the area
        if (!_agent.pathPending &&
          _agent.remainingDistance < 0.5f)
        {
            _agent.destination =
```

```
            _startPosition + _points[_currentPoint];
        _currentPoint = (_currentPoint + 1) % _points.Length;
        UpdateState(ActionType.Move);
    }
  }
}
```

The chase behavior was already in the Update method, but now we have different conditionals to decide what to execute in this method. The first part of the method checks whether the enemy is dead, and if that is the case, we can ignore the rest of the code.

Then, we check whether the targetToFollow variable is not null, which means that we need to follow the unit that was attacking and now is trying to escape.

However, if there is no valid targetToFollow object, we can patrol the area, which is the new code added to this method. We are using the agent variable, from the NavMeshAgent class, to check that there is no valid pathPending object and the remaining distance between the current position and the destination is less than 0.5, and if both are true, we can move to a different point.

The new destination will be the startPosition value plus the Vector3 value of the currentPoint property, and then we update the currentPoint property to be the next available point or the first point, using the mod operator, %. The last thing we do is to change the action to Move, so the walking animation is played while the enemy is moving toward the new destination point.

The last existing method that we are going to update is the OnStopAttacking method, which has also changed quite a bit, so we can replace all its content with the following code snippet:

```
private void OnStopAttacking(Transform target, bool opponentIsDead)
{
  if (IsDead)
  {
    return;
  }

  // Chase the unit
  if (!opponentIsDead)
  {
    transform.LookAt(target.position);
    _targetToFollow = target;
  }

  // Patrol the area (when not chasing)
  _agent.isStopped = false;
  _startPosition = transform.position;
```

```
    UpdateState(ActionType.Move);
}
```

This method is executed when the enemy is told to stop attacking because the unit is dead or is trying to escape and so must be chased. The first check is to ignore the method if the enemy is already dead, otherwise, we can proceed to the rest of the code.

If the value of the `opponentIsDead` parameter is false, this means that we stopped attacking because the collision is not happening anymore, and the enemy must look at the `target` parameter and follow the unit. The last lines of the code are executed in both cases when the enemy is chasing the unit or patrolling the area. For both behaviors, the enemy will start to move, so we will change the value of `isStopped` to `false`, update `startPosition` to the current position, and change the action to `Move` so the walking animation is played.

Note that, if the value of the `opponentIsDead` variable is true, the value of the `targetToFollow` variable will not be set and will have the null value, so the patrol behavior will start the next time the `Update` method is executed internally by the `MonoBehavior` class.

That is the last change to this class, and once we save the script changes, we can run the game on the Editor and see that the enemies are now moving horizontally, patrolling the area, as we can see in *Figure 11.10*.

Figure 11.10 – The enemies patrolling the area, moving in different directions

If we run the game multiple times, we can see the random directions of the enemies, and this is even more evident when we look at the larger enemy group that is on the top-right side of the map. We can send a unit in that direction to explore the map and remove the fog so we can see the enemies moving around or disable the Fog GameObject while the game is running on the Editor.

With a few code changes in the NavMeshEnemyComponent class, we were able to add a new behavior for the enemy, and now the player will always see enemies moving and patrolling an area on the map instead of being idle in the same spot.

Summary

This was a very fun chapter – well done for finishing it! We added more features that are fundamental to RTS games, and we have almost reached the point where we have a solid gameplay experience with many challenges for the player.

In this chapter, we learned how to create and set up groups of enemies, and how to add them to the map using configurable spawn points. An enemy group is a very powerful and flexible solution to rapidly add or reuse enemies on the map, and we are going to use it extensively as we progress toward the end of the book.

We also learned how to add the Fog of War, a must-have feature present in any RTS game that has a map covered by fog to make the player explore areas by sending units. Our fog nicely clears once the player moves the camera and the selected units to any area of the map, revealing enemies hidden and waiting to attack the units. Our enemies have gotten a bit smart and challenging now that we have learned how to implement the patrol behavior between two points on the map, adding a new skill to the enemies, besides attacking and chasing.

In *Chapter 12, Balancing the Game's Difficulty*, we are going to start reusing all that has been developed so far to balance the gameplay experience for the player as well as learning how to set different levels of difficulty by making enemies stronger or weaker. We are also going to learn how to add unit tests to our project and how they can help us test the difficulty and balance of the game.

Further reading

In this chapter, we saw a few new APIs and concepts from Unity. The following links will help you to understand the topics we covered in this chapter better, as well as helping you with doubts you might have regarding this content:

- *EditorGUILayout.PropertyField*: https://docs.unity3d.com/2023.1/Documentation/ScriptReference/EditorGUILayout.PropertyField.html

- *Vector3.sqrMagnitude*: https://docs.unity3d.com/2023.1/Documentation/ScriptReference/Vector3-sqrMagnitude.html

- *Physics.Raycast*: https://docs.unity3d.com/2023.1/Documentation/ScriptReference/Physics.Raycast.html

- *Nested Prefabs*: https://docs.unity3d.com/2023.1/Documentation/Manual/NestedPrefabs.html

- *Making an Agent Patrol Between a Set of Points*: https://docs.unity3d.com/2023.1/Documentation/Manual/nav-AgentPatrol.html

- *Fog of War*: https://en.wikipedia.org/wiki/Fog_of_war

Part 4:
The Gameplay

In this last part of the book, you will learn how to piece together the features we have developed so far to create the gameplay of our RTS game. You will start by learning how to balance the units and enemies so that they provide a fair and challenging battle for the player, using unit tests to simulate different battle scenarios. You will also learn how to generate resources automatically and implement the resource-gathering action for the units. Then, you will see how to add buildings on the map that can upgrade the resource generation and create a defense tower to defend the settlement, with ranged attacks.

After implementing everything that is needed to generate and gather resources, including upgrades, you will learn how to create objectives for the player and how to track progress in a level. Finally, you will see how to export your game to desktop platforms manually and automatically with a script, as well as how to expand the game, adding new content such as units, enemies, objectives, and levels.

This part includes the following chapters:

- *Chapter 12, Balancing the Game's Difficulty*
- *Chapter 13, Producing and Gathering Resources*
- *Chapter 14, Crafting Buildings and Defense Towers*
- *Chapter 15, Tracking Progression and Objectives*
- *Chapter 16, Exporting and Expanding Your Game*

12

Balancing the Game's Difficulty

We have created a lot of very cool features for our RTS game, and our *Dragoncraft* game is almost complete. However, there is one thing that could make any game be considered boring or unfair: the game's difficulty. Whether the game is too easy without real challenges to the player, or too hard to the point of not being an enjoyable experience, it all comes down to the balance of difficulty in the game.

In this chapter, we are going to learn how to set up our Unity project to use Assemble Definition files, and how to configure the Unity Test Framework so we can use unit tests to validate the enemy and unit configurations in our game. We will also learn how to create test scripts to simulate battles between both one unit and a single enemy, and one unit and multiple enemies. Plus, we will see how we can use test scripts to help find the right balance of the game's difficulty for the battles and the benefits that unit tests can add to game projects.

By the end of this chapter, you will learn how to create unit tests using the Unity Test Framework, how to create flexible test scripts that will use a list of parameters to run different test cases, and how we can use unit tests to create multiple battle configurations to test. You will also learn how we can create a formula to set up a level multiplier for our units to make them stronger as we level up, and how it can be used to balance our game difficulty.

In this chapter, we will cover the following topics:

- Writing unit tests
- Using unit tests to simulate battles
- Balancing the battle difficulty

Technical requirements

The project setup with the imported assets for this chapter can be found on GitHub at https://github.com/PacktPublishing/Creating-an-RTS-game-in-Unity-2023 in the Chapter12 folder inside the project.

Writing unit tests

We have developed a lot of features and created configurations for units and enemies, while retaining the flexibility to expand and add many more characters to our RTS game. However, with such flexibility comes the difficult task of testing all the configurations to make sure the units and enemies are balanced in strength. We could continue testing the units and enemies in battle, adding them using the **Debug** menu while running the game in the Unity Editor, but it would take a lot of time and effort to test multiple combinations.

So far, we have added the Warrior and Mage units, and the Orc, Golem, and Dragon enemies. Later, in *Chapter 14, Crafting Buildings and Defense Towers*, we are going to add a feature to train the units and level them up, making them stronger. After adding the unit level-up functionality, we will have to make sure that when the Warrior and Mage units each level up, they become stronger but not overpowered, as otherwise, we will make the game shallow and boring for players, without challenges to stimulate their strategic thinking.

For the game to work with the flexibility of having multiple units and enemies and unit level-ups, we need a tool to fully test and validate all battle combinations and their expected outcomes. Here we can use the **Unity Test Framework** (**UTF**) that comes with the Unity Editor and provides everything we need to validate and test our game.

The UTF is a framework included in the Unity Editor that enables projects to test code in Edit Mode (running tests in the Editor) and Play Mode (actually running the game to test the code). For our tests, which are character configurations and battle combinations, Edit Mode has everything we need. However, before we start creating the tests, we need to set up the project.

Setting up the UTF

To set up the UTF we are going to need an **Assembly Definition** file. The Assembly Definition is a configuration file placed on any folder that has scripts (including subfolders) and tells Unity to generate a **Dynamic Link Library** (**DLL**) for that specific folder and its scripts. By default, Unity will create a single DLL containing all the project code in the same library. However, projects with large code bases can benefit from having multiple Assembly Definition files in different script folders to create DLLs for each of them, thus taking a modular approach.

In our project, we do not need such modularity for our scripts, but we do need to create an Assembly Definition file because the UTF has its own DLL, and we need to create a reference for the project DLL so the test scripts can access and use the project classes. Let us first create the Assembly Definition file for the project:

1. In the **Project** view, right-click on the **Scripts** folder, select **Create | Assembly Definition**, and save it as `Dragoncraft`.

2. Left-click on the newly created `Dragoncraft.asmdef` file to select it. Then, in the **Inspector** view, under the **Assembly Definition References** section, click on the plus sign button (+) to add a new item to the list that was empty.

3. Click on the circle button on the right side of the new item and, in the new **Select AssemblyDefinitionAsset** window, select the file named `Unity.TextMeshPro`.

4. Scroll down to the bottom of the **Inspector** view and click on the **Apply** button to save the changes.

Once we click on the **Apply** button, the Unity Editor will recompile the scripts using the new Assembly Definition. Since this file groups together the assets in a DLL, we need to add a reference to any other DLLs that we are using but that are not part of the Unity engine. That is why we had to add a reference to the **TextMeshPro** Assembly Definition file, as shown in the following screenshot, so our code can still see and use the classes from that package:

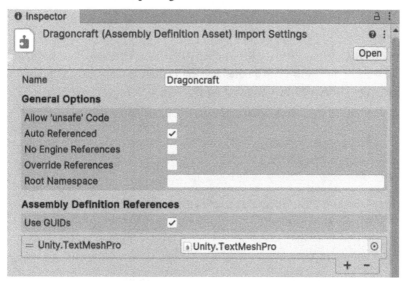

Figure 12.1 – The Dragoncraft Assembly Definition file

We can have multiple Assembly Definition files, however, every time a new one is created, we need to add a reference to any DLL that is used. All Unity packages that can be installed using the Package Manager must come with an Assembly Definition file. One important thing to keep in mind is that the project cannot have more than one Assembly Definition file with the same name, otherwise, a compilation error will occur.

The Edit Mode tests have a requirement that all test scripts must be inside an **Editor | Tests** folder structure. We can add this folder anywhere, but to keep our project organized we are going to create it inside the **Scripts** folder:

1. In the **Project** view, inside the **Scripts** folder, right-click on the existing **Editor** folder and select **Create | Testing | Tests Assembly Folder**. This menu option will create a folder named `Tests` and an Assembly Definition file named `Tests.asmdef` inside the new folder.

2. Double-click to open the new **Tests** folder, then right-click on the new `Tests.asmdef` file and select the **Rename** menu option. Change the filename to `Dragoncraft.Tests.asmdef`.

3. Left-click on the `Dragoncraft.Tests.asmdef` file to select it. Then, in the **Inspector** view, change the value of the **Name** property from **Tests** to `Dragoncraft.Tests`.

4. In the **Assembly Definition References** section, click on the plus sign button (+) to add a new item to the list. Click on the circle button on the right side of the new item and, in the new **Select AssemblyDefinitionAsset** window, select the file named `Dragoncraft`.

5. In the **Platforms** section, click on the first **Any Platform** checkbox to unselect it – this will select all platforms below it. Click on the **Deselect all** button and click on the checkbox on the right side of the **Editor** platform to select only this one.

6. Scroll down to the bottom of the **Inspector** view and click on **Apply** to save the changes.

A good practice is to name the Assembly Definition file the same as the namespace we are already using in our code, which is `Dragoncraft`, followed by the name of the folder, separated by a dot – for example, `Dragoncraft.Tests`, as we just did.

The new Assembly Definition file, which was created using the **Testing** menu option, comes with a couple of references to the **TestRunner** package, allowing us to use this package in our test scripts. We also included a reference to the `Dragoncraft` Assembly Definition file for the same reason, to use its existing code in our project.

The following screenshot shows the new Assembly Definition file after applying the changes we just discussed:

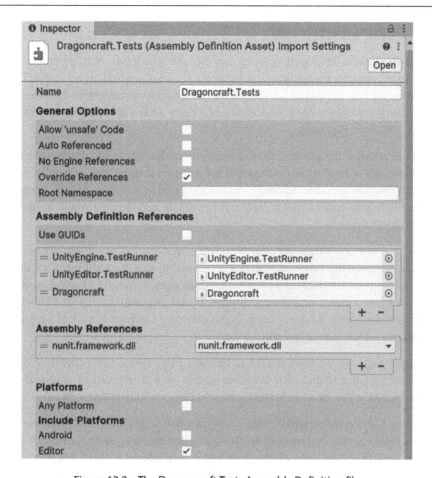

Figure 12.2 – The Dragoncraft.Tests Assembly Definition file

In our project, we are going to have only two Assembly Definition files: Dragoncraft.asmdef, which contains all the code we wrote for our RTS game, and Dragoncraft.Tests.asmdef, which will contain all the test scripts. Since we added only **Editor** as the platform in the **Dragoncraft. Tests.asmdef Platforms** section, Unity will take care to not include the test code in our game, and it will only work in the Editor.

Now that we have an Assembly Definition file for our project, one caveat to bear in mind is that we might need to add a reference there for any new package added to the project. This is not an issue; in fact, it is the industry standard when developing large projects on Unity since we can separate each module into a different Assembly Definition file.

Now that we have everything set up, we can move on to creating a few test scripts and using the Test Runner to run our tests in the Editor.

Creating the first test script

The UTF is a great solution developed by Unity that uses the **NUnit** unit testing framework, which is widely used to test code for C# applications. There are a few known limitations with UTF 1.3 (the current version included in Unity 2023), such as a lack of support for parameterized tests and repeated tests. However, such limitations are not an issue since Unity adds a lot of other features on top of NUnit to make it work on the engine.

It is also worth mentioning the software development process known as **Test-Driven Development (TDD)**, which consists of creating test case scripts before the code is written, instead of writing the code and then creating the tests later. Although this is not covered in this book, TDD is a great software development process with many benefits, but for it to work as intended, we would need to adopt it from the beginning of the project.

The TDD is not used as extensively in game development as in more traditional software development, probably due to the creativity involved in creating games, but this does not mean we should just ignore it. Perhaps instead of developing a whole game from scratch using TDD, which would require updating the test scripts every time a feature changes, a good idea would be to use TDD on isolated parts of the project that don't change as much as the gameplay mechanics, such as a login/account creation system or a matchmaking system.

Besides testing the code, we can also use UTF to test data, more specifically ScriptableObjects. This is exactly what we are going to do in our following scripts; we will add tests to validate the existing ScriptableObjects for the `UnitData`, `EnemyData`, and `EnemyGroupData` classes.

Before writing the first test script, let us create a new class that will be inherited by all test scripts so they can use the common methods without needlessly duplicating the code. In the **Project** view, open the **Scripts | Editor** folder, right-click on the **Tests** folder, then select **Create | C# Script** and name it `BaseTest`. Then open the new script in the IDE and replace the contents with the following code:

```
using UnityEditor;
namespace Dragoncraft.Tests
{
    public class BaseTest
    {
        protected UnitData LoadUnit(string unit)
        {
            return AssetDatabase.LoadAssetAtPath<UnitData>(
                $"Assets/Data/Unit/{unit}.asset");
        }
    }
}
```

We are going to use a particular namespace for all the test scripts, called `Dragoncraft.Tests`, to keep our code organized and match the name of the Assembly Definition file we have created for the test scripts. This new `BaseTest` class is not a test script itself and does not inherit from any other class because, as we will see in the next script, we will use an attribute in the methods declaration to let Unity know they are unit tests.

The first method of the `BaseTest` class is `LoadUnit`, which will load the ScriptableObject of the type `UnitData` using the `AssetDatabase` method `LoadAssetAtPath`. This method has a generic parameter type `<T>` that we use as `<UnitData>`, which indicates that the asset loaded by the `LoadAssetAtPath` method is converted to the type `UnitData` when returned.

The `LoadUnit` method has a `unit` string parameter, which is the name of the file. Using this filename, we create a string with the path of `UnitData` in our project, which is `Assets/ Data/ Unit`, and add the `.asset` extension at the end. For example, if the `unit` parameter has the value of `BasicOrc`, the string to load the ScriptableObject will have the value of `Assets/Data/Unity/BasicOrc.asset`.

The `LoadUnit` method is a very handy method that we are going to use in our test scripts. Now, let us add two more methods to load `EnemyData` and `EnemyGroupData`, respectively, using the same logic but with the corresponding path of each ScriptableObject in the project. Add the following methods in the new `BaseTest` class:

```
protected EnemyData LoadEnemy(string enemy)
{
    return AssetDatabase.LoadAssetAtPath<EnemyData>(
        $"Assets/Data/Enemy/{enemy}.asset");
}


protected EnemyGroupData LoadEnemyGroup(string enemyGroup)
{
    return AssetDatabase.LoadAssetAtPath<EnemyGroupData>(
        $"Assets/Data/EnemyGroup/{enemyGroup}.asset");
}
```

These two methods, `LoadEnemy` and `LoadEnemyGroup`, look similar but have different paths for each ScriptableObject type. The assets loaded and returned by the two methods also have their respective types as well.

Now that we can load our `UnitData`, `EnemyData`, and `EnemyGroupData` ScriptableObjects, we can move on to the first test script. In the **Project** view, open the **Scripts | Editor** folder and right-click on the **Tests** folder, then select the **Create | C# Script** menu option, and name it `UnitDataTest`. Open the new script in the IDE and replace the content with the following code:

```
using NUnit.Framework;
using System.Collections.Generic;
namespace Dragoncraft.Tests
{
    public class UnitDataTest : BaseTest
```

```
    {
      private static List<string> units = new List<string>()
      {
        "BasicWarrior",
        "BasicMage"
      };
    }
  }
```

The new UnitDataTest class is also in the Dragoncraft.Tests namespace and is inherited from the other class that we just created: the BaseTest class. Note that we specify the use of NUnit. Framework at the beginning of the class, so we can access the NUnit API in our test scripts.

In the previous code snippet, we added a new class variable called units, which is a List of strings and contains the name of the UnitData files we have created so far, BasicWarrior and BasicMage. This class variable is also static, and that is because we are going to use it as a parameter for our first test, which we'll see next.

The following method, TestWithUnits, will load a ScriptableObject of each UnitData asset on the list of units and will test each property to make sure they have valid values. So, add the following TestWithUnits method in the new UnitDataTest class:

```
[Test]
public void TestWithUnits([ValueSource("units")] string unit)
{
  UnitData data = LoadUnit(unit);
  Assert.IsNotNull(data, $"UnitData {unit} not found.");

  Assert.IsTrue(data.Health > 0);
  Assert.IsTrue(data.Attack > 0);
  Assert.IsTrue(data.Defense >= 0);
  Assert.IsTrue(data.WalkSpeed > 0);
  Assert.IsNotNull(data.SelectedColor);

  Assert.IsFalse(data.AnimationStateAttack01 == "");
  Assert.IsFalse(data.AnimationStateAttack02 == "");
  Assert.IsFalse(data.AnimationStateDefense == "");
  Assert.IsFalse(data.AnimationStateMove == "");
  Assert.IsFalse(data.AnimationStateIdle == "");
  Assert.IsFalse(data.AnimationStateCollect == "");
  Assert.IsFalse(data.AnimationStateDeath == "");

  Assert.IsTrue(data.AttackRange >= 0);
```

```
    Assert.IsTrue(data.ColliderSize > 0);
    Assert.IsTrue(data.Level >= 0);
    Assert.IsTrue(data.LevelMultiplier > 0);
    Assert.IsFalse(data.Actions == ActionType.None);
}
```

This new TestWithUnits method contains quite a lot of information, so let us go over each part individually. Remember that one of the limitations of NUnit in the UTF is the lack of parameterized tests? We cannot use parameterized tests from NUnit; however, we can use the ValueSource attribute, which allows the method to use each item of a static list as a parameter.

The first thing we need to do is to add the Test attribute on top of our method to tell Unity that this is a unit test method. Now that Unity knows it is a unit test, the next thing we do is add the ValueSource attribute before the first parameter of the method. The ValueSource attribute has a parameter as well, which is a string with the name of a static variable that contains the list of parameters for this unit test.

Next, we tell Unity that the TestWithUnits method is a unit test method by using the Test attribute. We also added the units list of strings to the ValueSource attribute and the parameter unit that we will use to access each item in the list of units. Since the units static variable is a list of strings, the method will receive one string in the unit parameter. This will all make sense in a moment, when we run our tests using the Test Runner. Unity will look at the units list of strings and execute the TestWithUnits method for each item on that list, which will also be used in this method as the parameter unit.

Now that the method is set up, we can move to the actual test. The first thing we do is load the ScriptableObject using the LoadUnit method and the unit parameter; this will return a UnitData object. Then, we use the Assert API to validate what we want. In this case, we use the IsNotNull method, where the first parameter is an object to verify that it is not null, and the second parameter is a message for when the validation fails.

The Assert IsNotNull test is considered a success if the first parameter, the data object, is not null. Otherwise, this test will fail, and the method execution will stop. If this test fails, we will see a message in the Test Runner saying that UnitData was not found, so we will know why the test failed. The second parameter is optional, but will help us to understand what caused the test to fail, especially when we have more than one Assert in the same method.

We are also using another two methods from the Assert API, IsTrue and IsFalse, that validate whether a condition is true or false, respectively. This is useful to test, for example, whether the Health property has a value greater than 0, to make sure we do not have an invalid value that would not work in our game. All the Asserts methods have a second parameter with an optional message to be displayed if the test fails, and it is always recommended to add this. In the TestWithUnits method here, the messages were omitted to keep the code in the book simple and clean.

Now that we have reviewed what is happening in the TestWithUnits method, we can finally run the tests to see it all in action:

1. In the Unity Editor, click **Window** | **General** | **Test Runner** to open the **Test Runner** window. You can drag the window and dock it anywhere in the Editor to keep it visible.

2. In the new **Test Runner** window, click on the **Run All** button, located in the top-left corner of the window.

The **Test Runner** window will list all the tests in the project. Since we are using ValueSource to create one test for each item in the list of units, we will see two tests on the list: **TestWithUnits(BasicWarrior)** and **TestWithUnits(BasicMage)**. Both tests will use the same TestWithUnits method, but with different parameter values. If we add or remove items to or from the list of units, Unity will update the **Test Runner** window to display the current list of units to test.

ValueSource is a very powerful attribute because by using one single method, we can test the same method with multiple values. Once the Test Runner finishes running all the tests, we should see two tests with green checks indicating that they were run successfully:

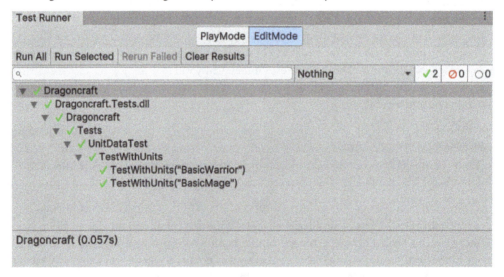

Figure 12.3 – Test Runner after running the tests for each UnitData ScriptableObject

The **Test Runner** window has more information and options for us to get used to. In the top right, we can see how many tests we have in the project and how many succeeded, failed, or were not run yet. We also have the **Run All**, **Run Selected** (which runs one or more selected tests), **Rerun Failed** (to retry failed tests), and **Clear Results** buttons. We can see the **PlayMode** button that switches to list all the tests that require the game to run to be executed, and the **EditMode** button that lists all the tests that can run in the Editor – this is the one we are using.

Now, let us add a test that we know will fail to see how the **Test Runner** window displays the results. In the `UnitDataTest` class that we just created, add a new string to the list of `units` with the `DarkMage` value, as shown in the following code block:

```
public class UnitDataTest : BaseTest
{
  private static List<string> units = new List<string>()
  {
    "BasicWarrior",
    "BasicMage",
    "DarkMage"
  };
}
```

This `UnitData` instance does not exist, so the test will fail when trying to load it.

Now save the test script and move back to the Unity Editor. This will refresh the **Test Runner** window and we will be able to see three tests now – a new one was added, called **TestWithUnits(DarkMage)**. Press the **Run All** button to execute all three tests; we should see two succeed (shown with a green checkmark) and one fail (the red circle), as shown in the following screenshot:

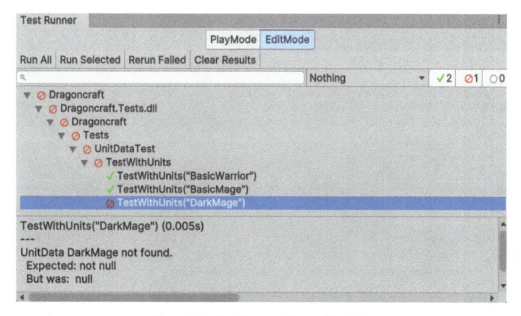

Figure 12.4 – Test Runner with one failed test

Left-click on the test that failed to select it and the **Test Runner** window will show the details in the panel below the list of tests, as we can see in the preceding screenshot. The details show which specific test and parameter failed and our custom message that says why the test failed: UnitData DarkMage was not found.

Remove DarkMage from the list of units, since it was only added there for demonstration purposes and we no longer need it. Now that we have learned how to set up, create, and run tests, we can move on to add more test scripts to our project.

Adding more test scripts

We have finished setting up our tests and adding the first test script to validate the ScriptableObjects of our both UnitData assets. Now we are going to add a couple more test scripts to validate the EnemyData and EnemyGroupData assets.

In the **Project** view, open the **Scripts | Editor** folder and right-click on the **Tests** folder, then select the **Create | C# Script** menu option, name it EnemyDataTest, and replace the content with the following code:

```
using NUnit.Framework;
using System.Collections.Generic;
namespace Dragoncraft.Tests
{
  public class EnemyDataTest : BaseTest
  {
    private static List<string> enemies = new List<string>()
    {
      "BasicOrc",
      "BasicGolem",
      "RedDragon"
    };
  }
}
```

The new EnemyDataTest class has the same base structure as the UnitDataTest class that we created previously. This class is also inherited from the BaseTest class and has a static variable, which is the list of strings containing all the asset names of enemies.

The TestWithEnemies method, defined next, is also similar to the TestWithUnits method from the previous class. However, here, we replace all the unit references with the enemy references. Add the following method to our new class:

```
[Test]
public void TestWithEnemies([ValueSource("enemies")] string enemy)
{
```

```
        EnemyData enemyData = LoadEnemy(enemy);
        Assert.IsNotNull(enemyData, $"EnemyData {enemy} not found.");

        Assert.IsTrue(enemyData.Health > 0);
        Assert.IsTrue(enemyData.Attack > 0);
        Assert.IsTrue(enemyData.Defense >= 0);
        Assert.IsTrue(enemyData.WalkSpeed > 0);
        Assert.IsNotNull(enemyData.SelectedColor);

        Assert.IsFalse(enemyData.AnimationStateAttack01 == "");
        Assert.IsFalse(enemyData.AnimationStateAttack02 == "");
        Assert.IsFalse(enemyData.AnimationStateDefense == "");
        Assert.IsFalse(enemyData.AnimationStateMove == "");
        Assert.IsFalse(enemyData.AnimationStateIdle == "");
        Assert.IsFalse(enemyData.AnimationStateCollect == "");
        Assert.IsFalse(enemyData.AnimationStateDeath == "");

        Assert.IsTrue(enemyData.AttackRange >= 0);
        Assert.IsTrue(enemyData.ColliderSize > 0);
    }
```

We are also using the `Test` and `ValueSource` attributes in the `TestWithEnemies` method so the test runs on each item in the list of `enemies` and this method will receive each item as an enemy parameter. As we just saw in the `TestWithUnits` method, the Test Runner will also create a test for each enemy in the list of `enemies` to execute this method.

Another difference in this method is that we are using the `EnemyData` class since this deals with the enemy tests, and we use the `LoadEnemy` method to get the ScriptableObject. After asserting that `enemyData` is not null, we check all properties of the ScriptableObject using the same asserts that we have in the `TestWithUnits` method. The only difference here is that the `EnemyData` class has fewer properties than the `UnitData` class.

Before running the new tests we have created for the `EnemyData` class, let us first create one more class to also test the `EnemyGroupData` class. In the **Project** view, open the **Scripts | Editor** folder and right-click on the **Tests** folder, then select the **Create | C# Script** menu option, name it `EnemyGroupDataTest`, and replace the content with the following code:

```
using NUnit.Framework;
using System.Collections.Generic;
namespace Dragoncraft.Tests
{
    public class EnemyGroupDataTest : BaseTest
    {
        private static List<string> groups = new List<string>()
```

```
    {
      "OrcDuo",
      "GolemAndOrcs"
    };
  }
}
```

The new `EnemyGroupDataTest` class is also like the two other classes, `UnitDataTest` and `EnemyDataTest`, inheriting as it does from the `BaseTest` class and having a `static` variable with a list of strings for each enemy group's asset name. The `groups` static variable is used by the `ValueSource` attribute in the new `TestWithGroups` method defined here:

```
[Test]
public void TestWithGroups([ValueSource("groups")] string group)
{
    EnemyGroupData enemyGroupData = LoadEnemyGroup(group);
    Assert.IsNotNull(enemyGroupData);

    Assert.IsFalse(enemyGroupData.Name == "");
    Assert.IsFalse(enemyGroupData.Enemies.Count == 0);
}
```

One more time, we add the `Test` and `ValueSource` attributes to a new `TestWithGroups` method, this time using the `groups` static variable that will make the Test Runner create a test for each group and execute this test using the `group` parameter.

This test is smaller than the previous two because the `EnemyGroupData` class has only two properties, `Name` and `Enemies`, so those are the asserts we need to validate, as well as checking that `enemyGroupData` is not null after loading the asset using the `LoadEnemyGroup` method.

Now that we have finished creating two new test scripts, we can go back to the Unity Editor and open the **Test Runner** window to see the new total of seven tests: two tests for `UnitData`, three for `EnemyData`, and two for `EnemyGroupData`. Once Unity finishes refreshing the new tests, click **Run All** in the **Test Runner** window to execute the tests. The result should be the same as shown in the following screenshot:

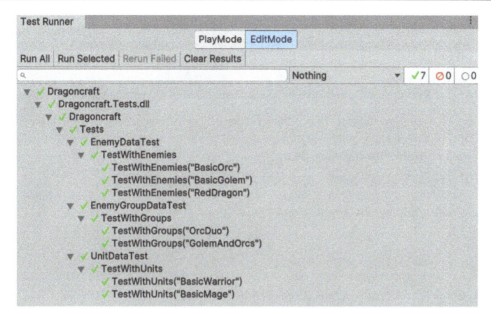

Figure 12.5 – Test Runner after running all the unit tests

The combination of the `ValueSource` attribute and the static list of strings makes the Test Runner dynamically generate our tests. When we create or remove a unit, enemy, or enemy group asset, we just need to update the respective list of strings and the Test Runner will take care of updating the tests for us. The test scripts we created are important to validate the integrity of each ScriptableObject and make sure that we are not adding a misconfigured asset in our game.

Although these three new test scripts are quite short and could have been added to the same script, it is good practice to have one test script for each context, in our case, unit, enemy, and enemy group. This also makes our tests better organized, grouping them by class and method, and making them easier to understand at a glance when we have many more assets in our game.

Now that we have set up the UTF and added test scripts to validate our asset data for units, enemies, and enemy groups, we can move to another cool usage of the Test Runner: battle simulations.

Using unit tests to simulate battles

So far, we have used the UTF to validate our existing ScriptableObjects containing the configuration for the units, enemies, and enemy groups. Now, we are going to add new tests using the flexibility provided by the `ValueSource` attribute to dynamically create tests for different battle configurations. For this test script, which we will reference as a single battle simulation, we will create a list of combat scenarios with one unit against one enemy, the level of the unit, and whether we expect the unit to win the battle.

Before we start creating the test script, we are going to introduce a new calculation for the attack and defense attributes of the units, which will consider the level of the unit and which will increase in strength as they level up. The unit level-up feature will be developed later in *Chapter 14, Crafting Buildings and Defense Towers*, but we need to add the logic now to be able to test it properly. The enemies do not have the level-up feature, but we can make stronger enemies by creating variations if needed.

Let us first add a couple methods to the existing script used by units and enemies. In the **Project** view, go to **Scripts | Character** and double-click on the `BaseCharacter.cs` script to open it, then add the following methods at the end of the class:

```
public virtual float GetAttack()
{
    throw new NotImplementedException();
}


public virtual float GetDefense()
{
    throw new NotImplementedException();
}
```

These two new methods, `GetAttack` and `GetDefense`, are `virtual` methods, which means that the classes that inherit from the `BaseCharacter` class will need to override these methods and add their own implementations, otherwise the code will throw an error saying that the methods were not implemented.

Next, let us add the implementation of the two methods we just created to one of the classes inherited from the `BaseCharacter` class. Open the script located at **Scripts | Enemy |** `EnemyComponent.cs` and add these two methods at the end of the class:

```
public override float GetAttack()
{
    return Attack;
}


public override float GetDefense()
{
    return Defense;
}
```

The implementation of the two `GetAttack` and `GetDefense` methods in the `EnemyComponent` class uses the `override` keyword to implement these two methods, but replaces their implementations by returning the `Attack` and `Defense` attributes, respectively.

The next class we are going to update with the new methods is the one responsible for the units. Open the script located at **Scripts | Unit |** Unit Component.cs and add the following two methods at the end of the class:

```
public override float GetAttack()
{
  return Mathf.Pow(Level, LevelMultiplier) + Attack;
}

public override float GetDefense()
{
  return Mathf.Pow(Level, LevelMultiplier) + Defense;
}
```

As we did in the previous EnemyComponent class, we also use the override keyword in this UnitComponent class to replace the method implementation from the BaseCharacter class. However, this time we do not simply return the values of Attack and Defense, respectively, but also add a small bonus to these values based on the Level property.

Using the Mathf API from Unity, and its Pow method, we calculate the value of the unit Level raised to the power of the LevelMultiplier, before adding this to the Attack and Defense values. This calculation will allow us to increase the bonus exponentially, so every time the unit is leveled up the bonus added to the Attack and Defense values will increase. The LevelMultiplier property is used to fine-tune the exponential growth of the bonus value in the ScriptableObject that contains the UnitData asset.

Now that we have added the new methods to get the Attack and Defense values, we need to update the code in a few places to use these methods. The first place to do so is in the UnitComponent class, in the existing UpdateAttack method. Replace Attack with GetAttack, as shown in the highlighted line in the following code block:

```
private void UpdateAttack()
{

  ...

  MessageQueueManager.Instance.SendMessage(
    new FireballSpawnMessage
  {
    Position = transform.position,
    Rotation = transform.rotation,
    Damage = GetAttack()
  });
  ...
}
```

The `UpdateAttack` method in the `UnitComponent` class has a unique code for ranged attacks that the `EnemyComponent` class does not have – that is why we are only updating the `UnitComponent` class. Besides the `UnitComponent` class, we have the `CollisionComponent` class that is indeed used by both unit and enemy to calculate the damage on every melee attack, and we need to update its code to also use the `GetAttack` and `GetDefense` methods.

Open the script located at **Scripts | Battle |** `CollisionComponent.cs` and, in the existing `TakeDamageOverTime` method, replace `Attack` and `Defense` in the damage variable with `GetAttack` and `GetDefense`, as we can see in the following code snippet:

```
private IEnumerator TakeDamageOverTime(BaseCharacter opponent)
{
  ...
  float damage =
    _character.GetAttack() - opponent.GetDefense();
  ...
}
```

Likewise, still in the `CollisionComponent` class, in the existing `TakeDamageFromProjectile` method, replace `Defense` with `GetDefense` in the damage variable, as shown here (note that we do not change the `opponentAttack` variable, only `Defense`):

```
private void TakeDamageFromProjectile(
   float opponentAttack, Transform target)
{
  ...
  float damage = opponentAttack - _character.GetDefense();
  ...
}
```

Those were the changes required to make sure the game will use the updated formula to calculate the bonus for both attack and defense when the unit is leveled up. Now, let us move to the first test script that will simulate the battle against a single enemy.

Simulating a battle against a single enemy

The test script to simulate a battle is quite simple, as we will see in a moment. However, we still need one extension class to make it easier to calculate the bonuses for Attack and Defense based on the unit level. The test script will not use the `UnitComponent` and `EnemyComponent` classes; instead, we are going to simulate the battle using the configuration of each unit and enemy from their respective ScriptableObjects. For this reason, we need to have an easy way to calculate the bonuses for Attack and Defense.

In the **Project** view, open the **Scripts | Editor** folder and right-click on the **Tests** folder, then select **Create | C# Script**, and name it `TestExtensions`. Now replace the content with the following code:

```
using UnityEngine;
namespace Dragoncraft.Tests
{
  public static class TestExtensions
  {
    public static float GetAttack(this UnitData data, int level)
    {
      return Mathf.Pow(level, data.LevelMultiplier) +
        data.Attack;
    }

    public static float GetDefense(this UnitData data, int level)
    {
      return Mathf.Pow(level, data.LevelMultiplier) +
        data.Defense;
    }
  }
}
```

As we learned in *Chapter 7, Attacking and Defending Units*, extension methods are very handy to extend one class with multiple methods. Both the class and method must be `static`, and the extension is applied to the class type of the first parameter in the method, with the `this` keyword before the parameter type and name. Here, we are extending the `UnitData` class to add two static methods, `GetAttack` and `GetDefense`.

One interesting thing to note is that the `TestExtensions` static class is in the `Dragoncraft.Tests` namespace, which means that these extension methods are only available for the test scripts in this namespace. All the other scripts in the project that are part of the game will not be able to use this extension method. This is a good example of how to extend classes but for specific purposes using the namespace.

Now that we have our extension methods, we can move on to our test script. We are going to split the script into three parts, starting from the class definition. Open the **Scripts | Editor** folder and right-click on the **Tests** folder, then select the **Create | C# Script** menu option, and name it `BattleSingleTest`. Now replace the content with the following code:

```
using NUnit.Framework;
using System.Collections;
using System.Collections.Generic;
using UnityEngine.TestTools;
using UnityEngine;
```

```
namespace Dragoncraft.Tests
{
  public class BattleSingleTest : BaseTest
  {
  }
}
```

The `BattleSingleTest` class is quite simple – we will have only one static class variable and one test method. We also inherit from the `BaseTest` class to use the methods defined there to load the ScriptableObjects required to simulate the battle between one unit and one enemy.

Next, we have the following big code snippet, which is the `static` list used by the test method. The UTF will create a test case for each item on this list and will use the test method that we are going to define in a moment. We also make use of a nice feature from C#, the **tuple**, which is a lightweight data structure to group multiple data elements. So, add the following static list to the class:

```
private static List<(string, string, int, bool)> battles =
  new List<(string, string, int, bool)>
{
  ("BasicWarrior", "BasicOrc",    0, false),
  ("BasicMage",    "BasicOrc",    0, false),
  ("BasicWarrior", "BasicGolem",  0, false),
  ("BasicMage",    "BasicGolem",  0, false),
  ("BasicWarrior", "RedDragon",   0, false),
  ("BasicMage",    "RedDragon",   0, false),
  ("BasicWarrior", "BasicOrc",    1, true),
  ("BasicMage",    "BasicOrc",    1, true),
  ("BasicWarrior", "BasicGolem",  1, false),
  ("BasicMage",    "BasicGolem",  1, false),
  ("BasicWarrior", "RedDragon",   1, false),
  ("BasicMage",    "RedDragon",   1, false),
  ("BasicWarrior", "BasicOrc",    2, true),
  ("BasicMage",    "BasicOrc",    2, true),
  ("BasicWarrior", "BasicGolem",  2, false),
  ("BasicMage",    "BasicGolem",  2, false),
  ("BasicWarrior", "RedDragon",   2, false),
  ("BasicMage",    "RedDragon",   2, false),
  ("BasicWarrior", "BasicOrc",    3, true),
  ("BasicMage",    "BasicOrc",    3, true),
  ("BasicWarrior", "BasicGolem",  3, false),
  ("BasicMage",    "BasicGolem",  3, false),
  ("BasicWarrior", "RedDragon",   3, false),
  ("BasicMage",    "RedDragon",   3, false),
  ("BasicWarrior", "BasicOrc",    4, true),
```

```
  ("BasicMage",    "BasicOrc",    4, true),
  ("BasicWarrior", "BasicGolem",  4, false),
  ("BasicMage",    "BasicGolem",  4, false),
  ("BasicWarrior", "RedDragon",   4, false),
  ("BasicMage",    "RedDragon",   4, false),
  ("BasicWarrior", "BasicOrc",    5, true),
  ("BasicMage",    "BasicOrc",    5, true),
  ("BasicWarrior", "BasicGolem",  5, true),
  ("BasicMage",    "BasicGolem",  5, true),
  ("BasicWarrior", "RedDragon",   5, false),
  ("BasicMage",    "RedDragon",   5, false)
};
```

The tuple is a data structure that can have multiple data elements. To use a tuple as a data type, we simply add the variable types we want between parentheses, just as we do in the battles static variable. The type of the items in the battles list is a tuple with (string, string, int, bool), which means that each item in the list will have a data group with four data elements: the unit name (string), the enemy name (string), the unit level (int), and a flag indicating whether the unit should win the battle (bool).

Following the definition of the data elements in the tuple, the first item in the list is a battle between BasicWarrior and BasicOrc, where BasicWarrior is at level 0 and should not win the battle (false). Each element of the list is a different scenario involving one unit facing off against one enemy, with the level of the unit provided (ranging from 0 to 5) along with whether the unit should win the battle simulation.

All of the battle scenarios in the battles static variable will give us 36 different battle simulations, which will execute the same test method. The TestWithSingleEnemy method is a bit long, but we will review the code in detail. So, add the following code to the test script:

```
[UnityTest]
public IEnumerator TestWithSingleEnemy(
  [ValueSource("battles")] (string unit, string enemy,
    int level, bool unitWin) battle)
{
  UnitData unitData = LoadUnit(battle.unit);
  float unitHealth = unitData.Health;

  EnemyData enemyData = LoadEnemy(battle.enemy);
  float enemyHealth = enemyData.Health;

  float timer = 0;
  while (unitHealth > 0 && enemyHealth > 0)
  {
```

```
    if (timer % unitData.AttackSpeed == 0)
    {
      enemyHealth -=
        Mathf.Max(unitData.GetAttack(battle.level) -
          enemyData.Defense, 0);
    }

    if (timer % enemyData.AttackSpeed == 0 &&
      enemyHealth > 0)
    {
      unitHealth -= Mathf.Max(enemyData.Attack -
        unitData.GetDefense(battle.level), 0);
    }

    timer += 0.5f;
    yield return null;
  }

  Assert.AreEqual(battle.unitWin, unitHealth > 0);
}
```

In this test method, we use the `UnityTest` attribute, which is part of the Unity testing API. The `Test` attribute, previously used in test methods, is part of NUnit, and the difference here is that using the `UnityTest` attribute from the UTF allows the test method to have `IEnumerator` as the return type.

We use the same tuple as defined in the `battles` static variable for the `ValueSource` of this method, however, this time we add names to each data element to make it easier to use in our code by giving meanings to the tuple data elements.

Each data element of the tuple will be used as a property of the `battle` parameter. The tuple data elements (`string, string, int, bool`) in this method correspond to the unit, enemy, `level`, and `unitWin` properties respectively, accessible through the `battle` parameter. The combination of `ValueSource` and the tuple makes our test method very flexible and customizable for our needs.

Since we are simulating a battle between a unit and an enemy, we need to load their respective asset data to use in the simulation using the `LoadUnit` and `LoadEnemy` methods. There is one caveat here: we had to define two float variables, `unitHealth` and `enemyHealth`, and use these variables as the life counters for our combatants, otherwise we would need to modify the values of the ScriptableObjects in the asset file and their respective lives in the asset file would change to the updated lives after the simulation, resulting in units and enemies with 0 value. To avoid modifying the values in their asset files, we instead use these two temporary variables during the battle simulation.

After loading the `unitData` and `enemyData`, we proceed to the main part of this test method: the `while` loop that simulates the battle. We are using a `timer` local variable to count the time and, at the end of the while loop, increment it by `0.5`. The loop will execute while the health values of both the unit and the enemy are greater than `0`, which means they are both alive and still battling. When the health of either of the two is reduced to `0` (or less), the loop will stop the execution.

Inside the `while` loop we have two `if` statements. The first validation is for the Unit, in which we do a modulus operation to get the remainder of the division between the current `timer` and `AttackSpeed`. If the remainder of the modulus operation is equal to `0`, this means that the value of the `timer` variable is a multiple of the `AttackSpeed` property and we can perform a new attack. For example, if `timer` is 3 and `AttackSpeed` is 1.5, the remainder of the division 3 ÷ 1.5 is 0, so we can perform an attack.

If the unit can attack, we must cause damage to the enemy by reducing the value of the `enemyHealth` variable. The damage formula is simple: we get the unit attack using the `GetAttack` extension method, and subtract the enemy `Defense`. After calculating the damage, we use the `Max` method from the `Mathf` API, which returns the larger value of the damage and `0` to avoid negative values.

The next `if` conditional is almost the same, but with a few differences. Besides the enemy `AttackSpeed`, we also check whether `enemyHealth` is greater than `0`. Since the unit attacks first, `enemyHealth` could be 0 and we need to skip the attack made back to the unit. If the enemy is still alive, we decrease `unitHealth` using the following damage calculation: we get the enemy `Attack` and subtract the unit defense using the `GetDefense` extension method. Once the damage has been calculated, we use the `Max` method to get the larger value between the damage and `0` to avoid negative values as well.

We keep increasing the `timer` and yielding the `IEnumerator` until the unit or the enemy is defeated and, after exiting the `while` loop, we assert whether the test was a success by comparing the values of `unitWin` and `unitHealth` using the `AreEqual` method. If the unit should win the battle and the value of `unitHealth` is positive at the end of the battle, we consider the test result a success.

Now that we have finished writing our test script, we can go back to the Unity Editor and open the **Test Runner** window to run our 36 new tests. Once Unity refreshes the **Test Runner** window, we can click on the **Run All** button and wait for the results. As we can see in the following screenshot, some of the tests will fail:

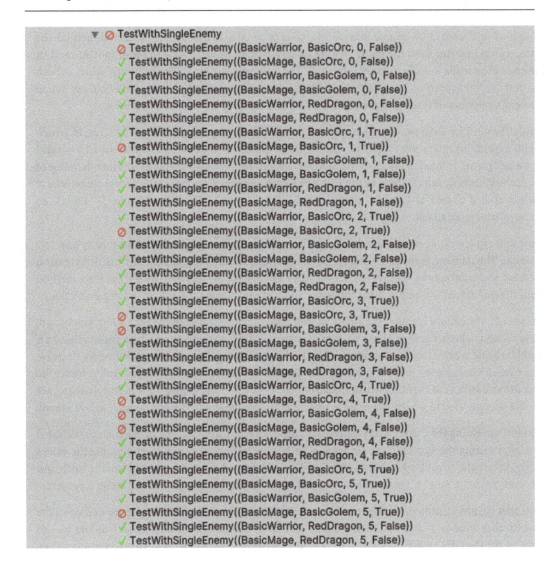

Figure 12.6 – Simulations of single enemy battles with some tests failing

It is completely fine that some of the tests fail at this point because we are writing the tests with the outcomes we desire, meaning that we provide a list of battles with each enemy and unit at different unit levels, and whether we expect the unit to win the battle. Next, we are going to update the units and enemies to attain a better balance and achieve our desired outcome without updating the tests in a moment. In short, the failing tests here are not wrong because they reflect what we expect – what needs to change is the data used to calculate the battle that seems to be a bit unbalanced.

Before moving on to balancing the unit and enemy, we are going to create another test script, this time to simulate battles between one unit and multiple enemies.

Simulating battles against multiple enemies

Now that we have finished the battle simulation against a single enemy, we can move on to a more advanced version of the previous test script where we will simulate battles between one unit and a list of enemies, still considering the level of the unit and the expected outcome of the battle as to whether the unit should win or not.

The new test script has a lot of similarities to the previous one, so we are going to focus on what is different in the new version that supports multiple enemies. Open the **Scripts | Editor** folder and right-click on the **Tests** folder, then select the **Create | C# Script** menu option, and name it `BattleMultiTest`. Now replace the content with the following code:

```
using NUnit.Framework;
using System.Collections;
using System.Collections.Generic;
using System.Linq;
using UnityEngine.TestTools;
using UnityEngine;

namespace Dragoncraft.Tests
{
  public class BattleMultiTest : BaseTest
  {
    private List<EnemyData> _enemyDataList =
      new List<EnemyData>();
    private List<float> _enemyHealthList =
      new List<float>();
  }
}
```

The new `BattleMultiTest` class has the same base structure as the previous `BattleSingleTest` class and we are still inheriting from the same `BaseTest` class. The two class variables enemyDataList and enemyHeathList are support lists that we are going to use during the test to help simulate the battles against the enemies and to keep track of their health during the combat with the unit.

Next, we are going to create a static variable for the list of tests. The `battles` static list, which contains each test case for the battle simulation, is defined in the following code block; we have a total of seven tests. Add the following code to the new `BattleMultiTest` class:

```
private static List<(string, List<string>, int, bool)> battles =
  new List<(string, List<string>, int, bool)>
{
  ("BasicWarrior", new List<string>{
    "BasicOrc", "BasicOrc"}, 1, false),
```

```
("BasicWarrior", new List<string>{
   "BasicOrc", "BasicOrc"}, 2, true),
("BasicWarrior", new List<string>{
   "BasicOrc", "BasicOrc", "BasicOrc"}, 2, false),
("BasicWarrior", new List<string>{
   "BasicOrc", "BasicGolem"}, 2, false),
("BasicWarrior", new List<string>{
   "BasicOrc", "BasicOrc", "BasicOrc"}, 3, false),
("BasicWarrior", new List<string>{
   "BasicOrc", "BasicGolem"}, 3, false),
("BasicWarrior", new List<string>{
   "BasicOrc", "BasicOrc", "BasicGolem"}, 3, false)
};
```

The main difference here, when compared to the previous `battles` static list, is that the tuple uses a list of strings in the second parameter, instead of a single string, to make it possible to add multiple enemies to the battle simulation.

The type of the items in the `battles` list is a tuple with (`string`, `List<string>`, `int`, `bool`), which means that each item in the list will have a data group with four data elements: the unit name (`string`), the list of enemy names (`List<string>`), the unit level (`int`), and a flag defining whether the unit should win the battle (`bool`).

Next is the main part of this test script, the `TestWithMultiEnemy` test method used to simulate the battle using the `battles` static variable. Add the following code snippet to the class:

```
[UnityTest]
public IEnumerator TestWithMultiEnemy(
   [ValueSource("battles")] (string unit,
     List<string> enemies, int level, bool unitWin) battle)
{
   UnitData unitData = LoadUnit(battle.unit);
   float unitHealth = unitData.Health;
   InitializeListsWithEnemies(battle.enemies);

   float timer = 0;
   bool battleEnded = false;
   while (unitHealth > 0 && !battleEnded)
   {
      AttackEnemy(timer, unitData, battle.level);
      AttackUnit(timer, unitData, battle.level,
        ref unitHealth);

      battleEnded = !_enemyHealthList.Any(i => i > 0);
```

```
    timer += 0.5f;
    yield return null;
}


Assert.AreEqual(battle.unitWin, unitHealth > 0);
}
```

The `TestWithMultiEnemy` method has a similar signature to the `TestWithSingleEnemy` method that we saw in the previous `BattleSingleTest` class, using the same `UnityTest` and `ValueSource` attributes to read the `battles` static list. The main difference in the new method is the precense of a list of `enemies` instead of one single enemy. This is reflected in the `battle` tuple that has four parameters: the `unit` name (`string`), the list of `enemies` (`List<string>`), the unit `level` (`int`), and `unitWin`, which indicates whether the unit should win the battle or not (`bool`).

We still have one `Unit`, and we continue to load the `UnitData` using the `LoadUnit` method and use the variable `unitHealth` to keep track of the unit `Health` property during the battle. However, since we now have a list of `enemies`, we need to load each of them into a list of `EnemyData` using the `LoadEnemy` method, as well as adding each enemy `Health` value to a separate list of `float` to keep track of their lives during the battle. This is very similar to what we did with our single enemy in the previous `TestWithSingleEnemy` method, but now we need the two lists, `enemyDataList` and `enemyHealthList`, since the Unit will battle against more than one enemy. Both of these lists are initialized in the `InitializeListsWithEnemies` method.

The `while` loop uses the same logic that we saw in the previous method. First, the unit will attack one enemy that is still alive in the list, and then all the enemies that are still alive will attack the unit. We control the `while` loop by checking that `unitHealth` is still greater than 0 and that the `battleEnded` variable is not true.

We have a sequence of attacks simulated by the `AttackEnemy` and `AttackUnit` methods, which we will define in a moment. After both methods have been executed, we check whether there are any values in the `enemyDataList` greater than 0 using the `Any` method and the predicate `i => i > 0`, which works as a short version of a loop to find any item in the list by attributing the item to the `i` variable and checking whether the value of `i` is greater than 0. The value of the `battleEnded` variable changes to `true` when there is no value in the `enemyHealthList` greater than 0, otherwise it is `false`.

Finally, at the end of the `while` loop, we increase the variable timer by `0.5` and `yield` to the next loop. Once the unit or all enemies are defeated, we exit the `while` loop and assert whether the test was a success by comparing the values of `unitWin` with `unitHealth` using the `AreEqual` method. If the unit wins the battle and the value of `unitHealth` is positive at the end of the battle, we consider the test a success.

Now, let us add the methods that we used inside the `TestWithMultiEnemy` method, starting with the following code snippet responsible for initializing the enemyDataList and enemyHeathList lists:

```
private void InitializeListsWithEnemies(List<string> enemies)
{
    _enemyDataList.Clear();
    _enemyHealthList.Clear();

    foreach (string enemy in enemies)
    {
        EnemyData enemyData = LoadEnemy(enemy);
        _enemyDataList.Add(enemyData);
        _enemyHealthList.Add(enemyData.Health);
    }
}
```

The `InitializeListsWithEnemies` method receives the list of enemies as a parameter, and uses this variable to load each `EnemyData` asset using the `LoadEnemy` method. Once loaded, we add the enemyData variable to the enemyDataList and the Health value to the enemyHeathList. Note that, before we loop through the enemies, we use the `Clear` method on both lists to clean it up before using them because it might have data from the previous test.

The next method we add to the script is the `AttackEnemy` method, defined in the following code block:

```
private void AttackEnemy(float timer, UnitData unitData, int level)
{
    if (timer % unitData.AttackSpeed == 0)
    {
        for (int i = 0; i < _enemyDataList.Count; i++)
        {
            if (_enemyHealthList[i] > 0)
            {
                _enemyHealthList[i] -= Mathf.Max(
                    unitData.GetAttack(level) -
                    _enemyDataList[i].Defense, 0);
                break;
            }
        }
    }
}
```

For the unit attack, we use the same modulus operation to check whether the remainder of the division between timer and AttackSpeed is 0, which indicates that the unit can attack. Then, we loop through each enemy in the enemyDataList and check in the corresponding index of enemyHealthList whether the current enemy is still alive. Since both lists are the same size, it is safe to use the same index on both. When we find the first enemy alive, we decrease its health using the same damage calculation as used in the previous TestWithSingleEnemy method. The important part here is the break keyword that will break out of the loop as soon as we damage one enemy to avoid the unit hitting multiple enemies at once, since only one can be attacked at a time.

The last method we are going to add to the BattleMultiTest class is the AttackUnit method, defined in the following code block:

```
private void AttackUnit(float timer, UnitData unitData,
    int level, ref float unitHealth)
{
    for (int i = 0; i < _enemyDataList.Count; i++)
    {
        if (_enemyHealthList[i] <= 0)
        {
            continue;
        }

        if (timer % _enemyDataList[i].AttackSpeed == 0)
        {
            unitHealth -= Mathf.Max(
                _enemyDataList[i].Attack -
                    unitData.GetDefense(level), 0);
        }
    }
}
```

We loop through each enemy in the enemyDataList again, but this time to attack the unit. Here we do a different validation by checking whether the corresponding enemy's health in the enemyHealthList variable is less than or equal to 0, and if it is, we use the continue keyword to skip this iteration and move to the next index. Then, we do the modulus validation to check whether the current enemy can attack and, if the attack is possible, we decrease the unitHealth value using the damage calculation we saw in the previous TestWithSingleEnemy method as well. Since all enemies can attack the same Unit, we do not break out of the loop but rather continue to the next enemy in the list to check whether it can also attack the unit.

That is everything we need to include in our BattleMultiTest class to simulate battles between one unit and multiple enemies. Now that we have finished writing our new test script, we can go back to the Unity Editor and open the **Test Runner** window to run our seven new test cases. Once Unity refreshes the **Test Runner** window, we can click on the **Run All** button and wait for the results. As we can see in the following screenshot, some of the tests will fail, but this is expected because we still need to balance the units and enemies:

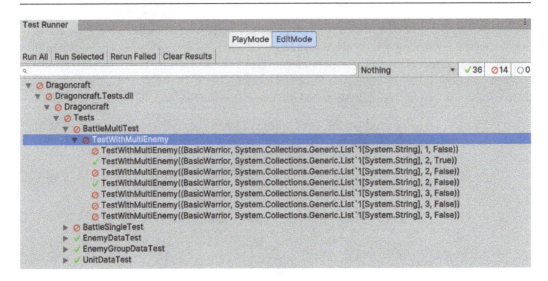

Figure 12.7 – Simulations of multi-enemy battles with some tests failing

We just finished creating a very flexible test script to simulate battles between one unit and multiple enemies. We can easily make the unit stronger by increasing its level in the test case, and adding information regarding whether it should win the battle or not. We have more tests failing now, but it is acceptable since we are in the process of creating our battle simulations and can tweak each unit and enemy to get a good balance and the expected outcome in each battle.

With these two test scripts done, we can run our test battles against single and multiple enemies. Let us start to analyze the specifics of each unit and enemy configuration and the changes we need to make to achieve a good balance of difficulty.

Balancing the battle difficulty

In game genres such as RTS, it is very important to have a well-balanced difficulty, otherwise the game will either be too easy and boring without challenges for the player, or too difficult, unfair, and unenjoyable. In both cases, the player will likely stop playing the game. We need to find the right balance that will make the game strategically challenging, but with a learning curve that will not punish the player right at the beginning of the session.

We could change the values in all ScriptableObjects for each unit and enemy and play all of our levels until we find the right configurations. However, that would take a long time and it is very likely that we will miss some cases when testing in-game. Fortunately, we now have a great tool to validate our configuration changes in the form of the test script that we just wrote, which can validate configurations and simulate battles in a few seconds. Indeed, we already have 50 tests that cover a lot of battle scenarios.

Besides using the test scripts to validate and simulate our configurations, we can also use them to help us find the balance we want. The following chart shows a battle between a level 4 `BasicWarrior` unit and a `BasicGolem` enemy using the `BattleSingleEnemy` test script. In this battle, we expect that the `BasicWarrior` at level 4 would be strong enough to defeat the `BasicGolem`; however, the following chart shows that the `BasicWarrior` health reaches 0 while the `BasicGolem` is still alive. The *y* axis indicates their respective lives and the *x* axis the time.

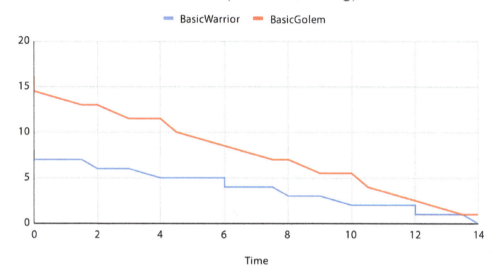

Figure 12.8 – The battle between the level 4 BasicWarrior and BasicGolem before balancing

This chart is a great visual representation of the test script, clearly illustrating how they attack each other over time. We can also see that we have almost achieved what we were looking for, which is our level 4 `BasicWarrior` defeating the `BasicGolem` enemy. In this case, we need to make the `BasicWarrior` a bit stronger, but not too much, otherwise other tests will fail when we expect the `BasicWarrior` to lose the battle.

To help with the unit attack and defense, we have the level and level multiplier to make it a bit more powerful on each level up. In this chapter, we changed the `UnitComponent` script to include the `Level` and `LevelMultiplier` properties in the `Attack` and `Defense` values using the formula: `Level ^ LevelMultiplier + Attack (or Defense)`. We are elevating the value of `Level` to the power of `LevelMultiplier` before adding it to `Attack` or `Defense`. This formula is also present in the `TestExtensions` script, which is used in our test scripts.

The following chart shows this formula applied to a unit, where the *y* axis is the value of Attack or Defense, and the *x* axis is the Level. Note that Attack starts with the value 2 and the Defense starts with the value 0. When the unit reaches level 4, Attack changes to the value 4 and Defense to the value 2. In this chart, we also consider the **Level Multiplier** as 0.5.

Unit Attack and Defense by Level with LevelMultiplier

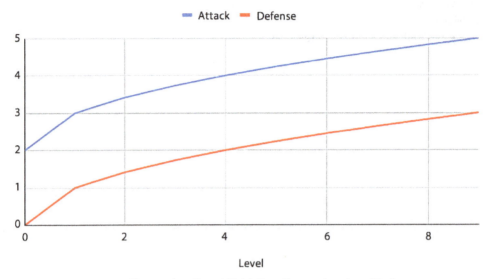

Figure 12.9 – The Level and Level Multiplier effect on Attack and Defense

The chart shows how the **Level** and the **Level Multiplier** affect the **Attack** and **Defense**, making the unit stronger with each level – but not too much. It also features a nice bump with level 1 so the player can feel a sense of reward from achieving the level up and will continue to upgrade the unit, even though the effect is lower after the first time.

Now that we understand how the level and level multiplier affect the attack and defense of a unit, let us update both `BasicWarrior` and `BasicMage` to give them similar curves:

1. In the **Project** view, left-click on the asset file located at **Data | Unit |** `BasicWarrior.asset`.

2. In the **Inspector** view, change the properties of **Health** to 8, **Attack** to 1.5, **Defense** to 0, **Attack Speed** to 1.5, **Level** to 0, and **Level Multiplier** to 0.5.

3. In the **Project** view, left-click on the asset file located at **Data | Unit |** `BasicMage.asset`.

4. In the **Inspector** view, change the properties of **Health** to 4, **Attack** to 2, **Defense** to 0, **Attack Speed** to 1, **Level** to 0, and **Level Multiplier** to 0.5.

5. Click **File | Save Project** to save the changes.

Once we finish updating the properties for both `BasicWarrior` and `BasicMage`, it should appear as in the following screenshot. Double-check that all the values are the same in your project, otherwise the test scripts will fail when simulating the battles.

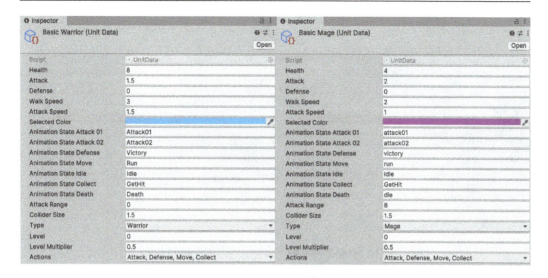

Figure 12.10 – BasicWarrior and BasicMage with updated values

Now that we updated both our `BasicWarrior` and `BasicMage` units, we can move on to the enemies. They are a bit simpler because we do not have Level and Level Multipliers to worry about, and we use their respective Attack and Defense values in the battles without applying any modifiers.

The enemies have fixed configurations, meaning that when we want a more powerful enemy, we need to create a new variation by adding a new ScriptableObject. Let us set the values for both `BasicOrc` and `BasicGolem` so they are consistent and balanced when battling the units:

1. In the **Project** view, left-click on the asset file located at **Data | Enemy |** `BasicOrc.asset`.

2. In the **Inspector** view, change the properties of **Health** to 8, **Attack** to 2.5, **Defense** to 0, and **Attack Speed** to 1.5.

3. In the **Project** view, left-click on the asset file located at **Data | Enemy |** `BasicGolem.asset`.

4. In the **Inspector** view, change the properties of **Health** to 16, **Attack** to 3, **Defense** to 2, and **Attack Speed** to 2.

5. Click **File | Save Project** to save the changes.

After updating the properties, we should have the two enemies `BasicOrc` and `BasicGolem` with the values shown in the following screenshot. Again, double-check these values in your project to make sure they are the same, otherwise the test scripts may fail.

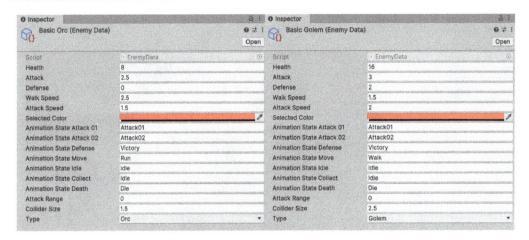

Figure 12.11 – BasicOrc and BasicGolem with updated values

The updated values for `BasicOrc` and `BasicGolem` are enough to balance the difficulty of the battles against the units. There is another enemy that we did not update here, `RedDragon`, but that one is already sufficiently well balanced to be a tough challenge for the player when their units are strong enough to fight our boss.

The following chart shows a battle between the level 4 `BasicWarrior` and `BasicGolem` enemy after updating their configurations to reflect the balance we want. We can see now that the `BasicWarrior`, when at level 4, can defeat the `BasicGolem` enemy with a bit more resistance.

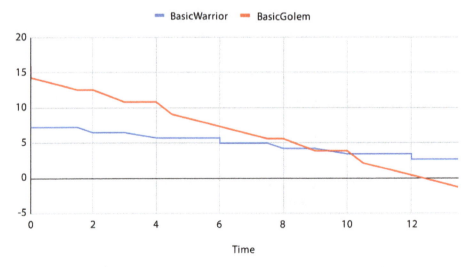

Figure 12.12 – The battle between the level 4 BasicWarrior and
BasicGolem enemy after balancing the configuration

It is good practice to adjust the balance of difficulty using resources such as charts and simulations. It gives a solid foundation to our game design decisions and allows us to see the outcome of the changes we make without having to add all the required assets and scripts into the game to test it live. It is like a prototype with data, but without random decisions or needing to guess the configuration values.

We can now open the **Test Runner** window and click on the **Run All** button. Using the updated configurations for the units and enemies, we should now see all 50 tests returning successful results, as shown in the following screenshot:

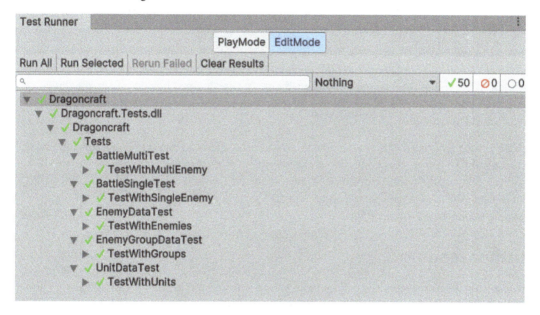

Figure 12.13 – Test Runner after running the 50 tests with successful result

The UTF is a powerful tool for writing unit tests, but also helps to validate configurations and simulate battles as we just did. The UTF and the 50 test cases we have will now be available to help us ensure the game remains balanced when we update any unit or enemy, as well as when new characters are introduced to the game and we want to quickly test their power in a simulated battle.

Test scripts are an easy way to test new enemies and units by creating or duplicating existing ScriptableObjects, and balance them even before adding any 3D models, Prefabs, and scripts into our project to test. Feel free to add or update your units and enemies, but do not forget to add new tests when creating new character configurations.

Summary

Well done for reaching the end of this chapter! Although we did not add new features to the gameplay of our RTS game, we now have a very flexible and powerful tool to validate our unit and enemy configurations, as well as simulate battles to check the game balance and make informed decisions regarding our game's difficulty.

In this chapter, we learned how to set up a Unity project using Assembly Definition files, how to set up the Unity Test Framework, and how to write unit tests that can be executed in the Unity Editor by the Test Runner.

We also learned how to create test scripts that use a list of parameters to create dynamic test cases and how to simulate a battle between both one unit and a single enemy, and one unit and multiple enemies. At the end of the chapter, we examined our data and learned how to update the difficulty curve, evaluating the impact of our changes on the balance of the game by using our test scripts to validate the configurations.

In *Chapter 13, Producing and Gathering Resources,* we are going to add one very important and useful feature of all RTS games: the production and gathering of resources. We will learn how to produce resources, upgrade the resource production, and command the units to gather the resources for us. That is one of the last features we will add to our RTS game, by which point we will be getting close to having a completed *Dragoncraft* game.

Further reading

The following links are recommended material to learn more about the Unity Test Framework and TDD, as well as to refresh your knowledge of the new C# features that we used in this chapter:

- *Test-driven development*: `https://en.wikipedia.org/wiki/Test-driven_development`

- *Getting started with Unity Test Framework*: `https://docs.unity3d.com/Packages/com.unity.test-framework@1.3/manual/getting-started.html`

- *Testing Test-Driven Development with the Unity Test Runner*: `https://blog.unity.com/technology/testing-test-driven-development-with-the-unity-test-runner`

- *Assert*: `https://docs.unity3d.com/ScriptReference/Assertions.Assert.html`

- *Tuple types*: `https://learn.microsoft.com/en-us/dotnet/csharp/language-reference/builtin-types/value-tuples`

13

Producing and Gathering Resources

One of the most underrated features of an RTS game is resource gathering and production because, while battle and map exploration are important, only resource management can help the player get stronger than the opponent faster. Some resources are automatically generated, while others require that the player command their units to collect them, making this a crucial part of the game strategy.

In this chapter, we are going to learn how to create a system that will generate resources each second, with an option to generate them automatically or manually, depending on the configuration we are going to have in ScriptableObjects. We will also learn how to implement the automatic generation of **Gold** resources in our game scenes and the manual generation of **Food** and **Wood** resources, which will only happen when the player sends units to a specific GameObject with the `Collect` command.

By the end of this chapter, you will have learned how to develop and control resource production using an inventory system that will keep track of how many resources the player has of each type, which is the foundation for resource management. You will have also learned how to create upgrades for resource generation that will increase our production, generating even more resources, and how to wire the resource generation to the UI elements to represent how much the player has of each type. Finally, you will have learned how to expand our debug menu and unit tests to help us test and validate the resource production in our game.

In this chapter, we will cover the following topics:

- Generating resources
- Gathering resources
- Testing the resources

Technical requirements

The project setup for this chapter, along with the imported assets, can be found on GitHub at `https://github.com/PacktPublishing/Creating-an-RTS-game-in-Unity-2023` in the `Chapter13` folder inside the project.

Generating resources

We are almost done developing all the features required for our RTS game, so now, we are going to add resource generation to the game. **Resource generation** is a feature that will automatically produce one specific resource type, and we will add scripts to make it possible to control how much of a resource is generated per second and a multiplier that will increase resource generation.

So far in our project, we have created a script named **ResourceType**, which has an enum with the resource types of **Gold**, **Wood**, and **Food**. The usage of the **ResourceType** enum is in the **ResourceUpdater** script, which is responsible for receiving a message of the **UpdateResourceMessage** type and updating each resource amount in the top-right corner of the UI. We also have a debug script, **ResourceDebugger**, to send the message of the **UpdateResourceMessage** type and test the UI update.

Now, we are going to add information about the production per second and the production level to each resource. To do that, we need a new script to create ScriptableObjects for each pair of resource types and the respective data. So, let's create a new script in the **Scripts | Resource** folder and name it `ResourceData`. Then, replace the content with the following code snippet:

```
using UnityEngine;
namespace Dragoncraft
{
    [CreateAssetMenu(menuName = "Dragoncraft/New Resource")]
    public class ResourceData : ScriptableObject
    {
        public int ProductionPerSecond;
        public int ProductionLevel;
        public ResourceType Type;
    }
}
```

The `ResourceData` class is a very simple `ScriptableObject`, with properties to define how many resources are generated per second (`ProductionPerSecond`), the production level that we will be using as a multiplier to increase the amount of resource production generated per second (`ProductionLevel`), and what the type of resource that this date is for (`Type`). We included the `CreateAssetMenu` class attribute to let Unity create a new menu option to create this type of asset.

Now that we have the `ResourceData` class, we can go ahead and create a ScriptableObject for each resource that we have in our game:

1. In the **Project** view, right-click on the existing **Data** folder, select **Create | Folder**, and name it `Resource`.

2. Right-click on the new **Resource** folder, select **Create | Dragoncraft | New Resource**, and name it `Gold`.

3. Left-click on the newly created asset, **Gold**, and, in the **Inspector** view, change **Production Per Second** to **1**, **Production Level** to **1**, and **Type** to **Gold**.

4. Right-click on the new **Resource** folder, select **Create | Dragoncraft | New Resource**, and name it **Food**.

5. Left-click on the newly created asset, **Food**, and, in the **Inspector** view, change **Production Per Second** to **2**, **Production Level** to **1**, and **Type** to **Food**.

6. Right-click on the new **Resource** folder, select **Create | Dragoncraft | New Resource**, and name it **Wood**.

7. Left-click on the newly created asset, **Wood**, and, in the **Inspector** view, change **Production Per Second** to **3**, **Production Level** to **1**, and **Type** to **Wood**.

8. Use the **File | Save Project** option to save the changes on the new assets.

We just created three assets, one for each type of resource: **Gold**, **Food**, and **Wood**. The following screenshot shows each of the new assets in the **Inspector** view:

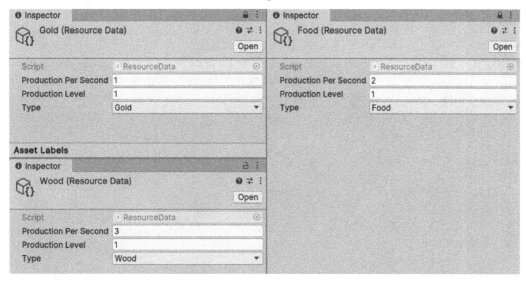

Figure 13.1 – The ResourceData ScriptableObjects for Gold, Wood, and Food

Note that each `ResourceData` file has a different value for the **Production Per Second** property, as well as the respective resource **Type**. The **Production Level** value is the same for all of them because the initial multiplier for the resource production is **1**.

Production Level will be used to multiply the **Production Per Second** value when an upgrade to the resource production is made by the player, so we can increase the production per second. Every time the player upgrades the source that produces a specific resource, we are going to increase the amount produced per second by multiplying the production level.

The player action to upgrade one resource production will be developed in *Chapter 14, Crafting Buildings and Defense Towers*; for now, we are going to prepare the foundation of this feature. We are going with a new message that will be sent when an upgrade happens. Create a new script in the **Scripts | MessageQueue | Messages | UI**, folder name it `UpgradeResourceMessage`, and add the following code to the script:

```
namespace Dragoncraft
{
  public class UpgradeResourceMessage : IMessage
  {
    public ResourceType Type;
  }
}
```

The `UpgradeResourceMessage` class implements the `IMessage` interface and has **only** one property, `ResourceType`, which will indicate whether the upgrade is for **Gold**, **Food**, or **Wood** production.

We also need to identify the type of production. The production can be `Manual` or `Automatic`, meaning that the units on the map, commanded by the player, will need to manually gather the resources to produce it or the resource will be automatically generated. To make it easier to set up the type of production, let's add a new enum with the two possibilities. Create a new script in the **Scripts | Resource** folder, name the file `ResourceProductionType`, and replace the content with the following code snippet:

```
namespace Dragoncraft
{
  public enum ResourceProductionType
  {
    Manual,
    Automatic
  }
}
```

The `ResourceProductionType` enum is very simple and contains only two possible options, which are to generate the resources automatically (`Automatic`) or to indicate that the resource is gathered manually (`Manual`).

Now that we've finished working on some base scripts for the resource data, message, and production type, we can move on to another part of resource generation, which is the player's inventory. This is where the amount of each resource will be kept and accessed by the game.

Managing the player's inventory

One of the features of an RTS game is the inventory, where the player can keep track of how many resources are available to use. Besides keeping an eye on the map to defend against enemy invasions or exploring regions covered by fog, the player also needs to pay attention to the inventory and manage the resource generation, gathering, and expenditure processes, and calculate what can be done with the amount available or if they need to wait a bit more before training more units or upgrading buildings.

The inventory will be accessed by different scripts in the game because some will need to consult if the player has the required resources while other scripts will increase or decrease the amount of one specific resource. For inventory management, we are going to create a couple of scripts that will contain the inventory data and the inventory manager.

So, inside the **Scripts** folder, create a new folder named `Inventory`. Inside, add a new script and name it `InventoryData`. Then, replace the content with the following code block:

```
using System;
namespace Dragoncraft
{
    [Serializable]
    public class InventoryData
    {
        public int Gold;
        public int Wood;
        public int Food;
    }
}
```

The `InventoryData` class will act as the data holder for our wallet, which is going to contain the updated amount of each resource type in the `Gold`, `Wood`, and `Food` variables. This class has the `Serializable` attribute because we are going to read and write into a file, and the data will be converted back and forth into an `InventoryData` object.

All scripts that need to get or set the resource value for each type will update this class, but not directly. We will create a management class that will be responsible for managing the inventory data. The manager class is a bit large, so we are going to create it method by method. Add a new script in the **Scripts | Inventory** folder, name it `InventoryManager`, and add the following code snippet:

```
using System;
using System.IO;
using System.Runtime.Serialization.Formatters.Binary;
using UnityEngine;
namespace Dragoncraft
{
  public class InventoryManager : IDisposable
  {
    private static string _fileName = "inventory.data";
    private string _filePath =
      $"{Application.persistentDataPath}/{_fileName}";
    private InventoryData _inventory;

    public InventoryManager()
    {
      LoadData();
    }
  }
}
```

The `InventoryManager` class will be responsible for loading and saving the data in the `InventoryData` object, as well as providing methods to access each resource type, updating their values, and cleaning up the inventory if needed. In the class constructor, only one method is executed – `LoadData`. This is going to be defined in a moment, but first, let's understand how we are going to store the inventory data serialized in a file.

We are going to serialize our data in a binary format, using the API from `Formatters.Binary`, in a file named `inventory.data`. This file will be read and written from the path in `persistentDataPath`, which is a property from the Unity `Application` API that returns a path that can be used as a persistent data directory for our game, no matter the operating system the game is installed on.

We will implement the methods that will read and write the serialized file shortly so that this process will make more sense. We're also going to delete the saved file. That is why this class implements the `Idisposable` interface, which requires us to add one method to our class, `Dispose`, as is defined here. This method is not automatically executed, so we can use it when we want our inventory to be disposable. Add the following method to our `InventoryManage` class:

```
public void Dispose()
{
```

```
    if (File.Exists(_filePath))
    {
        File.Delete(_filePath);
    }
}
```

The `Dispose` method is very simple: it checks whether a file exists, using the `File.Exists` method, in the path defined in the `filePath` class variable, and if the validation is true, we use the `File.Delete` method to remove the file from the persistent data directory.

Next, we are going to define one of the most important methods in the `InventoryManage` class – that is, `SaveData`:

```
private void SaveData()
{
    BinaryFormatter formatter = new BinaryFormatter();
    FileStream file = File.Create(_filePath);
    formatter.Serialize(file, _inventory);
    file.Close();
}
```

The `SaveData` method has four lines. Let's break down each of them to understand what is going on:

- First, we create a new `BinaryFormatter` object that is going to be used to serialize the `InventoryData` object into a binary file.

- In the second line, we create a new file using the `File.Create` method, in the path defined by the `filePath` class variable. The object that's returned by `File.Create` is a `FileStream` object that we will use to write our serialized object.

- In the following line, we have the `Serialize` method, from the `formatter` variable, which has two parameters: the `FileStream` object and the serializable object we want to serialize into `FileStream`. The `Serialize` method will convert our `InventoryData` object into a binary format and write it in the `file` object.

- The last thing we need to do is execute the `Close` method to save the changes to the file.

Note that we do not check whether the file exists before performing any operation. This is because when the `SaveData` method is called, we want to serialize and write the data in the file, so it is fine to override the current file since `InventoryData` is the most up-to-date version of the data. A nice feature of `File.Create` is that it overrides the existing file by default, which is another reason why we do not need to check if the file exists in this method.

Now that we've finished creating the SaveData method, let's implement its counter-part method, LoadData, which will be used to read the file created in the SaveData method and convert it into an object of the InventoryData type:

```
private void LoadData()
{
  if (File.Exists(_filePath))
  {
    BinaryFormatter formatter = new BinaryFormatter();
    FileStream file = File.Open(_filePath, FileMode.Open);

    _inventory = (InventoryData) formatter.Deserialize(file);
    file.Close();
  }
  else
  {
    _inventory = new InventoryData();
  }
}
```

The LoadData method has a couple of similarities to the SaveData method: we are using BinaryFormatter to create a new formatter object and FileStream to open the file located at filePath so that we can read its contents. Once the file is opened, we can use formatter to Deserialize the binary content into an InventoryData object, which is stored in the inventory class variable. As soon as we deserialize the file content, we can safely use the Close method to release the file from the memory.

Note that we only attempt to open the file to deserialize its contents into an InventoryData object if the file in filePath exists. If there is no file in filePath yet, we just initialize the inventory class variable by creating a new InventoryData object.

With that, we have created SaveData and LoadData so that we can manipulate the inventory in a persistent file. These two methods are private, which means that they can only be used on other methods in the InventoryManager class.

Next, we are going to create a couple of methods that can use these two methods to read and update the player's resources in the inventory, starting with the UpdateResource method:

```
public void UpdateResource(ResourceType type, int amount)
{
  switch (type)
  {
    case ResourceType.Gold:
      _inventory.Gold += amount;
      break;
```

```
    case ResourceType.Wood:
      _inventory.Wood += amount;
      break;
    case ResourceType.Food:
      _inventory.Food += amount;
      break;
    default:
      break;
  }
  SaveData();
}
```

The UpdateResource method has a couple of parameters: type and amount. These two parameters indicate what ResourceType we want to update and how many resources we want to add or remove from the inventory. Using the switch expression, we can evaluate the type parameter and determine what resource in the inventory needs to be updated.

The only operation we are doing with the inventory resources amount is addition because, even when we remove resources from the inventory, we can still add a negative value to reduce the resource amount. After the switch expression, we call the SaveData private method to persist our inventory changes in the binary file.

The last method we are going to add to the InventoryManager class is GetResource, which will return the number of resources the player currently has based on the specified ResourceType:

```
public int GetResource(ResourceType type)
{
  switch (type)
  {
    case ResourceType.Gold:
      return _inventory.Gold;
    case ResourceType.Wood:
      return _inventory.Wood;
    case ResourceType.Food:
      return _inventory.Food;
    default:
      return 0;
  }
}
```

The switch expression in the GetResource method is very similar to the one we just saw in the UpdateResource method. However, this time, we are just returning how many resources the player has based on the ResourceType parameter, type. No operation is required here, so we can return the value of the resource we want inside each case.

We do not need to use the LoadData method in GetResource because we are already using the LoadData method in the constructor of the InventoryManager class. So, when the class is instantiated, we read the binary file and create the InventoryData object.

When we execute the UpdateResource method, besides writing to the binary file, we also update the intentory class variable, which is always up to date with the latest values since the moment the InventoryManager class is created. For these reasons, we do not need to execute the LoadData method in any other place than the class constructor.

Now that we have finished our InventoryManager class, we can use it anywhere in our code to get the current amount of each resource type and update their values. Before we start using our inventory, however, we are going to create a script that will act as a trigger to update the resource value automatically, which is a component for generating resources automatically.

Producing resources automatically

We are going to have two different ways to produce resources: manually, where the player must command one or more units to gather the resources; or automatically, where we have a GameObject with a script attached to it that will generate resources over time. In this section, we will create a new script that is going to take care of producing resources automatically or manually.

The new script is a component that we can add to a GameObject and configure to produce the resources we want, with properties that will control how much to generate per second – the production is automatic. Create a new script in the **Scripts** | **Resource** folder, name it ResourceProductionComponent, and replace its content with the following code:

```
using UnityEngine;
namespace Dragoncraft
{
    public class ResourceProductionComponent : MonoBehaviour
    {
        [SerializeField] private ResourceData _resource;

        protected int _productionPerSecond;
        protected int _productionLevel;
        protected ResourceType _type;
        protected ResourceProductionType _productionType;

        private float _counter;
    }
}
```

The properties of the `ResourceProductionComponent` class are very important to make the whole script work as we expect:

- The `resource` property has a reference to the ScriptableObject's `ResourceData` asset. This will be the configuration for the resource we want to produce with this script.

- `productionPerSecond` is the number of resources that are going to be produced per second when `ResourceProductionType` is `Automatic`.

- `productionLevel`, as its name implies, specifies the level of production, which will be used in this class to multiply the `productionPerSecond` value and give us the real quantity produced.

- `type` is the `ResourceType` property we want to produce in this script.

- This script will have different behaviors depending on `productionType`, and most of the logic we are going to add here is for the `Automatic` production type.

- The last class variable in the code block is `counter`. It's a support variable that will help us count the seconds in the `Update` method, as we are going to see in a moment.

The `produtionPerSecond`, `productionLevel`, and `type` values come from the `resource` property, and we must assign each class variable to the proper value from `ResourceData` that is referenced in the GameObject that will have this script attached. We will do this in the `Start` method, as shown here:

```
protected virtual void Start()
{
  _productionPerSecond = _resource.ProductionPerSecond;
  _productionLevel = _resource.ProductionLevel;
  _type = _resource.Type;
  _productionType = ResourceProductionType.Automatic;
}
```

The `Start` method has the `virtual` keyword because later in this chapter, we are going to create another script called `ResourceCollectionComponent` that will be used when the resource is not generated over time but collected by the units when commanded by the player, overriding this method so that it's of the `Manual` production type.

After setting the values of the `productionPerSecond`, `productionLevel`, and `type` class variables from the `resource` property, we set the value of the `productionType` class variable to `Automatic` to ensure that this script works as an automatic resource generator when attached to a GameObject.

This script is going to generate resources automatically and increase the number of resources generated when an upgrade happens to the resource's source. In *Chapter 14*, *Crafting Buildings and Defense Towers*, we are going to implement the upgrade feature, but for now, we will prepare the ResourceProductionComponent class so that it receives a message that an upgrade happened by adding and removing a listener to/from the OnEnable and OnDisable methods, respectively:

```
private void OnEnable()
{
    MessageQueueManager.Instance.
     AddListener<UpgradeResourceMessage>(OnResourceUpgraded);
}

private void OnDisable()
{
    MessageQueueManager.Instance.
        RemoveListener<UpgradeResourceMessage>(OnResourceUpgraded);
}
```

The OnEnable and OnDisable methods are great places to use the MessageQueueManager singleton to add or remove a listener for the UpgradeResourceMessage message type. When the GameObject that has this script attached is enabled, the OnResourceUpgraded method is added to the list of listeners for UpgradeResourceMessage. Once the GameObject is disabled by Unity, the OnResourceUpgraded method is removed from the list of listeners for this message type.

Only when the GameObject is active in the game will this script receive the callback from the message system and execute the OnResourceUpgraded method, as shown in the following code snippet:

```
private void OnResourceUpgraded(UpgradeResourceMessage message)
{
    if (_type == message.Type)
    {
        _productionLevel++;
    }
}
```

The OnResourceUpgraded method is executed when UpgradeResourceMessage is received by the message system and forwarded to this method as the message parameter. The only validation that's done here is to make sure that the type of resource configured in this script is the same type of resource that's received in message. If that is the case, then we can increase our productionLevel class variable, which grows as we upgrade the source of a resource type.

Now that we have our message and method to upgrade the resource production, let's add the `Update` method, which will generate the resources per second automatically:

```
private void Update()
{
    if (_productionType == ResourceProductionType.Manual)
    {
        return;
    }

    _counter += Time.deltaTime;
    if (_counter > 1 )
    {
        _counter = 0;
        ProduceResource();
    }
}
```

The first thing we do in the `Update` method is check whether `productionType` is `Manual`; if that is true, we can exit the method by using the `return` keyword because we do not need to do anything here if `productionType` is not `Automatic`.

After the validation, we can increase our `counter` class variable by adding the `deltaTime` value, which specifies at the time between now and the last time the `Update` method was called internally by Unity. The difference between these two executions is in the `deltaTime` variable, which has a very small floating-point value, so it is safe to keep adding it to the counter until it reaches the value of 1 second.

When `counter` has a value greater than `1`, this means that 1 second has passed since we started adding the `deltaTime` value to the `counter` class variable. Then, we reset `counter` to `0` and call the `ProduceResource` method to generate a new amount of resources for the player. We are going to define this in the following code block.

The `Update` method we added to this class is very simple and aims to keep generating resources automatically. To do that, we must use the `ProduceResource` method:

```
private void ProduceResource()
{
    UpdateResourceMessage message = new UpdateResourceMessage
    {
        Amount = GetProducedAmount(),
        Type = _type
    };
    MessageQueueManager.Instance.SendMessage(message);
}
```

We already have a system in place that is listening for `UpdateResourceMessage` and increasing the number of resources of each type in the UI at the top right of the screen. Here, in the `ProduceResource` method, we are sending this message type, which is expected by the **ResourceUpdated** script, to the GameObject UI to update the values, with `Amount` calculated by the `GetProducedAmount` method and `Type` with the value of the `type` class variable, which specifies the resource type configured for this script.

Once we have created our `message` variable, we can use the `MessageQueueManager` singleton to send it to any script listening for the `UpdateResourceMessage` message type. In this `message`, we are going to send the resource amount produced in the `Amount` parameter, which is calculated by the `GetProducedAmount` method. This is defined in the following code block:

```
private int GetProducedAmount()
{
    return _productionPerSecond * _productionLevel;
}
```

The `GetProduceAmount` method does a simple calculation, which multiplies `productionPerSeconde` by `productionLevel` to get the actual produced amount. As we saw in the `OnResourceUpgraded` method, the `productionLevel` class variable is increased every time we receive a message of the `UpgradeResourceMessage` type. This is how we can increase the produced amount – by upgrading the source of the resource type that is generated automatically by this script.

This was the last method we needed to add to our **ResourceProductionComponent** script; now, we have a class that can generate resources automatically and be upgraded to produce more resources per second.

However, before we can start using this script, we need to update another couple of scripts, **LevelManager** and **ResourceUpdated**, so that they will be ready to generate resources for the player automatically.

Updating the LevelManager script

The existing **LevelManager** script is currently responsible for a few things, such as instantiating the mini-map Prefab, enabling fog on the map, and loading the scene that contains the UI. Now, we are going to expand it a bit so that we can take care of the player's inventory, adding a common way for the other parts of the game to access it.

Open the script located at **Scripts | Level** | `LevelManager.cs`. We will start by adding a new class variable called `inventory` that will be used by the script, as well as the variable creation in the `Awake` method:

```
private InventoryManager _inventory;
private void Awake()
{
```

```
...
    _inventory = new InventoryManager();
}
```

`inventory` can be added to the `LevelManager` class right before the declaration of the `Awake` method. Then, in the last line of `Awake`, we are creating the `inventory` object using the new keyword and the `InventoryManager` class type. The highlighted lines here are the only new additions to the code because the `Awake` method already exists and we are keeping all its code there, just adding the line at the end of the method.

Now that we have initialized the inventory variable, we can add a few methods to let other scripts use it and access it through the `LevelManager` singleton instance. The first method we are going to add to the script is `OnDestroy`, as defined here:

```
private void OnDestroy()
{
    _inventory.Dispose();
}
```

The `OnDestroy` method is a `MonoBehaviour` method that's automatically called by Unity when the GameObject that has this script attached to it is destroyed – for example, when the game level is uploaded or when the player closes the game. Here, we are calling the `Dispose` method of the `inventory` variable, which will clean up or inventory as soon as the GameObject is destroyed. Our intention here is to delete the player's inventory when a new game starts, so there is nothing in the inventory to give the player an advantage.

There are a couple more methods we need to add to this class. They will use the last couple of public methods available in the `InventoryManager` class. These are the `UpdateResource` and `GetResources` methods, as defined here:

```
public void UpdateResource(ResourceType type, int amount)
{
    _inventory.UpdateResource(type, amount);
}

public int GetResource(ResourceType type)
{
    return _inventory.GetResource(type);
}
```

In the first method, `UpdateResource`, we are simply sending both `type` and `amount` forward to the method in the `InventoryManager` class that has the same name. In the second method, `GetResource`, we are doing the same by passing the `type` parameter to the method with the same name in the `InventoryManager` class and returning the value from this method using the `return` keyword.

It might look like we are only hiding the `InventoryManager` class behind the `LevelManager` class – that is in fact what we are doing. The game code will have a single access to the inventory, via `LevelManager`, and if we need to change the usage or even update the `InventoryManager` implementation, there will be only one place to change rather than having to update every place that is using the class.

Now that we have finished adding the changes to the `LevelManager` class, we can change one script that will already use the new inventory to access the player's resources and update the UI to reflect the current resource values for each type.

Let's open the existing script located at **Scripts** | **UI** | `ResourceUpdater.cs` and replace the following highlight lines. We are replacing one line in the existing `OnResourceUpdated` method and updating another line in the existing `UpdateValue` method:

```
private void OnResourceUpdated(UpdateResourceMessage message)
{
    if (_type == message.Type)
    {
        LevelManager.Instance.UpdateResource(_type, message.Amount);
        UpdateValue();
    }
}

private void UpdateValue()
{
    _value.text =
        $"{_type}: {LevelManager.Instance.GetResource(_type)}";
}
```

`OnResourceUpdated` will only have one line changed, which is the first line inside the `if` statement – it is now using the `LevelManager` singleton, `instance`, to update the `type` and `Amount` parameters for the resources using the `UpdateResource` method. The `OnResourceUpdated` method is executed every time a new message of the `UpdateResourceMessage` type is received; we are using this to update the inventory as well.

The other existing method, `UpdateValue`, had only one line, so we are replacing its content with the new line. This method is responsible for updating the values of each resource type in the UI; now, we are doing so using the `GetResource` method with the `type` parameter, which is in this script already. As we did in the `OnResourceUpdate` method, we are accessing the inventory through the `LevelManager` singleton, `instance`.

These were the only changes that were required to make our new component work with our existing code and generate new resources automatically, updating the current values on both the UI and in the inventory data file.

Updating the LevelManager Prefab

Now that we have added the inventory to our game, let's also add the **ResourceProductionComponent** script that we created to the **LevelManager** Prefab so we can generate **Gold** automatically for the player when the game starts:

1. In the **Project** view, in the **Level** folder inside **Prefabs**, double-click to open the Prefab named **LevelManager**.

2. In the **Inspector** view, click on the **Add Component** button, search for the **Resource Production Component** script, and double-click to add it to the GameObject.

3. Regarding the **Resource Production Component** script that we just added, click on the circle button on the right-hand side of the **Resource** property and, in the **Select ResourceData** window, double-click on the **Gold** asset.

4. Use the **File | Save** menu option to save the changes.

Since the **LevelManager** Prefab is included in our scenes, it is a good place to add the script that's responsible for producing resources automatically. We set up this component to automatically generate the **Gold** resource, using the **ResourceData** asset that contains all the configuration we need to make our component work.

The following screenshot shows our **LevelManager** Prefab with the scripts we've added to it so far:

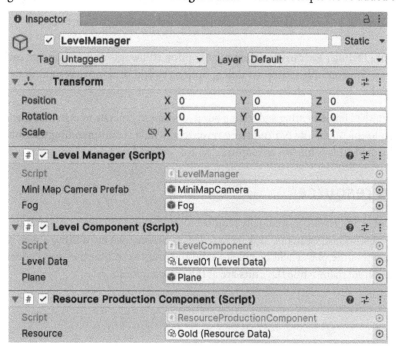

Figure 13.2 – The new Resource Production Component script with the Gold resource

After these last changes, we can go ahead and open our **Playground** scene. Run the game in the **Editor** area to see the number of gold resources increasing each second, in the top-right corner of the UI. As shown in the following screenshot, we have a value of **7** next to the **Gold** resource:

Figure 13.3 – The Gold resource being generated automatically each second

The **Dragoncraft | Debug | Resources** debug menu option still works and will update each resource on the UI, and we can continue to use the debug options to add or remove resources in the **Editor** area when needed.

Now that we've finished implementing and testing our new script to generate new resources automatically, we can move on to the other way of producing resources, which is commanding the units to gather resources on the map. This is known as manual resource generation.

Gathering resources

The feature for collecting resources manually requires some changes and new configurations regarding the project, Prefabs, scenes, cameras, and scripts. This is because we are adding a new interactable object to the map that the player will be able to send the units to so that they can collect resources and manually generate them while the units are at the resource location.

Adding the new Resource layer

The first thing we are going to add is a new layer. This will be used by the `Resource` object on the map so that it can collide with the unit, as well as a few other validations that we are going to do when we develop the next script. Open the **Project Settings** window by going to **Edit | Project Settings…**, select the **Tags and Layers** menu from the left panel, and add **Resource** as the value for **User Layer 11**. This should look as follows:

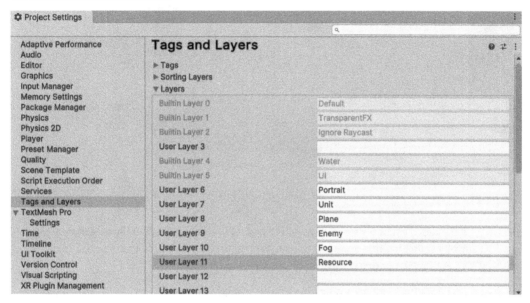

Figure 13.4 – The new Resource layer

We have one new layer now, but we also need to tweak the physics settings to make sure the objects within the **Resource** layer will only report a collision with the layers that we want.

In the same window, select **Physics** from the left panel and, under **Layer Collision Matrix**, uncheck the **Portrait**, **Plane**, **Enemy**, **Fog**, and **Resource** layers on the first column. As shown in the following screenshot, we disable the collision between the **Resource** layer and the layers where we do not want collision to be detected:

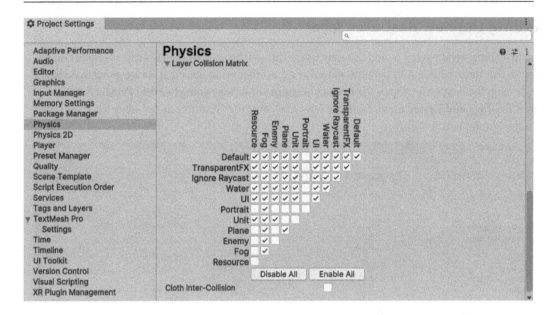

Figure 13.5 – The updated Layer Collision Matrix

Now that we have added a new layer, we also need to update the colling mask of the in-game cameras; otherwise, any object that has this new layer will not be rendered on either the map or mini-map. Let's update our camera Prefabs so that they have the new **Resource** layer:

1. In the **Project** view, in the **Level** folder inside **Prefabs**, double-click to open the **Main Camera** Prefab.

2. In the **Inspector** view, under the **Camera** component, click on the dropdown located on the right-hand side of the **Occlusion Culling** property and select the **Resource** option to also include this layer.

3. In the **Project** view, in the **UI** folder inside **Prefabs**, double-click to open the **MiniMapCamera** Prefab.

4. In the **Inspector** view, under the **Camera** component, click on the dropdown located on the right-hand side of the **Occlusion Culling** property and select the **Resource** option to also include this layer.

5. Use the **File | Save** menu option to save the changes.

After editing the two Prefabs, **Main Camera** and **MiniMapCamera**, our game will be ready to render any object that has the new **Resource** layer. We can now create a new script that will be used to collect resources manually from the units.

Producing resources manually

Before creating our new script to generate the resources manually, we need to update one of our scripts and add a new method that we will use to get the current action that the unit has. This will be used in the next script to validate whether the action is collected before producing a resource.

Open the script located in the **Scripts** | **Character** | BaseCharacter.cs folder and add the following GetActionType method to the class:

```
public ActionType GetActionType()
{
    return _action;
}
```

This method is very simple and is used to return the current action that the units are doing after receiving a command from the player. This is helpful for us to validate whether we should start manual resource generation or not in the next script.

We already have one script to generate resources automatically, with all the logic we need to be implemented there. The new script, **ResourceCollectionComponent**, will expand the **ResourceProductionComponent** script and add a few methods that will allow us to start and stop the manual resource generation based on the collision between the unit and an object with the **Resource** layer. So, create a new script in the **Scripts** | **Resource** folder, name it ResourceCollectionComponent, and replace its content with the following code block:

```
using UnityEngine;
namespace Dragoncraft
{
    public class ResourceCollectionComponent :
        ResourceProductionComponent
    {
        protected override void Start()
        {
            base.Start();
            _productionType = ResourceProductionType.Manual;
        }
    }
}
```

The ResourceCollectionComponent class is inherited from the ResourceProductionComponent class, so we can use all the code that is already there but with a few changes. Here, the Start method has the override keyword to indicate that we are replacing the virtual method's Start implementation from the parent class, ResourceProductionComponent, and reimplementing it in this class.

In the first line of the Start method, we are using the base class method, Start, which means that we want our override method to execute the virtual Start method before continuing to the next line. We are doing this because we need the virtual Start method to initialize the variables; then, we change one of the class variables, productionType, to Manual to make sure the script will not generate resources automatically.

Start is the only method that we are reusing here; the next couple of methods are new and will help us detect whether a unit is colliding with the GameObject that has this script attached and has the action to collect resources. Let's start with the OnCollisionEnter method:

```
private void OnCollisionEnter(Collision col)
{
    if (col.gameObject.TryGetComponent<UnitComponent>(out var unit))
    {
        if (unit.GetActionType() == ActionType.Collect)
        {
            _productionType = ResourceProductionType.Automatic;
        }
    }
}
```

The GameObject that will have this script attached to it will also have the **Resource** layer and, based on the collision matrix that we updated after adding **Resource** to the project, we are going to detect collisions between the **Resource** layer and the **Unit** layer. When a collision that meets this validation happens, the OnCollisionEnter method will be executed by Unity to let us know about the collision.

To make sure we have the correct object colliding with this class, we must use the TryGetComponent method with the UnitComponent type to get our unit object. If the unit object exists, then we enter the first statement and validate whether the unit action is Collect, using the GetActionType method. If the unit action is indeed Collect, we can change productionType to Automatic, meaning that the unit is in the GameObject collecting resources, so we can start generating it.

If the unit is moved to somewhere else by the player, the collision will stop, and we need to stop production as well. The OnCollisionExit method, defined here, is executed by Unity internally when the two objects that entered a collision previously are not colliding anymore:

```
private void OnCollisionExit(Collision col)
{
    if (col.gameObject.TryGetComponent<UnitComponent>(out var unit))
    {
        _productionType = ResourceProductionType.Manual;
    }
}
```

The OnCollisionExit method is very similar to the OnCollisionEnter method because we also use the TryGetComponent method with the UnitComponent type to get the unit variable if it was the object that exited the collision. After getting the unit variable, we do not need to check the unit action because it is not important now – the only thing we need to do is change the productionType variable to Manual so that we stop producing resources as soon as the unit stops the collision with this GameObject.

That is all we need to add to our new ResourceCollectionComponent class to generate the resources when the unit collides with the GameObject that has this script attached. We did this by changing productionType to Automatic when the collision starts and then changing back to Manual when the collision exists.

The last thing we need to do now is create a couple of Prefabs to represent a source of Wood and Food and have this script attached to it so that we can generate these resources when a unit collides and have the collect action. First, let's create our **Wood** Prefab:

1. Open the **Playground** scene and, in the **Hierarchy** view, right-click and select the **Create Empty** option to add a new GameObject. Name it Wood.

2. Left-click on the new **Wood** GameObject and, in the **Inspector** view, change **Layer** to **Resource**. Regarding the **Transform** settings, change **Position X** to **-5** and **Z** to **-5**.

3. Click on the **Add Component** button and add **Box Collider** to the GameObject. After adding it, from the **Box Collider** settings, change **Center Y** to **0.5** and **Size X** to **2**.

4. Click on the **Add Component** button one more time and add **Resource Collection Component** to the GameObject.

5. Regarding **Resource Collection Component**, click on the circle button on the right-hand side of the **Resource** property and, in the **Select ResourceData** window, double-click on the **Wood** asset.

6. In the **Project** view, search for the **rpgpp_lt_log_wood_01** Prefab, and then drag this Prefab from the **Project** view and drop it inside the new GameObject, **Wood**, in the **Hierarchy** view. **rpgpp_lt_log_wood_01** must be a child of the **Wood** GameObject.

7. In the **Project** view, create a new folder called Resource inside the **Prefabs** folder.

8. In the **Hierarchy** view, drag the **Wood** GameObject and drop it into the **Project** view, inside the **Prefabs | Resource** folder, to save it as a Prefab.

By completing these steps, we now have a Prefab that contains the 3D model of a pile of wood logs and, with the **ResourceCollectionComponent** script attached to it, can generate wood resources when a unit is commanded by the player to collect resources from it. We also need to create a Prefab for **Food**, which consists of similar steps but a few different parameters:

1. Open the **Playground** scene and, in the **Hierarchy** view, right-click and select the **Create Empty** option to add a new GameObject. Name it Food.

2. Left-click on the new **Food** GameObject and, in the **Inspector** view, change **Layer** to **Resource**. Regarding the **Transform** settings, change **Position X** to **5** and **Z** to **-5**.

3. Click on the **Add Component** button and add **Sphere Collider** to the GameObject. After adding it, go to the **Sphere Collider** settings and change **Center Y** to **1** and **Radius** to **1.5**.

4. Click on the **Add Component** button one more time and add **Resource Collection Component** to the GameObject.

5. Regarding **Resource Collection Component**, click on the circle button on the right-hand side of the **Resource** property and, in the **Select ResourceData** window, double-click on the **Food** asset.

6. In the **Project** view, search for the **rpgpp_lt_bush_01** Prefab, and then drag this Prefab from the **Project** view and drop it inside the new GameObject, **Food**, in the **Hierarchy** view. **rpgpp_lt_bush_01** must be a child of the **Food** GameObject.

7. In the **Hierarchy** view, drag the **Food** GameObject and drop it in the **Project** view, inside the **Prefabs | Resource** folder, to save it as a Prefab.

The **Food** Prefab uses a 3D model of a bush to represent the food we can gather in the wild. With the **ResourceCollectionComponent** script attached to it, it will generate **Food** while the unit is there to collect it. The **Food** Prefab has a different collider component than the **Wood** Prefab – a sphere instead of a box – because of the shape of the 3D model, but any collider would work on either Prefab.

The following screenshot shows the two Prefabs on the **Playground** scene. On the left-hand side, we can see the GameObjects and the respective 3D models we added as children of each:

Figure 13.6 – The new Wood and Food Prefabs

Now that we have all three resources configured in the game, we can run the **Playground** scene to test them. **Gold** is generated automatically by the **LevelManager** Prefab, while **Wood** and **Food** are manually collected when a unit has the **Collect** action and is colliding with the object in the scene. The following screenshot shows two units, each of them in one resource, collecting **Wood** and **Food**, while the game generates **Gold**:

Figure 13.7 – The units collecting the Wood and Food resources

The amount of each resource is updated on the top right-hand side of the UI, and we can also test that if we move the units away from the resources or change their actions, the respective resource will stop being collected and the values will not increase on the UI – apart from **Gold**, which is always generated.

We can also notice that, since each resource type has a different resource per second value, they are not generated evenly. We will always have more **Food** than **Wood**, and more **Wood** than **Gold**, unless we upgrade the resource type to generate more. Although we do not have the upgrade feature yet, we are going to add a few debug methods to help us test it.

Testing the resources

We already have a script with debug methods that can be used in the **Editor** area to add or remove resources of each type. Now, we are going to add a few more methods to this script so that we can also upgrade each of the resource types while the game is running in the **Editor** area.

Open the script located at **Scripts | Editor | Debug |** `ResourceDebugger.cs` and add the `UpgradeGold` method using the following code block:

```
[MenuItem("Dragoncraft/Debug/Resources/Upgrade Gold", priority = 6)]
private static void UpgradeGold()
{
  MessageQueueManager.Instance.SendMessage(
    new UpgradeResourceMessage { Type = ResourceType.Gold });
}
```

Here, we are adding an attribute called `MenuItem` with the `Upgrade Gold` menu option to the `UpgradeGold`. Method. This will send a new message of the `UpgradeResourceMessage` type with `ResourceType` set to `Gold`. When the game is running in the **Editor** area, we can use this menu option to send the upgrade message so that we can generate more `Gold` each second.

The other two methods, `UpgradeWood` and `UpgradeFood`, are very similar to `UpgradeGold`, with only naming changes and the respective resource types of `Wood` and `Food`:

```
[MenuItem("Dragoncraft/Debug/Resources/Upgrade Wood", priority = 7)]
private static void UpgradeWood()
{
  MessageQueueManager.Instance.SendMessage(
    new UpgradeResourceMessage { Type = ResourceType.Wood });
}

[MenuItem("Dragoncraft/Debug/Resources/Upgrade Food", priority = 8)]
private static void UpgradeFood()
{
  MessageQueueManager.Instance.SendMessage(
```

```
        new UpgradeResourceMessage { Type = ResourceType.Food });
}
```

We now have three new debug options to upgrade each resource type – Gold, Food, and Wood. Each time we use one of these menu options, the resource is upgraded one level. priority, in the MenuItem attribute, is there to order the menu options in the **Editor** area so that these three new options are the last ones from the menu.

Once we save the script changes, we can go back to the **Editor** area and run the game. While the game is running, we can use the **Dragoncraft | Debug | Resources | Upgrade Gold** menu option to see the game generate higher amounts of gold automatically each second in the UI. The same applies to the other resource upgrades, but since Food and Wood are not automatically generated, we need to spawn a couple of units and give them the action to collect the resources to see the upgrade working while they are gathering Food and Wood.

This debugging script is very helpful to use while the game is running to test different scenarios and will be a very important tool to have in the next chapter as we develop the buildings and upgrades for our resources.

Since the resources have configurations and data, we can also validate them using unit tests, as we did before for the enemies and units. Let's create a new script in the **Scripts | Editor | Tests** folder, name it ResourceDataTest, and replace its content with the following code:

```
using NUnit.Framework;
using System.Collections.Generic;
using UnityEditor;
namespace Dragoncraft.Tests
{
  public class ResourceDataTest
  {
    private static List<string> resources = new List<string>()
    {
      "Gold",
      "Food",
      "Wood"
    };
  }
}
```

In the ResourceDataTest class, we have a list of resources with the name of each ResourceData ScriptableObject that we have in the game. With this, we can test and validate their data and configuration when running our unit test.

To load each ScriptableObject we are going to use the LoadResource method, as defined here:

```
private ResourceData LoadResource(string resource)
{
    return AssetDatabase.LoadAssetAtPath<ResourceData>(
        $"Assets/Data/Resource/{resource}.asset");
}
```

This method is very similar to the one we used to load the enemy and unit ScriptableObjects but now, we are loading our resources and converting them into a ResourceData object.

We are going to use the LoadResource method in the following ResourceDataTestWithResources method to access the properties of the loaded ResourceData:

```
[Test]
public void ResourceDataTestWithResources(
    [ValueSource("resources")] string resource)
{
    ResourceData data = LoadResource(resource);

    Assert.IsNotNull(data, $"{resource} not found.");
    Assert.IsTrue(data.ProductionLevel > 0);
    Assert.IsTrue(data.ProductionPerSecond > 0);
}
```

The ResourceDataTestWithResources method has the Test attribute, which is there to tell Unity that this is a unit test and must be added to the test runner. This test method has one parameter resource, which is from the list of resources we defined in this class. We are going to have this test method executed for each item on the list of resources.

Inside the method, we use the parameter resource to load our ScriptableObject using the LoadResource method and add the return of the method to the variable data. After loading the resource, we can execute our tests, which are as follows:

- Checking if the resource exists using the IsNotNull assert
- Checking if ProductionLevel is greater than 0 using the IsTrue assert
- Checking if ProductionPerSecond is greater than 0 using the IsTrue assert

These are all basic validations, but without them, we cannot ensure the resources have the proper configuration. A resource with a zero value on ProductionLevel or ProductionPerSecond will not generate any resource, and this test is here to validate this.

After saving the changes to the new test script, we can go back to the **Editor** area and see the new unit tests appear in the **Test Runner** window:

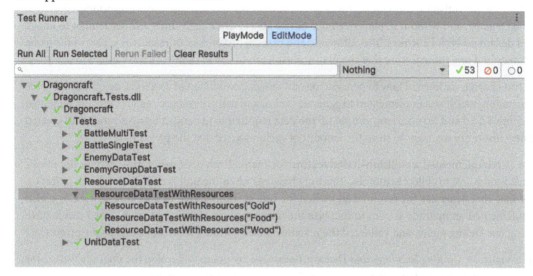

Figure 13.8 – The new tests for Gold, Food, and Wood

We can select **ResourceDataTest** and use the **Run Selected** or **Run All** button to execute the tests. Unity generated one test for each resource in our list, and since they were configured properly, all tests will pass and display the green check mark.

It is good practice to keep adding more tests to our project as we develop new features. Even basic tests, such as the ones we added here to check if the values are greater than zero, are good to keep the game working and add more changes to the code. It could save us from spending a good amount of time trying to understand why something is not working if the problem is just a ScriptableObject with a value that is not correct or expected in our game.

Debug scripts and unit tests are very powerful tools to have for any project, and they can help us test and validate features before we spend time developing them further or without waiting for an asset required for the feature. Now that we've covered resource generation, both automatic and manual, and have good testing tools, we can develop our buildings and upgrades in the next chapter.

Summary

Well done for finishing this chapter! Resource production is one of the most important features of an RTS game, and without it, players would not be able to grow their units and settlement to attack and defend powerful enemies. The automatic and manual production of resources is part of the game strategy and one more feature that the player must learn to use to dominate the game.

In this chapter, we learned how to generate resources using two different methods: automatic resource generation, which was implemented to generate Gold, and manual resource generation, implemented on both Food and Wood. Here, you set up the data required to generate these resources and created a new inventory manager to store the amount of each resource that the player has.

We also implemented a collision-based feature for manual resource generation, where the player must send a unit to collect resources. This only happens when the unit is colliding with the object on the map that generates the resource and when the unit is commanded to collect resources. You also implemented an upgrade system to increase the number of resources generated, tested this feature using our **Debug** menu, and validated the resource data by adding more unit tests to the project.

In *Chapter 14, Crafting Buildings and Defense Towers*, we are going to develop the upgrade feature for resource generation and unit training. Both will use our inventory system to spend the resources that we are generating and collecting. You will learn how to add buildings to the map, which will work as an upgrade for the resources and units, as well as the defense tower, which is a type of building that will help the player defend against enemies trying to invade the settlement. We are going to develop the last features of our *Dragoncraft* game before putting everything together and creating our game.

Further reading

In this chapter, we covered a lot of concepts and Unity features that are already familiar since we covered them in other chapters in this book. However, we saw something new, and it is worth reading more about it:

- *Application.persistentDataPath*: `https://docs.unity3d.com/2023.1/Documentation/ScriptReference/Application-persistentDataPath.html`

14
Crafting Buildings and Defense Towers

RTS games have many different features, and we have added most of the main functionalities to our *Dragoncraft* game through the book. Now, we are going to develop some new features that are a must-have for RTS games, which will allow the player to craft buildings and the Defense Tower.

In this chapter, we are going to learn how to create a customizable and flexible store for the player to purchase upgrades for resources and units, as well as how to craft buildings that can be moved around the map by the player. We will also learn how to implement the **Defense Tower**, a powerful unit that attacks enemies from a distance and, as a building, can be used to protect the settlement against invasions.

By the end of this chapter, you will learn how to create a flexible solution to add buildings to the map, increase resource production by upgrading the resource type, and increase the units' attack and defense attributes with upgrades that will level up each unit type. You will also learn how to create a robust and reusable system for the resources store that can be easily expanded to support the unit store with minimal changes. Plus, you will see how we can add a new unit to the game, the Defense Tower, that also has building characteristics, and how it can reuse the same systems that the units have for collision detection, ranged attack, and damage.

In this chapter, we will cover the following topics:

- Crafting and upgrading buildings
- Training units
- Defending with towers

Technical requirements

The project setup in this chapter, with the imported assets, can be found on GitHub at `https://github.com/PacktPublishing/Creating-an-RTS-game-in-Unity-2023` in the `Chapter 14` folder inside the project.

Crafting and upgrading buildings

Resource management is a big part of RTS gameplay, and the player must think about the strategy when commanding units to gather and expend resources to craft or upgrade new buildings. Crafting and upgrading buildings are the last features we are going to develop before putting together everything we have created so far, in *Chapter 15*, *Game Progression and Objectives*, then our RTS game, *Dragoncraft*, will be ready to play.

We will create new scripts and Prefabs to have a new store pop-up window in the UI so the player can choose which building or unit they want to craft or upgrade. Before creating the scripts for the store popup, let us create another script that is going to be used by all buildings after they are crafted, namely, the drag-and-drop functionality, so the player can position the buildings where they want on the map.

Dragging and dropping buildings

We are going to create a new script that will add the drag-and-drop feature to any GameObject that is placed on the map. Since we only want to add this feature to the buildings, we will create a new Tag for the buildings, and use it together with the `Resource` layer so the player can only move GameObjects that have this combination of tag and layer.

Let us add a new tag named `Building` to the project:

1. Open the **Project Settings** window using the **Edit** | **Project Settings…** menu and click on the **Tags and Layers** option on the left-side panel.

2. After selecting the **Tags and Layers** option, on the right-side panel, expand the tags group by clicking on **Tags** and then click on the + button.

3. The + button will display a **New Tag Name** text input field and a **Save** button.

4. Type the word `Building` in the text input field and click on the **Save** button, and the new **Building** tag will be added to the list.

The new **Building** tag will be included in the list of tags and layers, where we already have the `Plane` tag that was added there previously. Now that we have our new `Building` tag, we can move on to our script that will allow the player to drag and drop GameObjects that have both the `Building` tag and the `Resource` layer.

Add a new script to the **Scripts | Level** folder, name it DragAndDropComponent.cs, and replace the content with the following code block:

```
using UnityEngine;
using UnityEngine.AI;
using UnityEngine.EventSystems;
namespace Dragoncraft
{
  public class DragAndDropComponent : MonoBehaviour
  {
    [SerializeField] private LayerMask _layerMask;
    [SerializeField] private string _tag = "Building";

    private GameObject _selectedGameObject;
    private float _startPositionY;
    private float _positionYWhileMoving;
  }
}
```

The new DragAndDropComponent class is rather long, so we are going to review it piece by piece, starting with the private class variables:

- layerMask has the SerializeField attribute so we can set the layer we want for this script in the GameObject that will have the DragAndDropComponent script attached.

- tag also has the SerializeField attribute, but this time we are initializing it with a pre-defined value of Building.

- selectedGameObject is a private class variable that will have a reference to the GameObject that the player is dragging around the map, so we can update the position based on the mouse movement.

- The next class variable, startPositionY, as the name implies, is the initial value of the Y position, so we can set the Y position back to the original value once the player drops the GameObject.

- The last class variable, positionYWhileMoving, is a new value for the Y position while the player is dragging the GameObject on the map, so we can raise it slightly from the map floor to indicate that the building was picked up while dragging.

With the five private variables, we will be able to control the drag-and-drop action on a GameObject and make sure that only the GameObjects with the right combination of tag and layer can be moved by the player.

Now, we are going to add our first method to the DragAndDropComponent class, the Update method, with the following code block:

```
private void Update()
{
    if (EventSystem.current != null &&
        EventSystem.current.IsPointerOverGameObject())
    {
        return;
    }

    if (Input.GetKeyDown(KeyCode.Mouse0))
    {
        if (_selectedGameObject == null)
        {
            StartDragging();
        }
        else
        {
            StopDragging();
        }
    }

    if (_selectedGameObject != null)
    {
        DragObject();
    }
}
```

In the first few lines of the Update method, we have a validation to check whether EventSystem is not null and whether the mouse point is over a GameObject – we have already used both validations in our UnitSelectorComponent script in order to ignore GameObjects behind the UI elements when selecting the units on the map. Here, we are using the EventSystem validations in a similar way: avoid drag-and-drop GameObjects that are behind the UI and not visible to the player.

Then, we have another validation to check whether the player pressed the left button on the mouse, identified by the KeyCode.Mouse0 enum and using the Input.GetKeyDown method. Inside this validation, we check whether the selectedGameObject variable is null in order to call the StartDragging method; otherwise, we will call the StopDragging method. Here we are using the left mouse button to start and stop the dragging movement, so the player will not have to keep the button pressed – left-click on the GameObject once to move, and then left-click again to place the GameObject in the desired location.

The last thing we do in the Update method is check whether the selectedGameObject variable is not null, and then we call the DragObject method. Note that we only execute the StartDragging and StopDragging methods once per left-click on the GameObject. Meanwhile, the DragObject method is executed on every game loop when selectedGameObject is not null. This will make the GameObject follow the player's mouse movement until the player clicks on the map to position the GameObject.

Now, we are going to add the three methods that we executed in the previous code block: StartDragging, StopDragging, and DragObject.

Let us start with the StartDragging method, defined in the following code snippet:

```
private void StartDragging()
{
  RaycastHit hit = GetRaycastHit();
  if (hit.collider ==null ||
    !hit.collider.CompareTag(_tag))
  {
    return;
  }

  _selectedGameObject = hit.collider.gameObject;
  _startPositionY =
    _selectedGameObject.transform.position.y;
  _positionYWhileMoving = _startPositionY + 1;

  if (_selectedGameObject.TryGetComponent<NavMeshObstacle>(
    out var navMeshObstacle))
  {
    navMeshObstacle.enabled = false;
  }
  Cursor.visible = false;
}
```

At the beginning of the method, we get RaycastHit using the GetRaycastHit method, which we are going to create in a moment, and then we check whether the collider in the hit variable is null or does not have the tag value, which is verified using the CompareTag method. If any of these two validations are true, it means that the Raycast hit a GameObject that does not have a collider or does not have the tag we want, and then we can skip the execution using the keyword return.

Next, we have three class variables, selectedGameObject, startPositionY, and positionYWhileMoving, that are initialized when the player starts to drag the GameObject:

- selectedGameObject is the gameObject instance from the collider that the Raycast hit and has the tag we want

- startPositionY has the Y position of the object that the player is going to move, so we can change back the value of position Y when the object is dropped on the map by the player

- positionYWhileMoving has the value of startPositionY + 1 because we are lifting the GameObject just slightly while the player is moving it, to give feedback that the object was picked up when it started to drag

After initializing the variables, we check whether the selectedGameObject has a NavMeshObstacle component, and if that is the case, we disable it by setting the enabled property to false. We are disabling NavMeshObstacle to move the GameObject; otherwise, it would interact with units and enemies on the map while moving the GameObject around.

Finally, in the last line of the StartDragging method, we hide the mouse cursor and set the value of Cursor.visible to false. This will make the cursor invisible while the player is dragging the GameObject around the map and, as we will see in the StopDragging method in the next code block; we change it back to visible once the player drops the object on the map:

```
private void StopDragging()
{
    Vector3 worldPosition = GetWorldMousePosition();
    _selectedGameObject.transform.position =
        new Vector3(worldPosition.x, _startPositionY,
            worldPosition.z);

    if (_selectedGameObject.TryGetComponent<NavMeshObstacle>(
        out var navMeshObstacle))
    {
        navMeshObstacle.enabled = true;
    }

    _selectedGameObject = null;
    Cursor.visible = true;
}
```

In the StopDragging method, the first thing we do is get the current mouse position, so we know where the player clicks to drop the GameObject moving on the map. However, the mouse position alone is not enough, as we need to convert it to the 3D world coordinates before changing the GameObject position.

We are using the GetWorldMousePosition method, which we are going to create in a moment, to get Vector3 with worldPosition. Then, we are going to change the selectedGameObject position to a new Vector3 using X and Z from the worldPosition variable and the Y from the startPositionY variable, so we can drop the GameObject on the map.

Next, we check whether selectedGameObject has a NavMeshObstacle component to enable it, since we disabled it in the StartDragging method to avoid collision issues when the player was moving the GameObject on the map. After that, we do not need the selectedGameObject value anymore and the variable can be set to null.

In the last line of the StopDragging method, we are changing Cursor.visible to true, so the player can see the mouse cursor again on the screen after the drag-and-drop movement. Now that we have finished creating the methods to start and stop the drag movement, we can move on to the last method that is executed inside the Update method, which is the DragObject method, defined in this code block:

```
private void DragObject()
{
  Vector3 worldPosition = GetWorldMousePosition();
  _selectedGameObject.transform.position =
    new Vector3(worldPosition.x, _positionYWhileMoving,
      worldPosition.z);
}
```

The DragObject method is quite simple because we only need to get the mouse position and move selectedGameObject there. Since this method is used on every game loop by the Update method, we will keep selectedGameObject moving by following the player's mouse position.

As we did in the previous StopDragging method, we also need to get the mouse position and convert it into the 3D world coordinates before applying the values of the X and Z to the selectedGameObject position by creating a new Vector3 object. Note that for position Y, we are using the positionYWhileMoving variable, which is startPositionY + 1 that was set in the StartDragging method.

selectedGameObject will follow the mouse position after the player clicks with the left mouse button on the GameObject and will stop following the mouse position as soon as the player clicks with the left mouse button one more time. Both the StopDragging and DragObject methods use the following GetWorldMousePosition method to get the mouse position converted into the 3D world coordinates:

```
private Vector3 GetWorldMousePosition()
{
  Vector3 selectedPosition =
    Camera.main.WorldToScreenPoint(
```

```
      _selectedGameObject.transform.position);

   Vector3 mousePosition =
     new Vector3(Input.mousePosition.x,
       Input.mousePosition.y, selectedPosition.z);

   Vector3 worldPosition =
     Camera.main.ScreenToWorldPoint(mousePosition);
   return worldPosition;
 }
```

The GetWorldMousePosition method can be divided into three parts that we execute to convert the mouse position from the 2D screen to the 3D world:

- The first thing we do is get the selectedGameObject position from the Camera API, using the WorldToScreenPoint method to create a new Vector3 selectedPosition variable

- The next thing is to get mousePosition using the Input API and to create a new Vector3 object using the X and Y positions from mousePosition and position Z from selectedPosition

- Finally, we can convert mousePosition to our 3D world position using the ScreenToWorldPoint method, which is also from the Camera API but the other way around this time – screen to world instead of world to screen

Once we finish the conversion, we can return the worldPosition variable. While the GetWorldMousePosition method is used on every frame to move selectedGameObject around the map, the next and last method of the DragAndDropComponent class is used only when the player clicks on the screen to use a Raycast and tries to find a GameObject. The GetRaycastHit method is defined in the following code block:

```
private RaycastHit GetRaycastHit()
{
   Vector3 worldMousePositionMax =
     Camera.main.ScreenToWorldPoint(
       new Vector3(Input.mousePosition.x,
         Input.mousePosition.y,
           Camera.main.farClipPlane));

   Vector3 worldMousePositionMin =
     Camera.main.ScreenToWorldPoint(
       new Vector3(Input.mousePosition.x,
         Input.mousePosition.y,
           Camera.main.nearClipPlane));
```

```
    RaycastHit hit;
    Physics.Raycast(worldMousePositionMin,
      worldMousePositionMax - worldMousePositionMin, out hit,
        1000, _layerMask);
    return hit;
}
```

The `GetRaycastHit` method can also be divided into three parts where we define the minimum and maximum positions for the Raycast, and then we use the `Physics` API to get `RaycastHit` if any GameObject was hit by the Raycast:

- `worldMousePositionMax` is a `Vector3` position and, using the `Camera` API and the `ScreenToWorldPoint` method, we convert from the `mousePosition` values as positions X and Y and from the `Camera` `farClipPlane` value as position Z.

- `worldMousePositionMin` is very similar to the previous `worldMousePositionMax` variable, converted using the `mousePosition` values as positions X and Y but a `Camera` `nearClipPlane` value as position Z.

- The last part of the method is to use the `Physics` API and the `Raycast` method to get the `RaycastHit` variable hit using the `out` keyword in the method signature. `Raycast` uses `worldMousePositionMin` as the origin and the subtraction between `worldMousePositionMax` and `worldMousePositionMin` to find the direction vector. The last parameter, `layerMask`, is very important so we can ignore any GameObject that does not have the `layerMask` instance we want.

The `GetRaycastHit` method always returns a `RaycastHit` variable; however, we need to check whether there is a valid `collider` in this variable before using it; this is what we are currently doing on the `StartDragging` method, checking whether the `collider` value from `RaycastHit` is not `null` before using it.

With the last method completed, we can save our script and add it to the **LevelManager** Prefab so we can use it in our game:

1. In the **Project** view, inside the **Prefabs | Level** folder, double-click on the **LevelManager** Prefab to edit it.

2. In the **Hierarchy** view, left-click on the **LevelManager** GameObject to select it.

3. In the **Inspector** view, click on the **Add Component** button, search for the `DragAndDropComponent` script, and double-click to add it.

4. Under the new **Drag And Drop Component** component, left-click on the **Layer Mask** property and select the **Resource** value from the list of layer masks.

5. In the **Tag** property, type in the `Building` text for our tag name.

6. Save the changes to the Prefab.

The **LevelManager** Prefab is already included in our **Playground** and **Level01** scenes, so both will automatically have the new `DragAndDropComponent` script after it is added to the Prefab. *Figure 14.1* shows the **LevelManager** Prefab after adding our latest script to it:

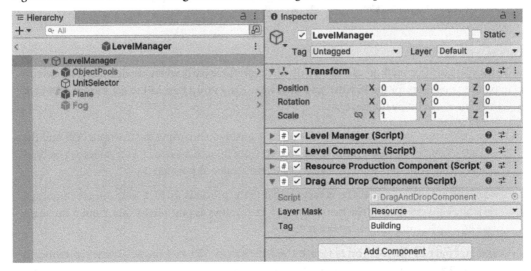

Figure 14.1 – DragAndDropComponent script added and configured on the LevelManager Prefab

Now that our component to drag and drop GameObjects on the map is ready and configured in our **LevelManager** Prefab, we can move on to the next part of our feature, which is configuring the store from which the player will be able to purchase and upgrade buildings (and use the drag-and-drop component to move them around the map).

Configuring the Resource Store

The Resource Store will be a pop-up window that the player will be able to open using a new menu button and choose to upgrade the Gold, Wood, or Food production levels. Each item available in the store will have a ScriptableObject to configure how much it costs to craft or upgrade the resource production building, and a build Prefab associated with it, which we are going to add to the map as soon as the player makes the purchase.

Most of the scripts and assets that we are going to create now will also be used in the Unit Store in the *Training Units* section of this chapter. To get started with the store, let us first create the class that will have the individual configuration of each store item:

1.   First, inside the **Scripts** folder, create a new folder, and name it `Store`.

2.   Then, in the new **Store** folder, add a new script and name it `StoreItem`.

3. Finally, replace the content with the following code block:

```
using System;
using UnityEngine;
namespace Dragoncraft
{
  [Serializable]
  public class StoreItem
  {
    public string Description;
    public int PriceGold;
    public int PriceResource;
    public ResourceType CurrencyResource;
    public GameObject Prefab;
    public Sprite Image;
  }
}
```

The StoreItem class is very simple, just a Serializable class with a few variables. This class will be used as a base for both resource crafting and unit upgrades:

- Description is a string that will be displayed on a given item in the store, so the player knows what this is about.

- PriceGold is how much Gold resource a given item costs. We will always charge an amount of Gold for the given item.

- PriceResource is interesting because it will make the player think about the strategy to manage the resource production. Besides Gold, we are also going to add another resource cost to the given item. The resource type is set in the CurrencyResource variable and can be Wood or Food. Typically, we are going to ask for Wood for buildings and Food for units.

- Prefab is a reference to the GameObject that we are going to instantiate on the map when a new building is added. If the build is already built, we will not add another copy, just upgrade the resource production instead.

- The last Image variable is a Sprite instance that will represent the item in the store window, like an icon.

Now that we have our StoreItem class, we will create another script that will have the information relating to the resource we are going to craft or upgrade for the player. In the same **Scripts | Store** folder, add a new script named ResourceStoreItem, and replace the content with the following code block:

```
using System;
using UnityEngine;
namespace Dragoncraft
```

```
{
  [Serializable]
  public class ResourceStoreItem : StoreItem
  {
    public ResourceType Resource;
  }
}
```

The ResourceStoreItem class is also a Serializable class, but in this case, we are inheriting it from the StoreItem class that we just created. We are creating a new ResourceStoreItem class because we need to have ResourceType to indicate which Resource the given item will craft or upgrade.

Since we are going to use StoreItem to upgrade units as well, we need a generic class that could be used as an inherited class to expand the properties, as we are doing with the ResourceStoreItem class by adding the Resource property.

Now that we have our specialized class for resources, we can create the ScriptableObject that is going to use the ResourceStoreItem class to configure our Resource Store.

In the same **Scripts | Store** folder, add a new script named ResourceStoreData, and replace the content with the following code block:

```
using System.Collections.Generic;
using UnityEngine;
namespace Dragoncraft
{
  [CreateAssetMenu(menuName = "Dragoncraft/New Resource Store")]
  public class ResourceStoreData : ScriptableObject
  {
    public List<ResourceStoreItem> Items =
      new List<ResourceStoreItem>();
  }
}
```

The ResourceStoreData class is a standard ScriptableObject class, with a CreateAssetMenu attribute that will add a new menu option to make it easier for us to add new configurations. This class has only one property, which is a ResourceStoreItem list that we will display for the player in the pop-up window that we are going to build in a moment.

Now that we have our first store scripts done, let us create a new `ResourceStoreData` ScriptableObject with upgrades for the Gold, Wood, and Food resource types:

1. In the **Project** view, create a new `Store` folder inside the **Data** folder. Then, right-click on the new **Data | Store** folder, select the **Create | Dragoncraft | New Resource Store** menu option, and save the new asset as `ResourceStore`.

2. Left-click on the new **ResourceStore** asset to select it and, in the **Inspector** view, change the value of **Items** to 3 – Unity will create three new items in the list for us.

3. In the first item on the list, add the text `Increases the Gold Production` in the **Description** property. Then, do as follows:

   I.   Set the **Price Gold** values and the **Price Resources** properties to `100`.

   II. Change the **Currency Resource** property to `Wood`.

   III. Change the **Resource** property to `Gold`.

   IV. Click on the circle button on the right side of the **Image** property and, in the new window, search for the `skill_icon_01` sprite and double-click on it to add it to the property.

   V.   Next, click on the circle button on the right side of the **Prefab** property and, in the new window, search for the `rpgpp_lt_building_01` Prefab and double-click on it to add it to the property.

4. In the second item on the list, add the text `Increases the Wood Production` in the **Description** property. Then, do as follows:

   I.   Set the **Price Gold** values and the **Price Resources** properties to `1000`.

   II. Change the **Currency Resource** property to `Wood`.

   III. Change the **Resource** property to `Wood`.

   IV. Click on the circle button on the right side of the **Image** property and, in the new window, search for the `skill_icon_04` sprite and double-click on it to add it to the property.

   V.   Next, click on the circle button on the right side of the **Prefab** property and, in the new window, search for the `rpgpp_lt_building_02` Prefab and double-click on it to add it to the property.

5. In the third item of the list, add the text `Increases the Food Production` in the **Description** property. Then, do as follows:

   I.   Set the **Price Gold** values and the **Price Resources** properties to `10000`.

   II. Change the **Currency Resource** property to `Wood`.

   III. Change the **Resource** property to `Food`.

IV.   Click on the circle button on the right side of the **Image** property and, in the new window, search for the `skill_icon_02` sprite and double-click on it to add it to the property.

V.   Next, click on the circle button on the right side of the **Prefab** property and, in the new window, search for the `rpgpp_lt_building_03` Prefab and double-click on it to add it to the property.

6.   Save the changes using the **File | Save Project** menu option.

Once we have finished setting up the `ResourceStore` ScriptableObject, we will have all we need to populate the data in the pop-up window, so the player can choose what to upgrade, based on the cost of each item.

Figure 14.2 shows the `ResourceStore` ScriptableObject after adding the configuration for the Gold, Wood, and Food resource types:

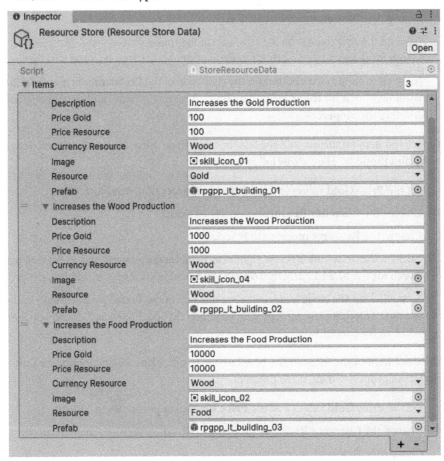

Figure 14.2 – Resource Store configured with Gold, Wood, and Food upgrades

That is all we need to create before starting to create scripts and Prefabs for the store interface and player interaction. Now, let us make some changes to the `LevelManager` class to add a starting Gold value for the player and to add a building to the map when purchased in the Resource Store.

Open the existing **LevelManager** script, located inside the **Scripts | Level** folder, and add the following highlighted line inside the existing `Awake` method, as the last line in the method:

```
private void Awake()
{
    ...

    _inventory.UpdateResource(ResourceType.Gold, 100);
}
```

We are adding 100 gold resources to the player inventory when the game starts, so the player will not need to purchase units in the Unit Store.

Next, let us add the new `AddBuilding` method to the `LevelManager` class, which will be used later to instantiate the building Prefab on the map after the player crafts the building in the Resource Store. The following method should be added to the `LevelManager` class:

```
public void AddBuilding(GameObject prefab)
{
    GameObject building = Instantiate(prefab);
    building.AddComponent<MeshCollider>();
    building.AddComponent<NavMeshObstacle>();
    building.transform.localScale = Vector3.one * 0.5f;
    building.layer = LayerMask.NameToLayer("Resource");
    building.tag = "Building";
}
```

The `AddBuilding` method will receive `Prefab` as a parameter, and it will be used to create the `building` GameObject through the `Instantiate` method. Once we instantiate the Prefab, we will use the `building` variable to change a few things in the new GameObject:

- We added a new `MeshCollider` component to the GameObject, so we can detect it using Raycast in the `DragAndDropComponent` class.

- We also added a new `NavMeshObstacle` component to the enemies, and units can avoid it while moving around the map. (Do not forget to add the line `using UnityEngine.AI;` at the beginning of the script so we can use the `NavMeshObstacle` class!)

- `localScale` has changed to half of the size, so the building is not too big on the game screen. `Vector3.one` is a shortcut to `(1, 1, 1)` and, once we multiply by `0.5`, `localScale` will be `(0.5, 0.5, 0.5)`.

- The last few things we need to do here is change layer to Resource and tag to Building, so our DragAndDropComponent class can use MeshCollider, plus these settings, to detect the GameObject when the player clicks on it, which allows the drag-and-drop movement on the map.

That is all we need to change in the LevelManager class. Now, we can move on to create the scripts and Prefabs for the Store UI, which we will use to display the available resource upgrades for the player.

Creating the Resource Store UI

To create our Resource Store UI, we will need to add a couple of scripts first: one to manage each store item and another to manage the list of store items. These two scripts will also be used and expanded in the *Training Units* section to create the Unit Store, but for now, let us focus on the Resource Store.

In the **Scripts | Store** folder, add a new script as StoreItemComponent, and replace the content with the following code block:

```
using System;
using TMPro;
using UnityEngine;
using UnityEngine.UI;
namespace Dragoncraft
{
    public class StoreItemComponent : MonoBehaviour
    {
        [SerializeField] private Image _image;
        [SerializeField] private TMP_Text _description;

        private StoreItem _storeItem;
        private Action<bool, string> _callback;
    }
}
```

The StoreItemComponent class is rather long, so we are going to create it in parts. Here, in the first code block, we are adding the class variables that we are going to expose in the **Inspector** view, through the SerializeField attribute, and the ones we will use in the class methods:

- The two properties with the SerializeField attribute are, respectively, the image and the description of the store item that we are going to display in the UI for the player.

- The storeItem is a reference to the ScriptableObject configuration that this item represents, and we are going to get all the data from there. Here, we are not using the ResourceStoreItem class to keep it generic, so we can use it for the Unit Store later as well.

- `callback` is an `Action` instance that we are going to invoke after the player attempts to purchase an upgrade. The first `bool` indicates whether or not the transaction was a success, and the second `string` is a message in case an error occurs, such as there being not enough resources to purchase the upgrade, so we can display it in the UI to let the player know what happened.

All these four class variables are initialized in the method defined in the following code block, which is called `Initialize` and has a couple of parameters, `storeItem` and `callback`:

```
public void Initialize(StoreItem storeItem,
  Action<bool, string> callback)
{
  _storeItem = storeItem;
  _callback = callback;
  _image.sprite = storeItem.Image;

  _description.text = $"{storeItem.Description}\n" +
    $"Price: {storeItem.PriceGold} Gold";

  if (storeItem.PriceResource >  0)
  {
    _description.text += $" + {storeItem.PriceResource} " +
      $"{storeItem.CurrencyResource}";
  }
}
```

Here we are setting the values of each class variable based on the `storeItem` data and the `callback` parameters. Both `image` and `description` are UI elements that will be set using the **Inspector** view later when we create the Prefab, so we consider that these two UI elements were set in the UI.

`description` has dynamic text that is created when the `Initialize` method is called. The text contains `Description`, `PriceGold`, `PriceResource`, and `CurrencyResource` information from `storeItem`, so we can create a nice text in the UI to show the player what the upgrade is for and the cost, including both gold resources and the other resources that are set up in the ScriptableObject.

The next method we are going to add to the `StoreItemComponent` class is `OnClick`, which will be assigned to a button in the UI when we create the Prefab in a moment. For now, let us define the `OnClick` method by adding the following code block to the class:

```
public void OnClick()
{
  if (LevelManager.Instance.GetResource(
    _storeItem.CurrencyResource) <
      _storeItem.PriceResource)
  {
    _callback(false, $"Low balance of " +
```

```
        $"{_storeItem.CurrencyResource}. " +
          $"Required: {_storeItem.PriceResource}");
      return;
    }

    if (LevelManager.Instance.GetResource(
      ResourceType.Gold) < _storeItem.PriceGold)
    {
      _callback(false, $"Low balance of Gold. " +
        $"Required: {_storeItem.PriceGold}");
      return;
    }

    UpdateInventory();
    UpgradeResource();
    _callback(true, null);
}
```

The OnClick method will make some validations before updating the inventory, which are there to make sure the player has enough resources of each required resource type; if this is not the case, callback will be invoked with a false value, meaning that it was not a successful transaction, and a message letting the player know which required resource does not have enough balance in the inventory will be sent. Here is the sequence of validations and method executions:

- First, we check whether the player has enough resources of the resource type set up in the CurrencyResource property by checking the inventory through the LevelManager instance. If the current amount of the resource is lower than PriceResource, we use callback to indicate that the transaction failed because of the low balance.

- The second validation is like the first validation; however, this time we are checking for the Gold resource type. If the player does not have enough Gold, we also invoke callback with the false value and the error message for the UI.

- If the player has enough resources of both types, we will execute the UpdateInventory method to deduct the resources paid for the upgrade, and then execute the UpgradeResource method to increase the upgrade level of the resource.

- The last thing here is callback with the true value. There is no message to indicate that everything worked well, and the player has a new upgrade in the game.

If the player has enough resources, we call the UpdateInventory method defined in the following code block to remove the amount paid for the upgrade from the inventory on each resource type, and to send a message to update the values on the UI as well:

```
private void UpdateInventory()
{
```

```
LevelManager.Instance.UpdateResource(
  _storeItem.CurrencyResource, -
    _storeItem.PriceResource);
MessageQueueManager.Instance.SendMessage(
  new UpdateResourceMessage {
    Type = _storeItem.CurrencyResource });

LevelManager.Instance.UpdateResource(
  ResourceType.Gold, -_storeItem.PriceGold);
MessageQueueManager.Instance.SendMessage(
  new UpdateResourceMessage { Type = ResourceType.Gold });
}
```

In the UpdateInventory method, we first update the resource amount in the inventory using the UpdateResource method from LevelManager to reduce the resource type in the CurrencyResource property by the value of the PriceResource property, with the minus signal, to indicate that we are removing this value. Then, we send a UpdateResourceMessage message with the Type parameter as the CurrencyResource property, to update the value in the UI as well.

The second part of the UpdateInventory method does the same things, but this time for the Gold resource. We reduce the amount of Gold by the value of the PriceGold property, using the UpdateResource method from LevelManager, and then send a new UpdateResourceMessage message with the Type parameter as Gold.

Once we finish updating the inventory to remove the resources the player spent to upgrade, we send a message to upgrade the resource type and add a new building to the map. We are going to do these things in the UpgradeResource method defined in the code block here:

```
private void UpgradeResource()
{
  if (_storeItem.GetType() != typeof(ResourceStoreItem))
  {
    return;
  }

  ResourceStoreItem item = (ResourceStoreItem)_storeItem;
  MessageQueueManager.Instance.SendMessage(
    new UpgradeResourceMessage { Type = item.Resource });
  LevelManager.Instance.AddBuilding(item.Prefab);
}
```

The UpgradeResource method, which is the last method in the StoreItemComponent class, performs a series of operations to update the upgrade level and add a new building to the map if it has not yet been added:

- At the beginning of the method, we check whether the type of the storeItem object is ResourceStoreItem, and if it is not the same type, we return to ignore the rest of the method. This validation is here to avoid entering this method when the store item is a unit.

- Since we already did the validation, we can safely create a new item variable and cast storeItem to ResourceStoreItem so we can use it here.

- Next, we send a new UpgradeResourceMessage message with the type of resource that the player upgraded. This message updates the UI and resource generation because after the upgrade, it should generate more resources of that type.

- In the last line of the UpgradeResource method we are calling the AddBuilding method, from the LevelManager class, to instantiate a new building model on the map, based on the Prefab property.

That is all we need to do in the StoreItemComponent class to update the resource upgrade level when the player has enough resources to upgrade one resource type. Now, we can move on to the next script, which is responsible for creating the list of items, based on the ResourceStoreData ScriptableObject, and update the UI with error messages from the callback instance that the previous StoreItemComponent class invokes when the player has an insufficient balance to upgrade.

In the existing **Scripts | Store** folder, create a new script and name it StoreComponent. We are going to start the StoreComponent class with the variable declarations, so replace the content of the new script with the following code block:

```
using System.Collections.Generic;
using TMPro;
using UnityEngine;
namespace Dragoncraft
{
  public class StoreComponent : MonoBehaviour
  {
    [SerializeField] private ResourceStoreData _resourceStoreData;
    [SerializeField] private TMP_Text _errorMessage;
    [SerializeField] private ObjectPoolComponent _objectPool;
    [SerializeField] private GameObject _content;
  }
}
```

The `StoreComponent` class is responsible for the `StoreItemComponent` GameObject initialization and displaying an error message, indicating whether the player is not able to purchase the upgrade. This class has the following serializable properties that we are going to set up in the *Creating the Store UI* section:

- `resourceStoreData` is a reference to the ScriptableObject that has the list of store items we configured with all the data we need to use here

- `errorMessage` is a text label that will only be used when the callback from the `StoreItemComponent` class returns an error message, so we can display it in the UI for the player

- `objectPool` is a reference to `ObjectPoolComponent` that will manage the list of `StoreItemComponent` GameObjects for us

- The last property, `content`, is a GameObject where we will add `StoreItemComponents` from `objectPool` as a child

The four properties have the `SerializeField` attribute, which means we will need to manually set them up in the UI once we add this script to a GameObject. We are going to work on that in a moment, but first, let us add a few more methods to this `StoreComponent` class. The next one is the `OnEnable` method, defined in the next code block:

```
private void OnEnable()
{
   if (_resourceStoreData != null)
   {
      for (int i = 0; i < _resourceStoreData.Items.Count; i++)
      {
         InitializeStoreItem(_objectPool.GetObject(),
           _resourceStoreData.Items[i]);
      }
   }
   _errorMessage.text = string.Empty;
}
```

The `OnEnable` method is executed automatically by Unity as soon as the GameObject that has the `StoreComponent` script attached to it is activated in the scene. We are going to do our initialization using this method so, after checking whether the `resourceStoreData` is not null, we are going to loop through each item in the `Items` list of this ScriptableObject and call the `InitializeStoreItem` method using two parameters: one `StoreItemComponent` item from the `objectPool` and one `StoreItem` item from the `resourceStoreData.Items` list.

For each item in the ScriptableObject, we will have a corresponding item on the UI to display the item information on the UI. The last thing we do in this method is initialize `errorMessage` with an empty text, so we make sure we remove the last message that could be there when the player closes and opens the interface again that contains the `StoreComponent` script.

The next method is the counterpart of the `OnEnable` method, which is the `OnDisable` method that is called every time the GameObject is deactivated in the scene. The code for the `OnDisable` method is quite simple, as we can see in the following code block:

```
private void OnDisable()
{
    for (int i = 0; i < _content.transform.childCount; i++)
    {
        _content.transform.GetChild(i).gameObject.SetActive(false);
    }
}
```

The `OnDisable` method will loop through each GameObject that is a child of the content GameObject and will disable all of them, using the `SetActive` method from each `gameObject` property. This will make sure the GameObjects are returned to `objectPool` to be reused.

The next `InitializeStoreItem` method is the one we just used inside the `OnEnable` method, and it is also simple because it only does a few things, as we can see in the following code snippet:

```
private void InitializeStoreItem(GameObject component,
    StoreItem storeItem)
{
    component.transform.SetParent(_content.transform, false);

    StoreItemComponent storeItemComponent =
        component.GetComponent<StoreItemComponent>();
    storeItemComponent.Initialize(storeItem,
        PurchaseCallback);
    storeItemComponent.gameObject.SetActive(true);
}
```

We are adding `component` as a child of the `content` GameObject and then getting `StoreItemComponent` from the component GameObject, so we can use it here. This method is responsible for calling the `Initialize` method on the `StoreItemComponent` component, with `storeItem` and `PurchaseCallback` as the parameters.

As we saw in the previous `StoreItemComponent` class, the `Initialize` method will take care of setting up the store item details, and `PurchaseCallback` will be executed when the player purchases an upgrade, either successfully or not. The last line in this method simply enables the GameObject that has the `StoreItemComponent` script, using the `SetActive` method, so we can see it in the UI.

The `PurchaseCallback` method that we are sending to `StoreItemComponent` is defined in the following code block:

```
private void PurchaseCallback(bool success, string errorMessage)
{
  if (success)
  {
    gameObject.SetActive(false);
  }
  else
  {
    _errorMessage.text = errorMessage;
  }
}
```

This method, which will be executed as a response to the player's action of purchasing an upgrade, checks the `success` parameter to hide this GameObject (the pop-up window with the list of items to purchase) so the player can see the building on the map and then upgrade. Otherwise, it will display the `errorMessage` text in the UI so the player can see the low balance message when the purchase fails.

That was the last script we needed to create before moving on to the interface. Now we can start creating the UI that will use the previous scripts.

Creating the Store UI

We are going to work on the **GameUI** scene now to set up the UI with the new scripts we have. We need to do a couple of things here: the first thing is to add a new button in the top-left corner of the UI to open the new store pop-up window, and the second thing is to create the store pop-up window.

Adding the new button

We will be beginning by adding the new button:

1. Open the **GameUI** scene and, in the **Hierarchy** view, expand the **Canvas** GameObject, and then expand the **Menu** GameObject.

2. Right-click on the existing **MenuButton** GameObject, which is under the **Menu** GameObject, and select the **Duplicate** option. Change the name of the new copy to **ResourceStoreButton**.

3. Left-click on the new **ResourceStoreButton** GameObject and, in the **Inspector** view, under the **Rect Transform** component, change the **Pos X** value to 120.

4. In the **Hierarchy** view, expand the **ResourceStoreButton** GameObject to see the child **Text (TMP)** GameObject. Left-click on the **Text (TMP)** GameObject to select it and, in the **Inspector** view, under the **TextMeshPro – Text (UI)** component, change the text to Resource.

5. Save the changes in the **GameUI** scene using the **File | Save** menu.

Now that we have the button, we can start to work on the pop-up window that this button will open. This window has quite a lot of changes, so we will do these in parts, starting with the background and then assigning the new GameObject to the button we just added, so we can display this pop-up window when the button is clicked.

Adding the pop-up window background

Let us create a new GameObject and add the pop-up window background:

1. In the **Hierarchy** view, expand the **Canvas** GameObject, and then expand the **Menu** GameObject. Right-click on the **Menu** GameObject, select the **UI | Image** option, and name it ResourceStorePopup.

2. Left-click on the new **ResourceStorePopup** GameObject to select it and, in the **Inspector** view, click on the checkbox on the left side of the GameObject name to uncheck it and disable this GameObject.

3. Under the **Rect Transform** component, change **Anchor Preset** to stretch on the top and stretch on the left and, after that, change the values of the **Left**, **Top**, **Right**, and **Bottom** properties to 0.

4. Under the **Image** component, click on the **Color** property and change the **R, G, B**, and **A** values to 0, 0, 0, and 155, respectively.

5. In the **Hierarchy** view, right-click on the **ResourceStorePopup** GameObject, select the **UI | Image** option, and name the new Background GameObject.

6. Left-click on the new **Background** GameObject and, in the **Inspector** view, under the **Rect Transform** component, change the **Width** value to 1024 and the **Height** value to 512.

7. Under the **Image** component, click on the **Source Image** property and, in the new window, search for the **barmid_ready** asset, and double-click to select it.

8. Save the changes in the **GameUI** scene using the **File | Save** menu.

Figure 14.3 shows both the **ResourceStorePopup** and the **Background** GameObjects that we just added to the **GameUI** scene:

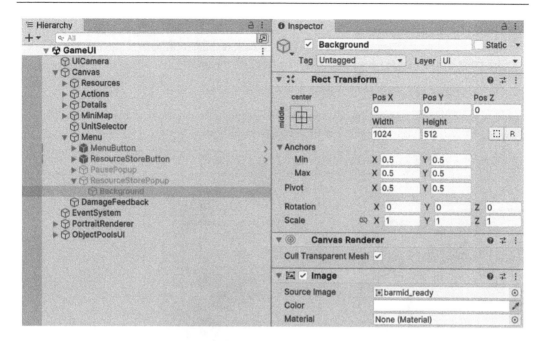

Figure 14.3 – The new Background GameObject

In the next step, we will add a close button to the background.

Adding a close button to the background

Now, let us add a close button to the **Background** GameObject that we just created:

1. In the **Hierarchy** view, right-click on the newly added **Background** GameObject, select the **UI | Button - TextMeshPro** option, and name the new GameObject CloseButton.

2. Left-click on the new **CloseButton** GameObject and, in the **Inspector** view, under the **Rect Transform** component, change **Anchor Preset** to right on the top and top on the left.

3. Then, change **Pivot X** to 1 and **Pivot Y** to 1.

4. After that, change **Pos X** to 0, **Pos Y** to 0, **Width** to 64, and **Height** to 64.

5. Under the **Image** component, click on the **Source Image** property and, in the new window, search for the **button_cancel** asset, and double-click to select it.

6. Under the **Button** component, click on the + button in the **On Click ()** property to add a new item.

7. Drag the **ResourceStorePopup** GameObject from the **Hierarchy** view, and drop it in the field that is below the **Runtime Only** option inside the new item added to **On Click ()**.

8. After adding **ResourceStorePopup** to the **On Click ()** item, select the **GameObject | SetActive (bool)** option in the drop-down menu on the right side of the **Runtime Only** option.

9. In the **Hierarchy** view, expand the new **CloseButton** GameObject, right-click on the child GameObject named **Text (TMP)**, and select the **Delete** menu option to remove it. We do not need text in this button because we have the close button sprite.

10. Save the changes in the **GameUI** scene using the **File | Save** menu.

In *Figure 14.4*, we can see the new **CloseButton** GameObject added under the **Background** GameObject and the **On Click ()** property set up to use the **SetActive** method:

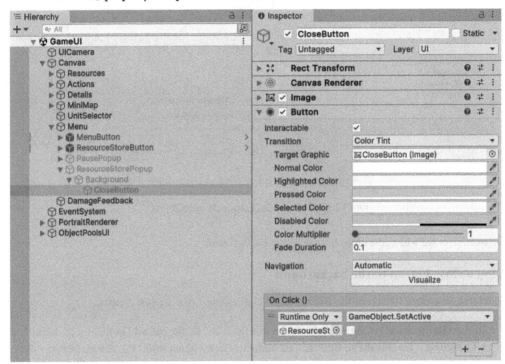

Figure 14.4 – The new CloseButton GameObject

Finally, we will add the last GameObject to the background – the error message. Let us see how in the next section.

Adding the error message to the background

After adding **CloseButton**, we need to add one more GameObject under the **Background** GameObject, which is the **ErrorMessage** text:

1. In the **Hierarchy** view, right-click on the newly added **Background** GameObject, select the **UI | Text - TextMeshPro** option, and name the new GameObject `ErrorMessage`.

2. Left-click on the new **ErrorMessage** GameObject and, in the **Inspector** view, under the **Rect Transform** component, change **Anchor Preset** to `stretch` on the top and `bottom` on the left.

3. Then, change **Left** to 0, **Pos Y** to 32, **Right** to 0, and **Left** to 0.

4. Under the **TextMeshPro – Text (UI)** component, change the text to Error Message.

5. In the **Main Settings** section, change **Font Style** to I (which means italic), **Font Size** to 14, and **Alignment** to Center.

6. Click on the **Vertex Color** property and, in the new **Color** window, change the **Hexadecimal** value to FFA300.

7. Save the changes in the **GameUI** scene using the **File | Save** menu.

Figure 14.5 shows the new **ErrorMessage** GameObject set up with the changes we made in the **TextMeshPro – Text (UI)** settings:

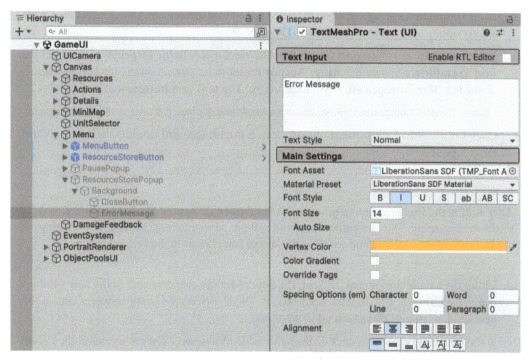

Figure 14.5 – The new ErrorMessage GameObject

That was the last item we needed to add to the background, and now we can move on to the main part of the pop-up window, which is the container that will have the list of store items with which the player can interact.

Adding the container for the items

Let us add the container for the items using these steps:

1. In the **Hierarchy** view, right-click on the newly added **ResourceStorePopup** GameObject, select the **Create Empty** option, and name the new GameObject `Container`.

2. Left-click on the new **Container** GameObject and, in the **Inspector** view, under the **Rect Transform** component, change **Width** to `1024` and **Height** to `512`.

3. Click on the **Add Component** button, search for **Scroll Rect**, and double-click on it to add the component. Under the **Scroll Rect** component, uncheck the checkbox on the right side of the **Vertical** property.

4. In the **Hierarchy** view, right-click on the newly added **Container** GameObject, select the **Create Empty** option, and name the new GameObject `Viewport`.

5. Left-click on the new **Viewport** GameObject to select it and, in the **Inspector** view, under the **Rect Transform** component, change **Anchor Preset** to `stretch` on the top and `stretch` on the left. Then, change **Left** to `50`, **Right** to `50`, **Top** to `0`, and **Bottom** to `0`.

6. Click on the **Add Component** button, search for **Mask**, and double-click on it to add the component.

7. Click on the **Add Component** button, search for **Image**, and double-click on it to add the component.

> **Note**
>
> After adding the **Image** component to the GameObject that has the **Mask** component, you might see a warning. This is a known issue and will not affect the functionality. The warning can be resolved by disabling and enabling the **Image** component in the **Inspector** view.

8. Under the **Image** component, click on the **Source Image** property and, in the new window, search for **UIMask**, and double-click to add it. Then, click on the **Color** property and, in the new window, change **Hexadecimal** to `000000`.

9. In the **Hierarchy** view, right-click on the newly added **Viewport** GameObject, select the **Create Empty** option, and name the new GameObject `Content`.

10. Left-click on the new **Content** GameObject to select it and, in the **Inspector** view, under the **Rect Transform** component, change **Width** to `0` and **Height** to `512`.

11. Click on the **Add Component** button, search for **Horizontal Layout Group**, and double-click on it to add the component. Under the **Horizontal Layout Group** component, change the **Spacing** value to `16`, and the **Child Alignment** value to the `Middle Center` option. Then, in the **Child Force Expand** property, uncheck both the **Width** and **Height** checkboxes.

12. Click on the **Add Component** button, search for **Content Size Fitter**, and double-click on it to add the component. Under the **Content Size Fitter** component, change the **Horizontal Fit** property to `Preferred Size`.

13. In the **Hierarchy** view, left-click on the newly added **Container** GameObject and, in the **Inspector** view, under the **Scroll Rect** component, left-click on the circle button on the right side of the **Content** property and, in the new window, double-click to select the **Content** GameObject. Left-click on the circle button on the right side of the **Viewport** property and, in the new window, double-click to select the **Viewport** GameObject.

14. Save the changes in the **GameUI** scene using the **File | Save** menu.

Figure 14.6 shows the new **Container** GameObject with the **Horizontal Layout Group** component set up to not expand based on **Width**, so we can keep the store items visible inside the container:

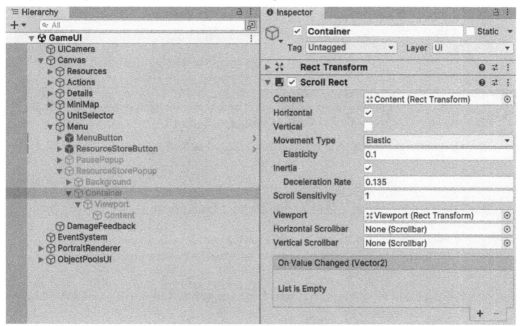

Figure 14.6 – The new Container GameObject

Now that we have the **Container** GameObject set up, we can add another GameObject for the store items.

Adding store items

Since each store item will have its own details, we are going to create a Prefab to make it easier to maintain the store items. The next steps are rather long, but let us add the store item GameObject first, and then create the Prefab:

1. In the **Hierarchy** view, right-click on the newly added **Container** GameObject, select the **UI | Image** option, and name the new GameObject StoreItem:

   I.   Left-click on the new **StoreItem** GameObject to select it and, in the **Inspector** view, under the **Rect Transform** component, change the **Anchor Preset** to left on the top and top on the left.

   II. After that, change the values of the properties, setting **Width** to 258 and **Height** to 385.

   III. Under the **Image** component, click on the **Color** property and change the **R, G, B**, and **A** values to 0, 0, 0, and 255, respectively.

2. In the **Hierarchy** view, right-click on the newly added **StoreItem** GameObject, select the **UI | Image** option, and name the new GameObject Image:

   I.   Left-click on the new **Image** GameObject and, in the **Inspector** view, under the **Rect Transform** component, change **Anchor Preset** to center on the top and top on the left.

   II. Then, change **Pivot Y** to 1 and, after that, change **Pos X** to 0, **Pos Y** to 0, **Width** to 256, and **Height** to 256.

3. In the **Hierarchy** view, right-click on the newly added **StoreItem** GameObject, select the **UI | Text - TextMeshPro** option, and name the new GameObject Description:

   I.   Left-click on the new **Description** GameObject and, in the **Inspector** view, under the **Rect Transform** component, change **Anchor Preset** to center on the top and bottom on the left.

   II. Then, change **Pivot Y** to 0 and, after that, change **Pos X** to 0, **Pos Y** to 64, **Width** to 256, and **Height** to 64.

   III. Under the **TextMeshPro – Text (UI)** component, change the text to Description.

   IV. In the **Main Settings** section, change **Font Size** to 14, and change **Alignment** to Center and Middle.

4. In the **Hierarchy** view, right-click on the newly added **StoreItem** GameObject, select the **UI |**
    Button - TextMeshPro option and name the new GameObject `PurchaseButton`:

    I.   Left-click on the new **PurchaseButton** GameObject and, in the **Inspector** view, under
          the **Rect Transform** component, change **Anchor Preset** to `center` on the top and
          `bottom` on the left.

    II.   Then, change **Pivot Y** to `0` and, after that, change **Pos X** to `0`, **Pos Y** to `0`, **Width** to `128`,
          and **Height** to `64`.

    III. Under the **Image** component, click on the **Source Image** property and, in the new window,
          search for the **button_ready_off** asset, and double-click to select it.

5. In the **Hierarchy** view, expand the new **PurchaseButton** GameObject to see the child **Text**
    (TMP) GameObject.

6. Left-click on the **Text (TMP)** GameObject to select it and, in the **Inspector** view, under the
    TextMeshPro – Text (UI) component, change the text to `Purchase`.

7. Save the changes in the **GameUI** scene using the **File | Save** menu.

After these lengthy steps, we finally have our **StoreItem** GameObject. Now we just need to set it up
and make a Prefab from it.

Adding and configuring the scripts

Let us see how we can add and configure the scripts:

1. In the **Hierarchy** view, left-click on the newly added **StoreItem** GameObject to select it.

2. In the **Inspector** view, click on the **Add Component** button, search for the `StoreItemComponent`
    script, and double-click on it to add it to the GameObject.

3. Under **Store Item Component**, there are two properties, **Image** and **Description**. Drag the
    Image GameObject that is a child of the **StoreItem** GameObject from the **Hierarchy** view, and
    drop it into the **Image** property, in the **Inspector** view.

4. Drag the **Description** GameObject that is a child of the **StoreItem** GameObject from the
    Hierarchy view, and drop it into the **Description** property, in the **Inspector** view.

5. In the **Hierarchy** view, left-click on the **PurchaseButton** GameObject, which is a child of the
    StoreItem GameObject, to select it.

6. In the **Inspector** view, under the **Button** component, click on the + button in the **On Click ()**
    property to add a new item.

7. Drag the **StoreItem** GameObject from the **Hierarchy** view and drop it in the field that is below
    the **Runtime Only** option, which has a value of **None (Object)**.

8. After adding **StoreItem** to the **On Click ()** item, select the **StoreItemComponent | OnClick ()** option in the drop-down menu on the right side of the **Runtime Only** option.

9. In the **Project** view, create a new folder called `Store` inside the **Prefabs** folder.

10. Drag the **StoreItem** GameObject from the **Hierarchy** view and drop it inside the new **Prefabs | Store** folder in the **Project** view, to create a new **StoreItem** Prefab.

11. In the **Hierarchy** view, delete the **StoreItem** GameObject – we will not need it anymore since we now have the Prefab.

12. Save the changes in the **GameUI** scene using the **File | Save** menu.

We now have a new **StoreItem** Prefab and we have added three of them to our UI, one for each resource type.

Figure 14.7 shows the **Container** GameObject with the three **StoreItem** Prefabs, and on the right side, we can see the **PurchaseButton** GameObject with the **Button** component configured with the **OnClick** action:

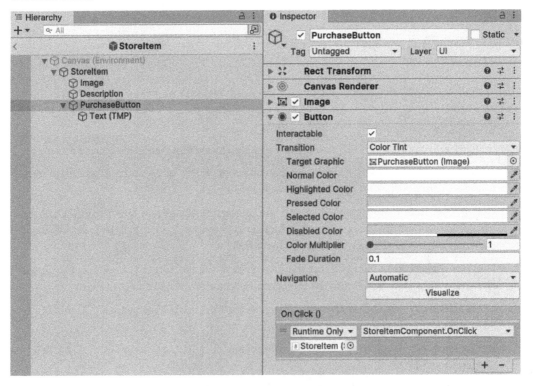

Figure 14.7 – The StoreItem Prefab

Before proceeding to the last part of the **ResourceStorePopup** setup, let us create a new Object Pool that will be required when adding the **StoreComponent** script:

1. In the **Hierarchy** view, right-click on the **ObjectPoolsUI** GameObject, select **Create Empty**, and name it **StoreItemObjectPool**.

2. Left-click on the newly created **StoreItemObjectPool** GameObject and, in the **Inspector** view, click on the **Add Component** button, search for **ObjectPoolComponent**, and double-click to add it.

3. Under the **Object Pool Component**, click on the circle button on the right side of the **Prefab** property and, in the new window, click on the **Assets** tab, search for **StoreItem**, and double-click to add it.

4. Change the **Pool Size** property to 5 and select the checkbox on the right side of the **Allow Creation** property to enable it.

5. Save the changes in the **GameUI** scene using the **File | Save** menu.

The last thing we need now is to add the **StoreComponent** script to our **ResourceStorePopup** GameObject and add the **StoreItem** Prefabs to this script, so we can fill them with the data from the ScriptableObject:

1. In the **Hierarchy** view, left-click on the **ResourceStorePopup** GameObject to select it.

2. In the **Inspector** view, click on the **Add Component** button, search for the `StoreComponent` script, and double-click on it to add it to the GameObject.

3. Under the **Store Component**, left-click on the circle button on the right side of the **Resource Store Data** property and, in the new window, double-click to select the **ResourceStore** asset.

4. Click on the circle button on the right side of the **Error Message** property and, in the new window, click on the **Scene** tab, and double-click to select the **ErrorMessage** GameObject.

5. Click on the circle button on the right side of the **Object Pool** property and, in the new window, click on the **Scene** tab, and double-click to select the **StoreItemObjectPool** GameObject.

6. Click on the circle button on the right side of the **Content** property and, in the new window, click on the **Scene** tab, and double-click to select the **Content** GameObject.

7. In the **Hierarchy** view, left-click on the **ResourceStoreButton** GameObject to select it.

8. In the **Inspector** view, under the **Button** component, click on the circle button inside the **On Click ()** section, located under **Runtime Only**.

9.   In the new window, search for **ResourceStorePopup** and double-click on it to add the GameObject in **On Click ()**.

10. Then, on the right side of **Runtime Only**, select the **GameObject | SetActive** option.

11. Save the changes in the **GameUI** scene using the **File | Save** menu.

So far, we have added the `StoreComponent` script to **ResourceStorePopup**, configured it with the **StoreItem** GameObjects in the **Container** GameObject, and set up the **ResourceStoreButton** action to activate the **ResourceStorePopup** GameObject and make it visible to the player.

Figure 14.8 shows the **ResourceStorePopup** GameObject configured:

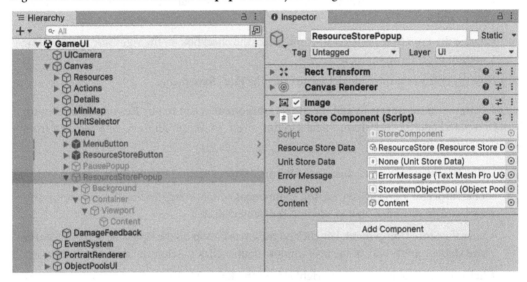

Figure 14.8 – The ResourceStorePopup GameObject configured

We have finally finished creating our Resource Store pop-up window, with all the data from the ScriptableObject to populate the details dynamically.

Now we can open the **Playground** scene and play the game on the Editor. When clicking on the **Resource** button in the upper-left corner of the screen, we will be able to see the new pop-up window with the three resources to upgrade:

Figure 14.9 – The Resource Store pop-up window in a game

This pop-up window is fully functional, and we can click on the **Upgrade** button that will display an error in orange at the bottom of the pop-up window if the balance is low for the required resource, as shown in *Figure 14.9*. Remember to use the **Dragoncraft | Debug | Resources** menu to add resources so that you can test each upgrade.

We will also see that, in the first upgrade, the building is added to the map and can be dragged and dropped where we want it. Play around with the Resource Store pop-up window, and keep in mind that you can add more debug options for the resources or change their values to give more resources to make it easier to test.

Now that we have finished the Resource Store, we can move on to the Unit Store, which will reuse everything we have done so far, besides expanding a few scripts to support the unit upgrades.

Training units

We have already done a lot of work to implement our Resource Store. As we create the Unit Store, we are going to reuse everything we have built, so it will be a bit easier and faster to have the Unit Store pop-up window configured and working.

Let us first start with the new scripts that are specific to the Unit Store configuration and then move on to the changes on the existing scripts and the UI. The upgrades on the Unit Store will be how we train the units to become stronger on the battlefield.

Configuring the Unit Store

The first script we are going to create is the message type that we will send when the unit has the upgrade level increased by the player. In the **Project** view, open the **Scripts | MessageQueue | Messages | UI** folder, and add a new script named `UpgradeUnitMessage`. Replace the content with the following code block:

```
namespace Dragoncraft
{
  public class UpgradeUnitMessage : IMessage
  {
    public UnitType Type;
  }
}
```

The `UpgradeUnitMessage` class is a very simple `IMessage` class that has only one property, `UnitType`. This message will be sent by the Unit Store when the player purchases the upgrade for each unit type.

Now that we have the upgrade message created, we can move on to the scripts that will configure the Unit Store. When we created the Resource Store scripts, we added the `StoreItem` class with the required properties for the store item configuration and created the `ResourceStoreItem` class to inherit from it and add `ResourceType`. We are going to do the same thing now for the Unit Store, so add a new script named **UnitStoreItem** inside the **Scripts | Store** folder, and replace the content with the following:

```
using System;
namespace Dragoncraft
{
  [Serializable]
  public class UnitStoreItem : StoreItem
  {
    public UnitType Unit;
    public bool IsUpgrade;
  }
}
```

`UnitStoreItem` is a `Serializable` class that inherits from the `StoreItem` class and adds the `UnitType` property, so we can select which unit we want when configuring the Unit Store. The `IsUpgrade` property will be used to identify whether a given item is an upgrade or a regular purchase.

In the Resource Store, we also created another script that was the ScriptableObject for us to configure each `ResourceStoreItem` and use it in the UI to set up the pop-up window. We will also create a similar ScriptableObject for the Unit Store, so add a new script named **UnitStoreData** inside the **Scripts | Store** folder, and add the following content to it:

```
using System.Collections.Generic;
using UnityEngine;
namespace Dragoncraft
{
    [CreateAssetMenu(menuName = "Dragoncraft/New Unit Store")]
    public class UnitStoreData : ScriptableObject
    {
        public List<UnitStoreItem> Items =
            new List<UnitStoreItem>();
    }
}
```

The `UnitStoreData` class has the `CreateAssetMenu` attribute to create a new menu item for us on Unity, so it is easy to create new assets from this ScriptableObject. It has only one property, the `UnitStoreItem` list that we are going to configure for all the unit types to which we want to add upgrade options in the game.

We can now create a new **UnitStoreData** ScriptableObject with upgrades for each unit type:

1. In the **Project** view, create a new `Store` folder inside the **Data** folder.

2. Then, right-click on the new **Data | Store** folder, select the **Create | Dragoncraft | New Unit Store** menu option, and save the new asset as `UnitStore`.

3. Left-click on the new **UnitStore** asset to select it and, in the **Inspector** view, change the value of **Items** to 4 – Unity will create four new items in the list for us.

4. In the first item on the list, add the text `Spawns a Warrior` in the **Description** property:

    I.   Set the value of the **Price Gold** property to `100`.

    II. Change the **Unit** property to `Warrior`.

    III. Click on the circle button on the right side of the **Image** property and, in the new window, search for the **warrior_silhouette_man** sprite, and double-click on it to add it to the property.

5. In the second item of the list, add the text `Spawns a Mage` in the **Description** property:

    I. Set the values of the **Price Gold** property to `500`.

    II. Change the **Unit** property to `Mage`.

    III. Click on the circle button on the right side of the **Image** property and, in the new window, search for the **warrior_silhouette_woman** sprite, and double-click on it to add it to the property.

6. In the third item of the list, add the text `Upgrades the Warrior Level` to the **Description** property:

    I. Set the values of the **Price Gold** and **Price Resources** properties to `5000`.

    II. Change the **Currency Resource** property to `Food`.

    III. Change the **Unit** property to `Warrior`.

    IV. Click on the checkbox on the right side of the **Is Upgrade** property to enable it.

    V. Click on the circle button on the right side of the **Image** property and, in the new window, search for the **warrior_silhouette_man** sprite, and double-click on it to add it to the property.

7. In the fourth item of the list, add the text `Upgrades the Mage Level` in the **Description** property:

    I. Set the values of the **Price Gold** and **Price Resources** properties to `10000`.

    II. Change the **Currency Resource** property to `Food`.

    III. Change the **Unit** property to `Mage`.

    IV. Click on the checkbox on the right side of the **Is Upgrade** property to enable it.

    V. Click on the circle button on the right side of the **Image** property and, in the new window, search for the **warrior_silhouette_woman** sprite, and double-click on it to add it to the property.

8. Save the changes using the **File** | **Save Project** menu option.

Once we have finished setting up the **UnitStore** ScriptableObject, we will have all we need to populate the data in the pop-up window for the player to choose what to upgrade, based on the cost of each item.

Figure 14.10 shows the **UnitStore** ScriptableObject after adding the configuration for the Warrior and Mage unit types:

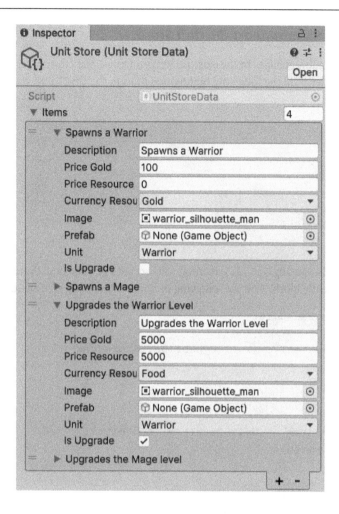

Figure 14.10 – The Unit Store configured with Warrior and Mage upgrades

Now that we have added the base scripts for the Unit Store configuration and the asset with the ScriptableObject data, we can move on to the Unit Store components and UI.

Creating the Unit Store UI

Now we are going to reuse everything we have already created for the Resource Store and expand it to support the Unit Store as well, with some minimal changes.

Updating the scripts to support the Unit Store

Let us start by adding a few changes to the existing StoreItemComponent script, located inside the **Scripts | Store** folder. Add the following highlighted lines to the OnClick method:

```
public void OnClick()
{
    ...
    UpdateInventory();
    UpgradeResource();
    UpgradeUnit();
    SpawnUnit();
    _callback(true, null);
}
```

The lines we added to the existing OnClick method will execute a new method, UpgradeUnit, which is defined in the next code block. Add the following method to the StoreItemComponent class:

```
private void UpgradeUnit()
{
    if (_storeItem.GetType() != typeof(UnitStoreItem))
    {
        return;
    }

    UnitStoreItem item = (UnitStoreItem)_storeItem;
    if (!item.IsUpgrade)
    {
        return;
    }
    MessageQueueManager.Instance.SendMessage(
        new UpgradeUnitMessage { Type = item.Unit });
}
```

The UpgradeUnit method is very similar to UpgradeResource, but with a few differences, since we are updating the unit upgrade levels:

- We first check whether the type of storeItem is UnitStoreItem, and if it is not, we just return from this method since there is nothing to do here for the other types.

- We create a new variable item by casting storeItem to the UnitStoreItem type. Since we already validated that storeItem has the UnitStoreItem type, we are safe to cast the variable.

- Next, we check whether the IsUpgrade property is not true, and return if we are not supposed to upgrade the given item.

- At the end of the method, we send a message using the new UpgradeUnitMessage message type to broadcast that the given Unit was upgraded.

After the UpgradeUnit method, let us add the following new SpawnUnit method:

```
private void SpawnUnit()
{
  if (_storeItem.GetType() != typeof(UnitStoreItem))
  {
    return;
  }

  UnitStoreItem item = (UnitStoreItem)_storeItem;
  if (item.IsUpgrade)
  {
    return;
  }

  if (item.Unit == UnitType.Warrior)
  {
    MessageQueueManager.Instance.SendMessage(new
      BasicWarriorSpawnMessage());
  }
  else if (item.Unit == UnitType.Mage)
  {
    MessageQueueManager.Instance.SendMessage(
      new BasicMageSpawnMessage());
  }
}
```

The first few validations in the SpawnUnit method are like the validations on the UpgradeUnit method: we ignore storeItem if it is not UnitStoreItem, and this time we return whether the configuration is to upgrade the given item, since here, we are spawning and not upgrading.

The last part of the method is a validation to check whether UnitType is Warrior or Mage, and to send the corresponding message to spawn a new unit, using BasicWarriorSpawnMessage or BasicMageSpawnMessage, respectively.

After adding the UpgradeUnit and SpawnUnit methods, our StoreItemComponent is also ready to support unit upgrades.

The only script left that we need to update, so that it also supports unit upgrades, is the `StoreComponent` class. Open the existing `StoreComponent` script, located inside the **Scripts | Store** folder, and add the following highlighted line at the beginning of the class:

```
public class StoreComponent : MonoBehaviour
{
    [SerializeField] private ResourceStoreData_resourceStoreData;
    [SerializeField] private UnitStoreData _unitStoreData;
    [SerializeField] private TMP_Text _errorMessage;
    [SerializeField] private List<StoreItemComponent>
        _storeItems = new List<StoreItemComponent>();

    ...
}
```

After adding the new `unitStoreData` property to the `StoreComponent` class, we can use it in the existing `OnEnable` method to read the data from the ScriptableObject and initialize the `StoreItemComponent` object using `unitStoreData`. We already added similar code to the `OnEnable` method for `resourceStoreData`, and now we are adding support for the unit data as well. Add the highlighted code inside the existing `OnEnable` method:

```
private void OnEnable()
{
    ...

    if (_unitStoreData != null)
    {
        for (int i = 0; i < _unitStoreData.Items.Count; i++)
        {
            InitializeStoreItem(_objectPool.GetObject(),
                _unitStoreData.Items[i]);
        }
    }
    _errorMessage.text = string.Empty;
}
```

The `OnEnable` method is executed automatically by Unity as soon as the GameObject that has the `StoreComponent` script attached is activated in the scene. We are going to carry out our initialization using this method so, after checking whether `unitStoreData` is not null, we are going to loop through each item in the Items list of this ScriptableObject and call the `InitializeStoreItem` method using two parameters: one `StoreItemComponent` item from `objectPool` and one `StoreItem` item from the `unitStoreData.Items` list.

Now that we have finished adding the Unit Store support to our scripts, we need to do the same for the UI elements.

Updating the UI elements to support the Unit Store

While updating our UI elements to support the Unit Store, the good news is that we can also reuse what we did for the Resource Store. So, with a few changes, we will have our Unit Store pop-up window ready.

Adding a new button

Let us first add a new button in the UI that we are going to use to open the Unit Store pop-up window:

1. Open the **GameUI** scene and, in the **Hierarchy** view, expand the **Canvas** GameObject, and then expand the **Menu** GameObject.

2. Right-click on the existing **ResourceStoreButton** GameObject that is under the **Menu** GameObject and select the **Duplicate** option. Change the name of the new copy to UnitStoreButton.

3. Left-click on the new **UnitStoreButton** GameObject and, in the **Inspector** view, under the **Rect Transform** component, change the **Pos X** value to 240.

4. In the **Hierarchy** view, expand the **UnitStoreButton** GameObject to see the child **Text (TMP)** GameObject.

5. Left-click on the **Text (TMP)** GameObject to select it and, in the **Inspector** view, under the **TextMeshPro – Text (UI)** component, change the text to Unit.

6. Save the changes in the **GameUI** scene using the **File | Save** menu.

Now that we have our button, we need to add the pop-up window as well.

Adding a pop-up window

This time, adding a pop-up window will be quite easy since we can duplicate **ResourceStorePopup** and then add a reference for the **UnitStoreData** asset that has the ScriptableObject with the configuration for the Unit Store:

1. In the **Hierarchy** view, right-click on the existing **ResourceStorePopup** GameObject that is under the **Menu** GameObject, and select the **Duplicate** option. Change the name of the new copy to UnitStorePopup.

2. In the **Hierarchy** view, left-click on the **UnitStorePopup** GameObject to select it.

3. In the **Inspector** view, under **Store Component**, left-click on the circle button on the right side of the **Resource Store Data** property and, in the new window, double-click to select the **None** item.

4. Left-click on the circle button on the right side of the **Unit Store Data** property and, in the new window, double-click to select the **UnitStore** asset.

5. Under the **Store Component** options, left-click on the circle button on the right side of the **Resource Store Data** property and, in the new window, double-click to select the **ResourceStore** asset.

6. In the **Hierarchy** view, left-click on the **UnitStoreButton** GameObject to select it.

7. In the **Inspector** view, under the **Button** option, click on the circle button inside the **On Click ()** section, located under **Runtime Only**.

8. In the new window, search for **UnitStorePopup** and double-click on it to add the GameObject to **On Click ()**.

9. Then, on the right side of **Runtime Only**, select **GameObject | SetActive**.

10. Save the changes in the **GameUI** scene using the **File | Save** menu.

Figure 14.11 shows our current GameObjects for the menu, including the two new buttons and the two new popups that we added in this chapter. Note that, in the **UnitStorePopup** configuration, we removed the references for **Resource Store Data** and added a reference for **Unit Store Data** – we cannot have both in the same popup:

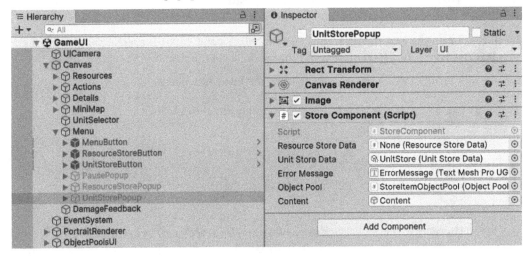

Figure 14.11 – The UnitStorePopup GameObject

We have just finished implementing the Unit Store and, as we can see, we reused all the scripts and GameObjects from the Resource Store, as well as adding a few new scripts that handle the unit upgrade data.

Adding and removing a listener

Before we can test the Unit Store on the Editor, we need to modify one more script to apply the upgrade level to the units. Open the `UnitComponent` script, located in the **Scripts | Unit** folder, and add the following highlighted line in the existing `OnEnable` method:

```
private void OnEnable()
{
   MessageQueueManager.Instance.
     AddListener<ActionCommandMessage>(
       OnActionCommandReceived);
   MessageQueueManager.Instance.
     AddListener<UpgradeUnitMessage>(OnUnitUpgradeReceived);
}
```

We are adding a listener to the new `UpgradeUnitMessage` message type when the GameObject is enabled, and assigning the listener to a new `OnUnitUpgradeReceived` method, which we are going to add to the `UnitComponent` class in a moment.

Let us also add the following highlighted line in the existing `OnDisable` method so we can remove the listener when the GameObject is disabled:

```
private void OnDisable()
{
   MessageQueueManager.Instance.
     RemoveListener<ActionCommandMessage>(
       OnActionCommandReceived);
   MessageQueueManager.Instance.
     RemoveListener<UpgradeUnitMessage>(
       OnUnitUpgradeReceived);
}
```

Now we are adding and removing the listener for the `UpgradeUnitMessage` message type and we need to add the new `OnUnitUpgradeReceived` method, which is called when a message is received:

```
private void OnUnitUpgradeReceived(UpgradeUnitMessage message)
{
   if (Type == message.Type)
   {
     Level++;
   }
}
```

When the `OnUnitUpgradeReceived` method is executed by the listener, we increase the value of the `Level` variable, if the `Type` from the message has the same value as `Type` in the `UnitComponent` class, so we can use it in the attack and damage calculations. That is the last change we need to make in the `UnitComponent` class, and now we are ready to test it.

Adding the upgrade message

Before testing, let us also update the debug script to include the unit upgrade message, so we can use it to speed up the unit upgrade test when playing on the Editor. Open the **UnitDebugger** script, located at **Scripts | Editor | Debug**, and add the following new `UpgradeWarrior` and `UpgradeMage` methods:

```
[MenuItem("Dragoncraft/Debug/Unit/Upgrade Warrior")]
private static void UpgradeWarrior()
{
    MessageQueueManager.Instance.SendMessage(
        new UpgradeUnitMessage { Type = UnitType.Warrior });
}

[MenuItem("Dragoncraft/Debug/Unit/Upgrade Mage")]
private static void UpgradeMage()
{
    MessageQueueManager.Instance.SendMessage(
        new UpgradeUnitMessage { Type = UnitType.Mage });
}
```

The two new `UpgradeWarrior` and `UpgradeMage` methods extend the debug menu by adding a couple of new options to upgrade the Warrior and Mage units, respectively, by sending a new `UpgradeUnitMessage` message type with `UnitType` as a parameter.

Testing the Unit Store

We can now run the **Playground** scene on the Editor and click on the **Unit** button, located in the top-left corner of the screen, to see the Unit Store pop-up window, with the two options to upgrade the **Warrior** or **Mage** levels, as we can see in the following figure:

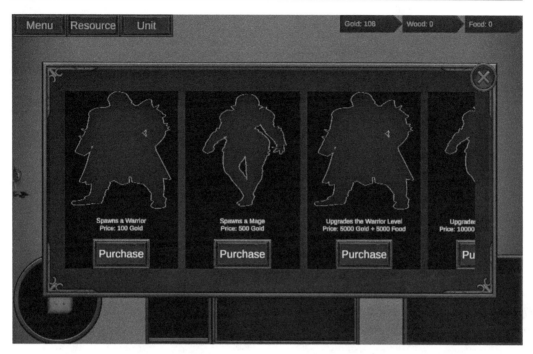

Figure 14.12 – The Unit Store pop-up window with the Warrior and Mage upgrade options

We can use the **Dragoncraft | Debug | Resources** debug menu to get **Gold** and **Food** resources to purchase the upgrade levels or use the **Dragoncraft | Debug | Unit** debug menu to upgrade both the **Warrior** and **Mage** directly. We can upgrade the unit levels and spawn new units to fight enemies and compare their increasing strength each time we upgrade their levels.

The Unit Store was much easier and faster to implement by reusing everything we created for the Resources Store; now we have both ready and working in our game.

The last thing we are going to do in this chapter is add one more unit, which is also a building, the **Defense Tower**.

Defending with towers

The Defense Tower is a powerful unit that is placed as a building but can attack enemies that are within its range. This is great when defending the base while sending units to explore the map and is a must-have feature for any RTS game.

We are going to reuse many different scripts and GameObjects to add the Defense Tower as a new unit. This section also works as a guide in case you want to add different units to the game later.

Adding the Tower model

We need to import a new asset package from the Unity Asset Store to include a Tower 3D model and Prefab in our project. Go through the following instructions to add the **Canon Tower** package to your assets and view it in **Package Manager** so you can import it into the project:

1. Open the link `https://assetstore.unity.com/lists/creating-a-rts-game-5773122416647` (you might need to log in to your Unity Asset Store account created in *Chapter 2*).

2. In the list of assets, find the one called **Canon Tower** and click on **Add to My Assets**.

3. Go back to the Unity Editor and click on **Window | Package Manager**.

4. Left-click on **Packages** and select the **My Assets** option to see all assets in your account.

5. Click on the **Download** button and then, once it has finished, click on the **Import** button.

6. Once the package is imported into your project, move the **CanonTower** folder into the **ThirdParty** folder to keep all assets organized.

Figure 14.13 shows the **Canon Tower** package after adding it to your assets, and the **Import** button to include the package in the project:

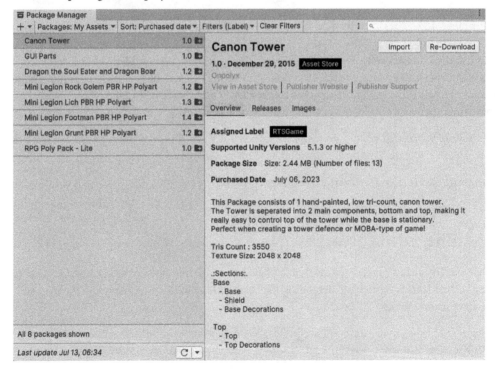

Figure 14.13 – The Canon Tower asset package

Now that we have added the new package to the project and have the Tower model and Prefab available, we also need to include a new layer in the project that we will use to set up the physics and the collisions.

Adding the new Tower layer

The new `Tower` layer will be used in the Tower Prefab to set up the physics and collision interaction between the Tower, enemies, and units.

Open the **Project Settings** window using the **Edit | Project Settings** menu option…, select the **Tags and Layers** menu on the left panel, and add **Tower** as the value for the **User Layer 12**. We have one new layer now, and we also need to tweak the physics settings to make sure that the objects with the **Tower** layer will report collisions only with the layers that we want.

In the same window, select the **Physics** menu option on the left panel and, under **Layer Collision Matrix**, uncheck the following layers on the first column: **Portrait**, **Unit**, **Resource**, and **Tower**. As we can see in the following figure, we have disabled the collision between the new `Tower` layer and the layers that we do not want to detect the collision:

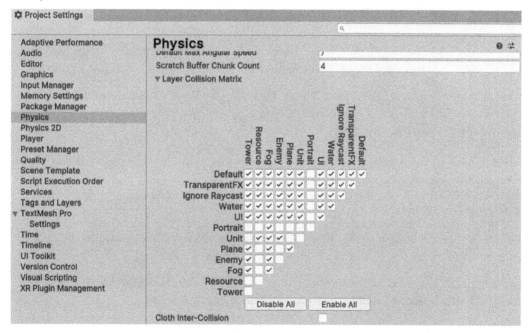

Figure 14.14 – The Layer Collision Matrix including the new layer, Tower

Now that we have added the new `Tower` layer, we also need to update the Occlusion Culling of the in-game cameras, otherwise, any object that has the new layer will not be rendered on either the map or the mini map. We also need to include the new `Tower` layer in the `DragAndDropComponent` script so the player can move the Tower GameObject on the map after building it.

Let us update our camera Prefabs to include the new **Tower** layer:

1. In the **Project** view, in the **Level** folder inside **Prefabs**, double-click to open the Prefab named `Main Camera`.

2. In the **Inspector** view, under the **Camera** component, click on the dropdown located on the right side of the **Occlusion Culling** property, and select the **Tower** option to also include this layer.

3. In the **Project** view, in the **UI** folder inside **Prefabs**, double-click to open the Prefab named `MiniMapCamera`.

4. In the **Inspector** view, under the **Camera** component, click on the dropdown located on the right side of the **Occlusion Culling** property and select the **Tower** option to also include this layer.

5. In the **Project** view, in the **Level** folder inside **Prefabs**, double-click to open the Prefab named `LevelManager`.

6. In the **Inspector** view, under the **Drag And Drop Component**, click on the dropdown located on the right side of the **Layer Mask** property and select the **Tower** option to also include this layer. We should now have both **Resource** and **Tower** layers selected here.

7. Use the **File | Save** menu option to save the changes.

After editing the **Main Camera**, **MiniMapCamera**, and **LevelManager** Prefabs, our game will be ready to render any object that has the new `Tower` layer.

We can now move on to a new script that will be used to control the Tower collision detection and attack.

Adding the Tower as a unit

Some of the changes we are going to make now are more specific to the Tower as a unit, whereas some are not required when adding a new unit to the game.

Modifying the scripts

One example is the following change that we are going to make by adding a new property to the `FireballSpawnMessage` class, that is specific to the Tower as a unit.

Open the existing `FireballSpawnMessage` script, located inside the **Scripts | MessageQueue | Messages | Battle** folder, and add the following highlighted line to the class:

```
public class FireballSpawnMessage : IMessage
{
    ...
    public bool IsTower;
}
```

The new `IsTower` property will allow us to use the `FireballSpawnMessage` message type to let the Tower attack enemies from a distance, and we are going to use it in the next script.

Now, we are going to create a new script that will take care of the Tower collision detection and attack. Inside the **Scripts | Battle** folder, add a new script named `TowerComponent`, and replace the content with the following code block:

```
using System.Collections.Generic;
using UnityEngine;
namespace Dragoncraft
{
  public class TowerComponent : MonoBehaviour
  {
    [SerializeField] private float _attack = 1;
    [SerializeField] private float _cooldown = 3;

    private List<GameObject> _targets = new List<GameObject>();
    private float _cooldownCounter;
  }
}
```

The `TowerComponent` class has a few properties that we are going to use as a setup to control the Tower attack cooldown, as well as to keep track of all the targets in range:

- The `attack` property is the default damage to the enemy in range, caused by the projectile shot from the Tower

- The `cooldown` property is the attack speed and will determine how many seconds we wait between attacks

- The `targets` variable is a list of all the GameObjects from the enemies that are in range and can be attacked

- The `attackCooldown` property is how we are going to count down to the next time we attack a target from the list of targets

The projectile does not consider the enemy's defense value, so it will always cause damage. For that reason, we are not applying an upgrade level to the Tower, otherwise, it would be too powerful and would unbalance the game.

Even though we have a list of `targets`, the Tower can only attack one target at once. This list is used to keep track of all enemies that entered the range, always attacking the last one to enter the range. If there are more than two enemies in the range, and one leaves or dies, the Tower will attack the next one from the list.

All the action in the TowerComponent class will happen in the Update method, which is in the following code snippet:

```
private void Update()
{
    if (_targets.Count == 0)
    {
        return;
    }

    _cooldownCounter -= Time.deltaTime;
    if (_cooldownCounter < 0)
    {
        transform.LookAt(
          _targets[_targets.Count - 1].transform.position);

        MessageQueueManager.Instance.SendMessage(
          new FireballSpawnMessage
        {
            Position = transform.position,
            Rotation = transform.rotation,
            Damage = _attack,
            IsTower = true
        });
        _cooldownCounter = _cooldown;
    }
}
```

Here in the Update method, we are performing a series of actions that will check whether the Tower can attack, and then send a message to attack with a fireball.

Let us see in detail what each part of the Update method is doing:

- First, we check whether there is any target in the list of targets, otherwise, we do not need to continue, and we can just return to ignore the rest of the execution.

- We subtract the cooldownCounter variable from the value of the current deltaTime property and, if the value of cooldownCounter is less than 0, we can proceed and attack one target.

- Using the LookAt method, we rotate the Tower cannon to face the enemy that is the last in the list of targets. We find the last target of the list using Count - 1 and then get the position of this target for the LookAt method.

- Now that we have the Tower cannon facing the enemy, we can send a new `FireballSpawnMessage` message type using the current `position` and `rotation` of the Tower, as well as the `attack` value. We also set the new `IsTower` property to `true`.

- The last thing to do is reset the `cooldownCounter` in order to have the `cooldown` value, so it can count down in the next game loop again until it is time to attack one more time.

The `Update` method is the core of the `TowerComponent` class, but there are a couple more methods that we need to include in the `TowerComponent` script for it to work properly. Let us add the following two methods, `OnCollisionEnter` and `OnCollisionExit` to the `TowerComponent` class:

```
private void OnCollisionEnter(Collision collision)
{
    _targets.Add(collision.gameObject);
}

private void OnCollisionExit(Collision collision)
{
    _targets.Remove(collision.gameObject);
}
```

These two methods, `OnCollisionEnter` and `OnCollisionExit`, are responsible for managing the list of targets, so every time an enemy collides with the Tower range, we add the enemy `gameObject` to the list. When an enemy moves out of the Tower range, it also moves out of the collision, and we remove the `gameObject` instance from the list.

We do not need to check whether the `gameObject` instance is an enemy because we have already configured the Layer Collision Matrix to optimize the collision detection and ignore the layers with which we do not want to collide.

That is all we need for the Tower to attack the enemies. Now we need to update a few more scripts, so we can reuse all the collision and battle systems to cause damage to the enemies.

Open the existing `ProjectileComponent` script, located inside the **Scripts | Battle** folder, and add the following highlighted line to the class:

```
public class ProjectileComponent : MonoBehaviour
{
    ...
    public bool IsTower;
}
```

The new `IsTower` public variable, added to `ProjectileComponent`, is set when we use the `Setup` method, so we also need to include it as a new parameter. Add the following two highlighted lines to the existing `Setup` method:

```
public void Setup(Vector3 position, Quaternion rotation,
    float damage, bool isTower)
{
    ...
    IsTower = isTower;
}
```

This is a simple change to add a new parameter to the `Setup` method, and then assign the value of the `isTower` parameter to the new `IsTower` public variable.

Now, we need to update the script that is using the `Setup` method as well as adding the `IsTower` parameter. Open the existing `FireballSpawner` script, located inside the **Scripts | Spawner** folder, and make the following highlighted change:

```
private void OnFireballSpawned(FireballSpawnMessage message)
{
    ...
    projectile.Setup(message.Position, message.Rotation,
        message.Damage, message.IsTower);
}
```

We have already added the `IsTower` property to the `FireballSpawnMessage` class, so here, we only need to add it as the last parameter of the `Setup` method.

Now, we need to update the script that is responsible for the collisions between enemies, units, and projectiles. Open the `CollisionComponent` script, located inside the **Scripts | Battle** folder, and add the following highlighted `isTower` parameter to the existing `TakeDamageFromProjectile` method, as well as its usage in the last line of the method, instead of the current `false` value:

```
private void TakeDamageFromProjectile(float opponentAttack,
    Transform target, bool isTower)
{
    ...
    StopAttacking(target, isTower);

}
```

The new `isTower` parameter is used to tell the `StopAttacking` method whether the enemy should follow the attacker or not. Using the value of the `isTower` parameter, we can make sure that the enemy will continue to follow the Mage unit, which is the one that also has the ranged attack, and will not follow the Tower since it is not possible to cause damage to it.

After adding the `isTower` parameter to the `TakeDamageFromProjectile` method, we also need to change the `OnCollisionEnter` method. Make the following highlighted change to the existing `OnCollisionEnter` method:

```
private void OnCollisionEnter(Collision collision)
{
    ...
    {
        collision.transform.gameObject.SetActive(false);
        TakeDamageFromProjectile(projectile.Damage,
            collision.transform, projectile.IsTower);
    }
}
```

With this change, we are sending the value of the new `IsTower` public variable from the `ProjectileComponent` class to the updated `TakeDamageFromProjectile` method. Now, we have all the damage and collision methods updated to work with the Tower, and we only need to update three more scripts, so the player will be able to build the Tower.

The Tower is a new unit type, so we must add it to the `UnitType` enum to use it. Open the existing **UnitType** script, located inside the **Scripts | Unit** folder, and make the following highlighted change as the last item in the enum:

```
public enum UnitType
{
    Warrior,
    Mage,
    Tower
}
```

Adding the Tower as the last item of the `UnitType` enum will ensure that we do not break GameObjects that are using `UnitType` by changing the index of the existing elements.

Now, let us update the `LevelManager` class by adding a new method to instantiate the Tower Prefab when we want to add it to the map. Open the existing script `LevelManager` and add the following method:

```
public void AddTower(GameObject prefab)
{
    Instantiate(prefab);
}
```

The new `AddTower` method will make sure we instantiate the Tower Prefab on the map, in the active level scene, such as **Level01** or **Playground**, for example. If it is instantiated in a different scene, for example, the **GameUI** scene, the Tower GameObject will not be visible on the map. That is why the `AddTower` method is simple, as we just need to instantiate the Tower Prefab on the map.

The last script we are going to update is `StoreItemComponent`, so when the player purchases a Tower, we also add the Tower GameObject on the map. Open the `StoreItemComponent` script, located inside the **Scripts | Store** folder, and add the highlighted lines at the end of the existing `SpawnUnit` method:

```
private void SpawnUnit()
{
    ...
    else if (item.Unit == UnitType.Tower)
    {
        LevelManager.Instance.AddTower(item.Prefab);
    }
}
```

Here, in the highlighted lines, we are using the last two changes we added to our code: the new `UnitType.Tower` and the new `AddTower` method. As we can see, the method is adding the Tower `Prefab` if the `Unit` type is `Tower`.

We are all done with the script changes for the Tower, and now we can move on to the last part, which is creating the Tower Prefab and updating the Unit Store data to have it as an option in the Unit Store.

Creating the Tower Prefab

We are going to create a new GameObject for the Tower, and then make a Prefab from it that can be used in the Unit Store data.

Let us start by creating the GameObject and the Prefab:

1. Open the **Playground** scene and, in the **Hierarchy** view, right-click and select the **Create Empty** menu option to add a new GameObject, and name it `DefenseTower`.

2. Left-click on the **DefenseTower** GameObject and, in the **Inspector** view, change **Tag** to `Building` and **Layer** to `Tower`.

3. Click on the **Add Component** button, search for **SphereCollider**, and double-click to add it.

4. Under the **Sphere Collider** component, change the **Radius** property to 8.

5. Click on the **Add Component** button, search for **NavMeshObstacle**, and double-click to add it.

6. Under the **Nav Mesh Obstacle** component, change the **Shape** property to `Capsule`, and then change the **Radius** property to 3.

7. Click on the **Add Component** button, search for **TowerComponent**, and double-click to add it.

8. Drag the existing **Tower** Prefab, located in the **ThirdParty | CanonTower | Prefabs** folder, from the **Project** view, and drop it on the **DefenseTower** GameObject, in the **Hierarchy** view. The **Tower** Prefab must be a child of the **DefenseTower** GameObject.

9. Now, drag the **DefenseTower** GameObject from the **Hierarchy** view, and drop it inside the existing **Prefabs | Level** folder in the **Project** view, to create a new Prefab.

10. Delete the **DefenseTower** GameObject from the **Hierarchy** view; we do not need it in the **Playground** scene.

Now that we have the **DefenseTower** Prefab, we can add it to the **UnitStore** ScriptableObject and have it automatically added as a new store item in the Unit Store pop-up window:

1. In the **Project** view, open the **Data | Store** folder, and left-click to select the **UnitStore** asset.

2. In the **Inspector** view, expand the **Items** property and click on the plus button located in the bottom-right corner of the list of items.

3. In the third item of the list, which we just added, change the text to **Builds a Defense Tower** in the **Description** property:

   I.   Set the values of **Price Gold** and **Price Resources** properties to 20000.

   II.   Change the **Currency Resource** property to Wood.

   III. Change the **Unit** property to Tower.

4. Click on the circle button on the right side of the **Image** property and, in the new window, search for the **weapon_icon** sprite, and double-click on it to add it to the property.

5. Next, click on the circle button on the right side of the **Prefab** property and, in the new window, click on the **Asset** tab, search for the **DefenseTower** Prefab, and double-click on it to add it to the property.

6. Use the **File | Save Project** menu option to save the changes.

We now have three units listed in the **UnitStore** ScriptableObject, with the upgrade available for the **Warrior** and **Mage** units, and the new **Tower** unit with the option to build the GameObject on the map.

These are all the scripts, Prefabs, and configurations we need to add our Defense Tower to the game to make it available in the Unit Store for the player to purchase and build, besides being able to drag and drop it anywhere on the map to the best position for defense.

Testing the Defense Tower

Open either the **Playground** or the **Level01** scene and run the game on the Editor. When clicking on the **Unit** button, the Unit Store pop-up window will be shown with the new Tower unit available to be built on the map, as we can see in the following figure:

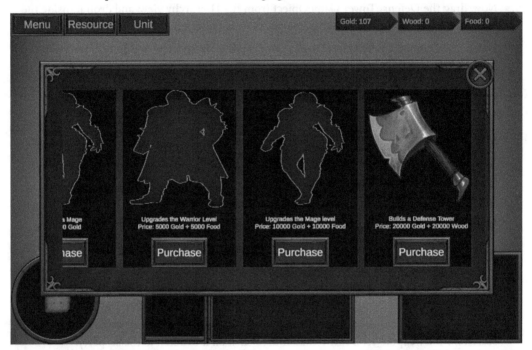

Figure 14.15 – The Unit Store pop-up window updated with the Defense Tower

With a few changes, we were able to add the Defense Tower as a new unit in the game, and most of the changes were due to the uniqueness of the Tower being both a building and a unit with a ranged attack.

Including the Tower on the Unit Store was even easier, since we only needed to add it to the ScriptableObject that has the Unit Store data, and after that, it was immediately available in the game to be purchased.

Use the **Debug** menu to add resources and build the Defense Tower on the map. You will be able to drag and drop it by clicking on the Tower to move it and then clicking one more time to place it back on the map. Move it closer to the enemies and watch it rotate and attack them using the fireball projectile, as shown in the following figure:

Figure 14.16 – The Defense Tower attacking an Orc that is within its range

The Defense Tower is a great unit to have in a RTS game because it helps the player protect the settlement while sending units to explore the map or invade the enemy's base. It has a higher cost than the regular upgrades for the units because of its great offensive and strategic advantages, making it a must-have feature in our RTS game. The same logic we implemented here also works for Tower Defense games, when the player needs to build Towers to defend against waves of enemies.

Summary

Well done on reaching the end of this chapter! It was a very long chapter, with lots of content and new features, and now we have finally added the last features to our RTS game. We will soon have our *Dragoncraft* game completed and ready to play.

In this chapter, we learned how to implement a drag-and-drop feature that can be used to move GameObjects on the map, and we applied it to buildings and the Defense Tower. We also learned how to create a customizable and flexible store to list resource upgrades, which also add new buildings on the map as well as increasing the amount of each resource that is generated.

We also implemented the unit upgrade store, reusing everything we created for the Resource Store, and adding specific scripts to upgrade the unit levels, based on the unit type.

Besides the craft and upgrade buildings to generate more resources over time, and the increase in the upgrade level of the training units, we also added a new unit type that works both as a building and a unit, the Defense Tower, a must-have feature for any RTS game that attacks enemies from a distance when in range.

In *Chapter 15, Game Progression and Objectives*, we are going to put together all the features we have developed so far in this book to create a game progression with objectives for the player, and we will use this system to evaluate whether the player wins or loses the game. We will also add a few UI elements to help communicate with the player, as well as a **Menu** screen and a **Game Over** screen, and we will complete the last details of our *Dragoncraft* game.

15

Tracking Progression and Objectives

Many RTS games have a list of objectives that the player needs to complete to win the level – usually, these objectives need to be completed within a time constraint, forcing the player to think fast and come up with an effective strategy quickly.

In this chapter, we are going to develop the last feature of our RTS game, the game progression, which will use objectives such as collecting resources and defeating enemies before the time is up. We will learn how to create a configuration file to define the objectives, how to track the player progressions on each objective, and how to show the current progression in the UI.

By the end of this chapter, you will learn how to create a flexible and powerful system to configure, track, and display the game objectives and the current player progression. You will also learn how to evaluate whether the player won or lost the game, based on whether they've completed the objectives and, if so, how long it took to complete them. Plus, we will see how to pause the game and display a game over message when the game finishes, with different feedback depending on the outcome.

In this chapter, we will cover the following topics:

- Setting up objectives
- Tracking objectives
- Winning or losing the game

Technical requirements

The project setup in this chapter, with the imported assets, can be found on GitHub at https://github.com/PacktPublishing/Creating-an-RTS-game-in-Unity-2023 in the Chapter15 folder.

Setting up objectives

In our RTS game, the player will have a list of objectives on each level to complete before running out of time. The objectives are a mix of collecting resources and killing enemies, as well as completing them before the time is up. We are going to create a few scripts and a ScriptableObject to help us configure the objectives and include them in the level so that the player knows what needs to be done in the current level to win the game.

Let's start with our first script, which will be a base for the following couple of scripts, which will list objectives related to resources and enemies. In the **Scripts** folder, create a new folder called Objective, and add a new script with the name BaseObjective. Then replace the content with the following code:

```
using System;
namespace Dragoncraft
{
    [Serializable]
    public class BaseObjective
    {
        public int Quantity;
    }
}
```

The BaseObjective class is very simple and contains only one property, Quantity, and the Serializable attribute that is required to use this class as data in a ScriptableObject. As the name suggests, this is a base for another couple of classes that we are going to create next.

The first class we are going to create to use the BaseObjective class is the ResourceObjective class. Add a new script to the **Objective** folder, name it ResourceObjective, and replace the content with the following code snippet:

```
using System;
namespace Dragoncraft
{
    [Serializable]
    public class ResourceObjective : BaseObjective
    {
        public ResourceType Type;
    }
}
```

The ResourceObjective class is inherited from the BaseObjective class and has a single property, Type, that allows us to set up the ResourceType of this objective. Both the Type and the Quantity (from the BaseObjective class) properties will define how many resources the player must collect to complete the objective.

Besides collecting resources, the other type of objective will be killing enemies. For this, we are going to add a new script that will work similarly to the previous ResourceObjective class, but configured to objectives with the enemy in mind. Create a new script in the **Objective** folder, name it EnemyObjective, and replace the content with the following code block:

```
using System;
namespace Dragoncraft
{
    [Serializable]
    public class EnemyObjective : BaseObjective
    {
        public EnemyType Type;
    }
}
```

The EnemyObjective class is very similar to the previous ResourceObjective class as it also inherits from the BaseObjective class and has one Type property that is used to set the EnemyType of this objective. We are going to use the EnemyObjective class to create objectives for the player to kill a certain quantity of enemies, using the Quantity property from the BaseObjective class, and from the specific EnemyType.

Now that have both the ResourceObjective and EnemyObjective classes to define objectives for resources and enemies, respectively, we can move on to the ScriptableObject that will use them as a configuration. In the **Objective** folder, add a new script named ObjectiveData, and replace the content with the following code snippet:

```
using System.Collections.Generic;
using UnityEngine;
namespace Dragoncraft
{
    [CreateAssetMenu(menuName = "Dragoncraft/New Objective")]
    public class ObjectiveData : ScriptableObject
    {
        public string Description;
        public int TimeInSeconds;
        public List<ResourceObjective> Resources =
            new List<ResourceObjective>();
        public List<EnemyObjective> Enemies =
            new List<EnemyObjective>();
    }
}
```

The `ObjectiveData` class has a few properties that we are going to use to define the objectives for the player in one specific level, and we will use this `ScriptableObject` to show the player the objectives in the UI and track their progress on each of them. Here is a quick look at the properties:

- `Description` will be used to display a short message of what the objectives are about

- `TimeInSeconds` is how much time the player has to complete the objectives, and we are going to show this in the UI as a countdown but in time format instead of just the number of seconds

- `Resources` is a list of `ResourceObjective` items, and each item has a quantity and type of resource that the player needs to collect

- The last property is `Enemies`, which is a list of `EnemyObjective` items, where each item has the quantity of a type of enemy that the player must eliminate to complete it

With the `CreateMenuAsset` attribute in the `ObjectiveData` class, we can easily create a new ScriptableObject and define a set of objectives that we will attach to one level. Let's create one asset now with a few objectives:

1. In the **Project** view, right-click on the **Data** folder and select **Create | Folder** to add a new folder. Name it `Objective`.

2. Right-click on the newly created **Objective** folder and select **Create | Dragoncraft | New Objective**. Name it `Objective01`.

3. Left-click on the new **Objective01** asset and, in the **Inspector** view, change the **Description** property to `Defeat and Collect` and change the **Time In Seconds** property to `900` (this is 15 minutes).

4. Expand the **Resources** property and add three new elements. Then, expand **Element 0** and change **Quantity** to `10000` and **Type** to **Gold**; expand **Element 1** and change **Quantity** to `1000` and **Type** to **Wood**; and expand **Element 2** and change **Quantity** to `1000` and **Type** to **Food**.

5. Expand the **Enemies** property and add two new elements. Then, expand **Element 0** and change **Quantity** to `10` and **Type** to **Orc**; and expand **Element 1** and change **Quantity** to `2` and **Type** to **Golem**.

6. Save the changes using the **File | Save Project** menu option.

We just created a new configuration that has three objectives for the player to collect resources, and two objectives for killing enemies on the map, besides the time to complete the objectives being set to 15 minutes (900 seconds). The following screenshot shows **Objective01** after our changes:

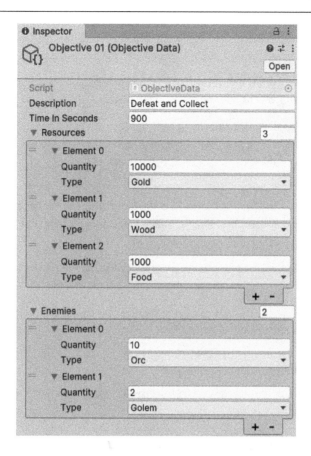

Figure 15.1 – The ObjectiveData ScriptableObject setup

Now that we have our ScriptableObject with a set of objectives configured, we can start using it in our game. The goal of having this asset is to create one configuration for each level, so it makes sense to be part of the level configuration; for this reason, we are going to add a reference to this asset to the LevelManager script and Prefab.

Let's start with the code changes. Open the existing LevelManager script, located inside the **Scripts | Level** folder, and add the following highlighted line inside the class:

```
public class LevelManager : MonoBehaviour
{
    ...

    [SerializeField] private ObjectiveData _objectiveData;

    ...
}
```

ObjectiveData will have a reference to the ScriptableObject that has the objectives for the current level, and we are using SerializeField so we can add the reference to the asset in the Prefab.

Before moving on to the Prefab changes, there is another modification we need to do in the LevelManager class, which is to add the following method:

```
public ObjectiveData GetObjectiveData()
{
  return _objectiveData;
}
```

The GetObjectiveData method, as the name suggests, returns the objectiveData reference that was added to the GameObject that has the LevelManager script attached, which is in the **LevelManager** Prefab.

After saving the changes in the LevelManager class, let's add the reference to the ScriptableObject in the **LevelManager** Prefab:

1. In the **Project** view, locate the existing **LevelManager** Prefab inside the **Prefabs | Level** folder, and double-click it to open it for editing.
2. In the **Inspector** view, under the **Level Manager** component, click on the circle button on the right side of the **Objective Data** property.
3. In the new window, double-click on the **Objective01** asset to select it and add to the Prefab.
4. Save the changes using **File | Save Project**.

Our **LevelManager** Prefab has the reference to the **Objective01** asset, which means that our existing scenes, **Playground** and **Level01**, also have it and can start using it.

Now we can move on to the scripts that will use the ObjectiveData class to display and track the player's progress to complete each objective while playing the game.

Tracking objectives

The classes we created so far in this chapter will help us to define the objectives on each level, and now we are going to implement a new class that is going to take care of showing the updated progress to the player, as well as keeping track of the progress.

To track the progress of how many resources the player collected, we will listen to an existing message, the UpdateResourceMessage class, which is triggered when a new number of resources is produced and is already used by our UI to display the values in the top-right corner of the screen.

We are ready to go with the resource collection progress; however, we do not have a similar message that we can use to track how many enemies were killed by the player. So, before moving on to the script that will keep track of the objectives, we first need to add a new message that should be triggered every time an enemy is defeated.

Let's create a new script in the **Scripts | MessageQueue | Messages | Enemy** folder, name it EnemyKilledMessage, and replace the content with the following code block:

```
namespace Dragoncraft
{
   public class EnemyKilledMessage : IMessage
   {
      public EnemyType Type;
   }
}
```

The EnemyKilledMessage class has only one property, Type, which will have the information related to the EnemyType that the player killed. Also, this class inherits from the IMessage interface so we can send and receive messages of the EnemyKilledMessage type using the MessageQueueManager class.

Now that we have our new message class, we need to send the EnemyKilledMessage in one of our existing classes that has information about the enemy that the player defeated. The best place to send the message is in the DeadComponent class, which is a script added to both unit and enemy GameObjects when they die in combat. Let's open the DeadComponent script, located inside the **Scripts | Battle** folder, and add the following highlighted method inside the class:

```
public class DeadComponent : MonoBehaviour
{
   ...
   public void Start()
   {
      UpdateObjective();
   }
   ...
}
```

This Start method will execute the UpdateObjective method as soon as the DeadComponent class is added to the GameObject as a component. However, this class is added to both units and enemies when they are killed.

In the following `UpdateObjective` method, we are going to add a validation to ensure the message is sent only when the GameObject is an enemy and not a unit. Add the following method to the `DeadComponent` class:

```
private void UpdateObjective()
{
    if (TryGetComponent<EnemyComponent>(out var enemy))
    {
        MessageQueueManager.Instance.SendMessage(
            new EnemyKilledMessage { Type = enemy.Type });
    }
}
```

The `UpdateObjective` method is using the `TryGetComponent` method from the `MonoBehaviour` class to get the `EnemyComponent` class from the GameObject. If the `EnemyComponent` class is present in this GameObject, the `TryGetComponent` method returns `true` and we enter the `if` statement with a valid enemy object, and then we use the `MessageQueueManager` singleton to send the `EnemyKilledMessage` object with the `Type` from the enemy variable.

Now that we have the two distinct messages for tracking the resource collection and enemies defeated, we can move on to the new `ObjectioveComponent` script that will use the `EnemyKilledMessage` and `UpdateResourceMessage` messages to track and update the player progress on each objective and create a new GameObject in the UI to display the current progress on each objective. We will also see how to pause the game when the player opens the pause pop-up window by manipulating the time scale from the Unity Engine.

Creating the Objective component

The next script is responsible for listening to the messages that will update the number of resources collected and the enemies that were killed in combat, and then use this information to update a new UI element to display the progress on each objective. We will first create the script, and then update the **GameUI** scene to have the new UI element that will display the objective progress.

Let's add a new script to the **Scripts | Objective** folder, name it `ObjectiveComponent`, and replace the content with the following code snippet:

```
using System;
using System.Collections.Generic;
using System.Text;
using TMPro;
using UnityEngine;
namespace Dragoncraft
{
    public class ObjectiveComponent : MonoBehaviour
```

```
    {
        [SerializeField] private TMP_Text _objectiveText;

        private Dictionary<ResourceType, int>
            _resourceCounter = new Dictionary<ResourceType, int>();
        private Dictionary<EnemyType, int> _enemyCounter =
            new Dictionary<EnemyType, int>();
        private ObjectiveData _objectiveData;
        private float _timeCounter;
        private bool _playerWin;
    }
}
```

The ObjectiveComponent class is a bit long, so we will see each part of it in the next code blocks. Now, let's look at each property that we defined in this class:

- objectiveText is a reference to a TextMeshPro text GameObject (TMP_Text) that we are going to use to display a formatted message with the objectives and the progress of each, as well as the time left and a short description – all the information will come from the ObjectiveData object and the messages we will use to track the progress.

- resourceCounter is a Dictionary that has the ResourceType as the key and the amount of resources as the value. We will use it to track how much of each ResourceType was collected or generated by the player.

- enemyCounter is also a Dictionary, but this time we are using the EnemyType as a key to count how many enemies the player killed of each type.

- objectiveData is a reference to the ScriptableObject that has the settings for the objectives in the current level, and we are going to get this reference using the LevelManager class.

- timeCounter is a simple float variable that we will use to count the elapsed time and check whether the player still has time left or it is game over.

- The last property, playerWin, is going to be used to validate whether the player completes all the objectives and, therefore, wins the game.

The properties we declared in the ObjectiveComponent class will be used in the subsequent methods that we are going to add to this class, starting with the following OnEnable method:

```
private void OnEnable()
{
  MessageQueueManager.Instance.
    AddListener<EnemyKilledMessage>(OnEnemyKilled);

  MessageQueueManager.Instance.
```

```
        AddListener<UpdateResourceMessage>(OnResourceUpdated);
}
```

The OnEnable method is executed every time the GameObject that has this script attached becomes active. Here we are adding a couple of listeners to the two messages we need to receive to update the progress: EnemyKilledMessage and UpdateResourceMessage. These two messages will execute the OnEnemyKilled and OnResourceUpdated methods, respectively.

Since we are adding listeners, we also need to remove them when they are not needed. We do that in the following OnDisable method:

```
private void OnDisable()
{
  MessageQueueManager.Instance.
    RemoveListener<EnemyKilledMessage>(
      OnEnemyKilled);

  MessageQueueManager.Instance.
    RemoveListener<UpdateResourceMessage>(
      OnResourceUpdated);
}
```

When the GameObject is deactivated, the OnDisable method is executed, and here, we are removing the listeners from EnemyKilledMessage and UpdateResourceMessage, as well as their respective callback methods, OnEnemyKilled and OnResourceUpdated. Using the OnEnable and OnDisable methods to add and remove the message listeners, we ensure that we are waiting for the messages when the object is active.

The next method, Start, is also called when the GameObject is active, but only once, which makes it ideal for initialization code. Add the following Start method to the class:

```
private void Start()
{
  _objectiveData = LevelManager.Instance.GetObjectiveData();
  _timeCounter = _objectiveData.TimeInSeconds;
}
```

In the Start method, we are doing a couple of things: it gets the objectiveData from the LevelManager and initializes the timeCounter property with the TimeInSeconds value. The timeCounter property will be decreased on each game loop, starting from TimeInSeconds; the player loses the game if the objectives are not completed once the timeCounter reaches zero.

The code that will reduce the timeCounter value on each game loop, as well as updating the objective progress in the UI and validating whether the player has reached the end of the game, is in the Update method, as we can see in the following code block:

```
private void Update()
{
    if (_timeCounter > 0)
    {
        _timeCounter -= Time.deltaTime;
        UpdateObjectives();
        CheckGameOver();
    }
}
```

The first thing we check in the Update method is whether the timeCounter is greater than 0, and only execute the next lines if this statement is true. Then, as mentioned prior to the code block, the timeCounter is decreased by subtracting the deltaTime on each game loop.

The last couple of lines in this method are executing the UpdateObjectives and CheckGameOver methods, where we update the UI information with the current progress and check whether the game is over, respectively. We are going to see these two methods in a moment, but first, let's define other methods, such as the OnResourceUpdated method, which we can see in the following code snippet:

```
private void OnResourceUpdated(UpdateResourceMessage message)
{
    if (message.Amount < 0)
    {
        return;
    }

    if (_resourceCounter.TryGetValue(message.Type, out int counter))
    {
        _resourceCounter[message.Type] =
            counter + message.Amount;
    }
    else
    {
        _resourceCounter.Add(message.Type, message.Amount);
    }
}
```

The OnResourceUpdated method is executed when we receive a message of the UpdateResourceMessage type, which indicates that a new amount of one specific ResourceType has been collected or produced by the player. If the Amount is less than 0, which is a negative value, the message was sent to consume resources, so we can ignore the rest of the method using the return keyword to exit the method execution because we will only count the resources that have been collected.

Since we are using a Dictionary to keep track of the resource objective progress, in the resourceCounter property, we first try to find a value using the message Type as the key. If we find an item using the TryGetValue method, we will have the current value in the counter variable and then add the value from the message Amount before updating it in the resourceCounter. If the Type is not found in the resourceCounter, we can safely add both Type and Amount as a new key and value, respectively, using the Add method.

This is how we are going to keep track of the number of resources collected by the player of each ResourceType. The next method defined in the following code block, the OnEnemyKilled method, is doing a similar thing, but instead counts the enemies defeated by the player:

```
private void OnEnemyKilled(EnemyKilledMessage message)
{
    if (_enemyCounter.TryGetValue(message.Type, out int counter))
    {
        _enemyCounter[message.Type] = counter + 1;
    }
    else
    {
        _enemyCounter.Add(message.Type, 1);
    }
}
```

In the OnEnemyKilled method, which is executed when we receive a message of the EnemyKilledMessage type, we are trying to find the current number of enemies of the Type received in the enemyCounter, through the TryGetValue method. If we find one item that has the enemy type, we increase the counter by 1 because this message is received once per enemy that is defeated. Then, if the item is not found, we add a new item to the enemyCounter with the Type key and a value of 1 using the Add method.

After the methods to update the resourceCounter and enemyCounter, we have the UpdateObjective method, which is executed in the Update method, and is defined in the following code snippet:

```
private void UpdateObjectives()
{
    StringBuilder objectives = new StringBuilder();
```

```
objectives.Append(
  $"{_objectiveData.Description}{Environment.NewLine}");

objectives.Append(GetTimeObjectiveText(_timeCounter));

foreach (EnemyObjective enemy in
  _objectiveData.Enemies)
{
  objectives.Append(GetEnemyObjectiveText(enemy));
}

foreach (ResourceObjective resource in
  _objectiveData.Resources)
{
  objectives.Append(GetResourceObjectiveText(resource));
}

  _objectiveText.text = objectives.ToString();
}
```

The UpdateObjectives method is responsible for creating a string with text that contains the Description for each objective, the updated timeCounter, the progress of each EnemyObjective, and the progress of each ResourceObjective. This is how the method creates the string:

- We are using a StringBuilder to create our string, which is perfect for manipulating a string because, being a mutable string, we can modify it without creating a new instance of the string on each change, which is fine; but here, we can use the extra performance that StringBuilder provides.

- The first string we append to the objectives variable is Description, and we are adding the Environment.NewLine constant after the string so we can have a line breaker at the end of the string.

- Next, we have the current time elapsed string, which is created using GetTimeObjectiveText.

- Then, we have a couple of foreach loops:

  - The first foreach loop goes through all the EnemyObjective items in objectiveData, represented by the Enemies property. Here, we are getting each individual EnemyObjective from the list and using the GetEnemyObjectiveText method to create the string with the updated progress.

  - The second foreach loop is similar, but now we are accessing each ResourceObjective in the Objectives property, and then creating the string using the GetResourceObjectiveText method.

- At the end of the method, we convert the `StringBuilder` to a regular string by using the `ToString` method and then setting this value to the `objectiveText` in the UI.

The `StringBuilder` is a great class to use when we are manipulating an unknown number of strings constantly, which is exactly what we are doing in the `UpdateObjectives` method because it is executed on every game loop by the `Update` method.

Now, let's see each individual method that is responsible for creating the strings, starting with `GetTimeObjectiveText` in the following code block:

```
private string GetTimeObjectiveText(float seconds)
{
    if (seconds < 0)
    {
        return $"<color=red>Time left:" +
            $" 00:00{Environment.NewLine}</color>";
    }

    TimeSpan time = TimeSpan.FromSeconds(seconds);
    return $"Time left: {time:mm\\:ss}{Environment.NewLine}";
}
```

The `GetTimeObjectiveText` method is responsible for creating a string with the current time left for the player in the level to complete the objectives. The `seconds` parameter has the information of how much time the player has left, but we are going to display this information to the player in minutes and seconds format, for example, 10:25.

However, before converting the time, we check whether the `seconds` variable is less than 0, which indicates that there is no time left for the player, and in this case, we return a string with the time 00:00. Here, we are using a nice feature of TextMeshPro, which is the `<color=red>` and `</color>` tags that change the color of the text in between these tags to red in the UI, which will give feedback to the player that they have missed this objective.

If the player still has time left, we convert the `seconds` variable to a `TimeSpan` variable using the `FromSeconds` method, and then we use the mm\\:ss string format to convert the `time` variable into a string with the format we want to display, which is minutes and seconds.

The next method, `GetEnemyObjectiveText`, is also creating a script with a format we want to display, but this time it shows how many enemies the player has defeated. Add the following method to the class:

```
private string GetEnemyObjectiveText(EnemyObjective enemy)
{
    _enemyCounter.TryGetValue(enemy.Type, out int counter);
```

```
if (counter >= enemy.Quantity)
{
    return $"<color=green>Kill {enemy.Type}: " +
        $"{enemy.Quantity}/{enemy.Quantity}" +
        $"{Environment.NewLine}</color>";
}

return $"Kill {enemy.Type}: " +
    $"{counter}/{enemy.Quantity}{Environment.NewLine}";
}
```

We are creating a string in the GetEnemyObjectiveText method that has the information on the Type of the enemy, the number of enemies of that type already killed by the player, and the expected Quantity that is in the EnemyObjective. Using the TryGetValue method, we look at the enemyCounter dictionary for the current quantity based on the Type key, and the updated value is returned in the counter variable.

Since the counter variable is an int, even if no value is found, the default value will be zero, which is fine for what we want. As soon as we have the counter variable, we check whether it is equal to or greater than the Quantity that we expect in this objective, and if that is true, we return a string with the <color=green> and </color> tags to give visual feedback that the objective was completed with some green text, and we stop modifying this string.

If this objective is still incomplete, we return a string that has the enemy Type, the counter, and the expected Quantity, so the player can have updated information about how many enemies of this type need to be defeated to complete the objective.

The following GetResourceObjectiveText method is very similar to the previous GetEnemyObjectiveText method, and we are doing the same things but using different variables and text. Add the following code snippet to the class:

```
private string GetResourceObjectiveText(ResourceObjective resource)
{
    _resourceCounter.TryGetValue(resource.Type,
        out int counter);

    if (counter >= resource.Quantity)
    {
        return $"<color=green>Collect {resource.Type}: " +
            $"{resource.Quantity}/{resource.Quantity}" +
            $"{Environment.NewLine}</color>";
    }

    return $"Collect {resource.Type}: " +
        $"{counter}/{resource.Quantity}{Environment.NewLine}";
}
```

The `GetResourceObjectiveText` method uses the `resource` parameter to find the current `counter` for the resource `Type` in the `resourceCounter` dictionary. Then, we check whether the value of the `counter` variable is greater than or equal to the `Quantity` we are expecting to complete this objective. If the player completes this objective, we return the object in green text to give visual feedback that it is completed and we are no longer counting this resource.

If the objective is not complete yet, we return a string with the `Type`, the current `counter` for this resource type, and the `Quantity` the player needs to collect to complete this objective. As we can see, the enemy and resource objective information is very similar but use their own variables and text.

The last method we are going to add to the `ObjectiveComponent` class is the `CheckGameOver` method. However, as we can see in the following code block, this method is empty now:

```
private void CheckGameOver()
{

}
```

The `CheckGameOver` method is empty because we still need to add a few scripts to the project before we can implement it, but we are going to do it in a moment, just after we finish the UI changes for the objective progression.

Creating the Objectives UI

Now that we have finished the `ObjectiveComponent` class, we can move on to the UI. We are going to add this script and a new text to display the string we are building and updating with the objective progress:

1. Open the **GameUI** scene and, in the **Hierarchy** view, right-click on the **Canvas** GameObject. Select the **Create Empty** option and name it `Objectives`. The new **Objectives** GameObject will be added as the last item under the **Canvas** GameObject – drag and drop it right below the **Canvas** GameObject to make it the first GameObject on the **Canvas**.

2. Left-click on the new **Objectives** GameObject and, in the **Inspector** view, in the **Rect Transform** component, change the **Anchor Preset** values to **right** on the right rectangle edge and **top** on the top rectangle edge. Then, change **Pivot X** to **1** and **Pivot Y** to **1**. After that, change **Pos X** to **0**, **Pos Y** to **-50**, **Width** to **200**, and **Height** to **150**.

3. In the **Hierarchy** view, right-click on the **Objectives** GameObject, select the **UI | Image** option, and name it `Background`.

4. Left-click on the newly created **Background** GameObject and, in the **Inspector** view, in the **Rect Transform** component, change the **Anchor Preset** values to **stretch** horizontally and **stretch** vertically. Then, change **Pos X** to **0**, **Pos Y** to **0**, **Right** to **0**, and **Bottom** to **0**.

5. In the **Image** component, click on the **Color** property and, in the new **Color** window, change the values of **R** to **0**, **G** to **0**, **B** to **0**, and **A** to **200**. This will change the image color to black with transparency.

6. In the **Hierarchy** view, right-click on the **Objectives** GameObject, select the **UI | Text - TextMeshPro** option, and name it Text.

7. Left-click on the newly created **Text** GameObject and, in the **Inspector** view, in the **Rect Transform** component, change **Anchor Preset** to **stretch** horizontally and **stretch** vertically. Then, change **Pos X** to **10**, **Pos Y** to **10**, **Right** to **0**, and **Bottom** to **0**.

8. In the **TextMeshPro – Text (UI)** component, click on the **Vertex Color** property and, in the new **Color** window, change the value of **R** to **255**, **G** to **255**, **B** to **0**, and **A** to **255**. This will change the text color to yellow. In the same component, change the **Font Size** property to **14** and click on the **SC** button in the **Font Style** property.

9. In the **Hierarchy** view, left-click on the **Objectives** GameObject to select it. In the **Inspector** view, click on the **Add Component** button, search for ObjectiveComponent, and double-click to add it. Drag the **Text** GameObject we just created from the **Hierarchy** view and drop it into the **Objective Text** property in the **Inspector** view, in the **Objective Component**.

10. Save the scene changes using the **File | Save** menu option.

The **Objectives** GameObject in the UI will display a black panel, with a bit of transparency, on the right side of the screen, with the list of objectives and their current progress in yellow. The following screenshot shows the **Objectives** GameObject after setting it up:

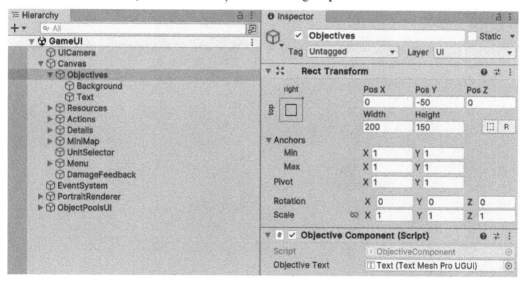

Figure 15.2 – The Objectives GameObject with ObjectiveComponent

The `ObjectiveComponent` script is going to use the **Text** GameObject to display the string we build in the code, with the description, time lapsed, resource collection progress, and enemy defeated progress. The whole text will be displayed in yellow but, as we saw in the `ObjectiveComponent` class, the time text changes to red when there is no more time left, and the resources and enemy-related objectives text changes to green when the objectives are completed.

We now have script, asset, and Prefab changes that enable us to create a configuration with objectives for the player, add it to a level, and display the player's progress toward the current objectives in the UI. The only thing left to do now is to check whether the player wins or loses the game, but before that, let's add one more script to pause the game when the menu is opened.

Pausing the game

We already have a **Menu** button in the top-left corner of the screen that opens the pause popup, with the options to **Resume** or **Exit** the game. However, when this popup is open, the game is not actually paused, and the **Exit** button does nothing at the moment.

Let's change that by adding a new script that will allow us to pause the game when the popup is open, and to resume the game when it is dismissed. Add a new script in the **Scripts | UI** folder, name it `PauseComponent`, and replace the content with the following code block:

```
using UnityEngine;
namespace Dragoncraft
{
  public class PauseComponent : MonoBehaviour
  {
    private void OnEnable()
    {
      Time.timeScale = 0;
    }

    private void OnDisable()
    {
      Time.timeScale = 1;
    }

    public void ExitGame()
    {
      Application.Quit();
    }
  }
}
```

The `PauseComponent` class is quite simple and has only three methods that have one line on each of them. The `OnEnable` and `OnDisable` methods are changing the value of the `Time.timeScale` property between `0` to pause the game and `1` to resume it. `timeScale` is used by the Unity Engine to set up things such as animation, physics, and the game loop, for example; by changing it to `0`, we are telling the engine to stop everything, and the game stays paused until we change it back to `1`.

The last method in this class, the `ExitGame` method, uses the `Quit` method from the `Application` API to close the game. It is worth mentioning that it does not work in the Editor, but it will close the game when we are running it on a desktop PC, for example.

Now that we have our new script for pausing and resuming the game, let's add it to the **PausePopup** GameObject in the UI, so the game is automatically paused and resumed when the pause popup is opened and closed, respectively:

1. Open the **GameUI** scene and, in the **Hierarchy** view, expand the **Canvas** GameObject, then expand the **Menu** GameObject and left-click on the **PausePopup** GameObject to select it.

2. In the **Inspector** view, click on the **Add Component** button, search for `PauseComponent`, and double-click to add it.

3. Save the scene changes using the **File** | **Save** menu option.

With a simple script, we can now pause and resume the game when the pause popup is opened or closed. We are going to reuse this script in the next part of this chapter, where we will create a new popup to display a message saying that the player has won or lost the game.

Winning or losing the game

We have added a nice new feature to our RTS game, the objectives, and now we can configure the level objectives, track the player's progress, and display the progress in the UI. However, there is still one thing left to implement here: the condition to win or lose the game with visual feedback to the player.

We are going to create a couple of new scripts with a new message type for the game over condition, a new component to listen to this message, and a new pop-up message that will display the game outcome to the player.

Let's start by adding the new message type. Create a new script in the **Scripts** | **MessageQueue** | **Messages** | **UI** folder, name it `GameOverMessage`, and replace the content with the following code block:

```
namespace Dragoncraft
{
    public class GameOverMessage : IMessage
    {
        public bool PlayerWin;
    }
}
```

The GameOverMessage class has only the PlayerWin property, which indicates whether the player has won or lost the game. This message will be sent only once during a gameplay session, when the game reaches the end.

To ensure that this message will be sent at the right moment, we are going to add its trigger to the ObjectiveComponent class, which is already taking care of the objective progression and has everything we need to validate whether the player wins or loses the game. We have already added one method, named CheckGameOver, to this class, but it was empty – now we are going to add the proper implementation there.

Open the ObjectiveComponent script, located inside the Scripts/Objective folder, and replace the existing CheckGameOver method with the following code:

```
private void CheckGameOver()
{
  if (_timeCounter <= 0)
  {
    MessageQueueManager.Instance.SendMessage(
      new GameOverMessage { PlayerWin = false });
    return;
  }

  _playerWin = IsEnemyObjectiveCompleted();
  if (!_playerWin)
  {
    return;
  }

  _playerWin = IsResourceObjectiveCompleted();
  if (_playerWin)
  {
    MessageQueueManager.Instance.SendMessage(
      new GameOverMessage { PlayerWin = true });
  }
}
```

The CheckGameOver method can be split into three different parts:

- The first part is a validation to check whether the player does not have time available to complete the objectives – this is done by checking whether the timeCounter is less than or equal to 0. If the validation is true and the time is up, we send a new message of the type GameOverMessage with the PlayerWin property set to false, then use the return keyword to exit this method execution.

- The second part is one of the objective validations. We use the `IsEnemyObjectiveCompleted` method to check whether all the objectives related to enemies being defeated are complete, and then we return the information to the `playerWin` property. If the player has not won yet, we use the `return` keyword to exit this method execution.

- The last part is to check whether the player has completed all the resource-related objectives by using the `IsResourceObjectiveCompleted` method. If `return` is `true` here, it indicates that the player completed both resource and enemy objectives, and we can send a new message of the type `GameOverMessage` with the `PlayerWin` property set to `true`.

It does not matter whether we check whether the player has completed the resource or enemy objectives first, but we need to make sure both are completed before we send the `GameOverMessage`.

Now, we need to add the two methods used in the previous code block to validate whether the objectives were completed or not. Let's start with the `IsEnemyObjectiveCompleted` method in the following code snippet:

```
private bool IsEnemyObjectiveCompleted()
{
    foreach (EnemyObjective enemy in _objectiveData.Enemies)
    {
        _enemyCounter.TryGetValue(enemy.Type, out int counter);
        if (counter < enemy.Quantity)
        {
            return false;
        }
    }
    return true;
}
```

In the `IsEnemyObjectiveCompleted` method, we loop over all the `EnemyObjective` items in `objectiveData` and use the `Type` to get the current `counter` for that enemy type from the `enemyCounter` Dictionary. If the `counter` is less than the `Quantity` required to complete the enemy objective, we return `false`. We do not need to continue the loop if we find one enemy objective that is not completed. Otherwise, if we do not find any incomplete enemy objectives, we return `true` in the last line of the method.

The `IsResourceObjectiveCompleted` method is very similar to the one we saw in the previous code block, but it uses different objects to validate the resource objectives, as we can see in the following code snippet:

```
private bool IsResourceObjectiveCompleted()
{
    foreach (ResourceObjective resource in
        _objectiveData.Resources)
```

```
    {
      _resourceCounter.TryGetValue(resource.Type,
        out int counter);

      if (counter < resource.Quantity)
      {
        return false;
      }
    }
    return true;
  }
```

The `IsResourceObjectiveCompleted` method will execute the same steps that we just saw in the `IsEnemyObjectiveCompleted` method, but this time using the `ResourceObjective` from `objectiveData`. We are checking how much of each resource `Type` was collected by the player in the `resourceCounter` dictionary, and then checking whether the `counter` is less than what the `Quantity` required is. As we did for the enemy objectives, as soon as we find a resource objective that is not complete, we return `false`. If all the resource objectives have the required `Quantity`, we return `true` in the last line of the method.

Our `ObjectiveComponent` class is ready to check whether the player has won or lost the game and sends the appropriate message. Now, the only thing left to do is to create the script to listen to the `GameOverMessage` and create the game over popup that we are going to display when the game ends.

Adding the Game Over popup

The Game Over popup is the last thing that will be displayed to the player during gameplay since it indicates whether they have won or lost the game. We updated our `ObjectiveComponent` script to check whether the game over condition has been met, and we send a message when that happens, including a property that has the information about whether the player won or lost the game.

We are going to create a new component that will listen to the `GameOverMessage` and show the Game Over popup when the message is received, with the correct text to feed back to the player about the outcome. Add a new script in the **Scripts | UI** folder, name it `GameOverComponent`, and replace the content with the following code block:

```
using System;
using TMPro;
using UnityEngine;
using UnityEngine.SceneManagement;
namespace Dragoncraft
{
  public class GameOverComponent : MonoBehaviour
  {
```

```
    [SerializeField] private GameObject _gameOverPopup;
    [SerializeField] private TMP_Text _gameOverText;
  }
}
```

The GameOverComponent class has a couple of properties with the SerializeField attribute, and we are going to add a reference to both gameOverPopup and gameOverText after adding this script to the game over popup that we will create in a moment. The gameOverPopup is the GameObject that we are going to show onscreen when the game is over, and the gameOverText is the description of the game outcome to the player.

The next couple of methods, OnEnable and OnDisable, are where we are going to add and remove the GameOverMessage listener, respectively. Add the following methods to the class:

```
private void OnEnable()
{
  MessageQueueManager.Instance.
    AddListener<GameOverMessage>(OnGameOver);
}

private void OnDisable()
{
  MessageQueueManager.Instance.
    RemoveListener<GameOverMessage>(OnGameOver);
}
```

We are adding the listener to the message type GameOverMessage in the OnEnable method and assigning the callback to the OnGameOver method. In the OnDisable method, we are doing the opposite action, removing the listener to the message type GameOverMessage and unassigning the callback from the OnGameOver method.

The OnGameOver method is defined in the following code block, and it is the main part of the GameOverComponent class because this is where the Game Over popup is displayed with the feedback text for the player:

```
private void OnGameOver(GameOverMessage message)
{
  _gameOverPopup.SetActive(true);

  if (message.PlayerWin)
  {
    _gameOverText.text =
      $"Game Over{Environment.NewLine}" +
        $"<color=green>You Win!</color>";
  }
```

```
    else
    {
      _gameOverText.text =
        $"Game Over{Environment.NewLine}" +
          $"<color=red>You Lose</color>";
    }
  }
}
```

Right after the `GameOverMessage` message type is sent by the `ObjectiveComponent` class, it is received by the `GameOverComponent` class, and the `OnGameOver` method is executed. Here, in the first line, we are making the Game Over popup visible by using the `SetActive` method with a value of `true`. For this to work properly, we are going to disable the Game Over popup GameObject by default. Then, we check whether the player has won the game with the `PlayerWin` property from the `message` parameter. If the player has won, we change the text to `You Win` in green; otherwise, the text will be `You Lose` in red.

Now, we can give feedback to the player about the game's outcome, but we also need to give an option to play again in the Game Over popup. The following `RestartLevel` method will be added to a button, so we can reload the current scene to start again:

```
public void RestartLevel()
{
  SceneManager.LoadScene(
    SceneManager.GetActiveScene().name);
}
```

The `RestartLevel` method will reload the current scene using the `LoadScene` method from the `SceneManager` API and get the current scene with the `GetActiveScene` method. We only need the scene name to load it, but in this case, we can avoid hardcoding the name of the scene by getting the active scene name, which will make this script work on any level you might create.

We just finished the `GameOverComponent` class after adding this last method, and we are ready to start creating our game over popup in the `GameUI` scene. We are going to duplicate the current pause popup and modify it a bit to create the new Game Over popup:

1. Open the **GameUI** scene and, in the **Hierarchy** view, expand the **Canvas** GameObject, and then expand the **Menu** GameObject. Right-click on the **PausePopup** GameObject, select the **Duplicate** option, and rename the duplicated GameObject `GameOverPopup`.

2. Expand the new **GameOverPopup** GameObject, right-click on the **ResumeButton** GameObject, and select the **Delete** option to remove it.

3. Rename the **ExitButton** GameObject `RestartButton` and left-click on it to select. Then, in the **Inspector** view, change **Pos Y** to **-64**. In the **Hierarchy** view, expand **RestartButton** and left-click on the **Text (TMP)** GameObject to select it and, in the **Inspector** view, change the text to `Restart`.

4.   Back in the **Hierarchy** view, right-click on the **GameOverPopup** GameObject, select the **UI |
     Text – TextMeshPro** option, and name it GameOverText. Left-click on the new **GameOverText**
     GameObject and, in the **Inspector** view, in the **Rect Transform** component, change **Pos Y** to **64**.

5.   In the **TextMeshPro – Text (UI)** component, change the **Font Size** property to **32** and click
     on the **SC** button in the **Font Style** property.

6.   In the **Hierarchy** view, click on the **Menu** GameObject and, in the **Inspector** view, click on
     the **Add Component** button, search for GameOverComponent, and double-click to add it.

7.   In the **Game Over Component**, click on the circle button on the right side of the **Game
     Over Popup** property and, in the new window, search for **GameOverPopup** in the **Scene** tab.
     Double-click to select it. Now, click on the circle button on the right side of the **Game Over
     Text** property and, in the new window, search for GameOverText in the **Scene** tab and
     double-click to select it.

8.   In the **Hierarchy** view, left-click on the **RestartButton** GameObject to select it. In the **Inspector**
     view, under **Button**, click on the circle button inside the **On Click ()** section, located under
     Runtime Only. In the new window, search for **Menu** and double-click on it to add the GameObject
     in the **On Click ()**. Then, on the right side of **Runtime Only**, select the **GameOverComponent
     | RestartLevel** option.

9.   Save the changes in the **GameUI** scene using **File | Save**.

After adding the new **GameOverPopup** GameObject to the **SceneUI** we will have the structure in the
Hierarchy view as shown in the following screenshot, as well as the GameOverComponent script
set up in the **Menu** GameObject, in the **Inspector** view:

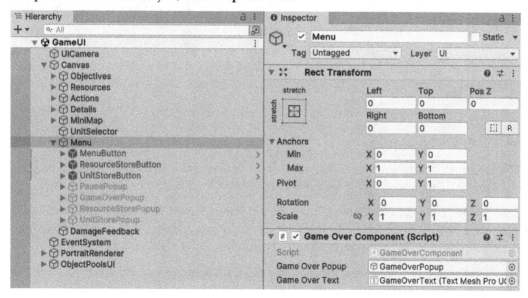

Figure 15.3 – The Menu GameObject with GameOverComponent

We can finally run the game in the Editor, using either the **Playground** or the **Level01** scene, and see the objective progression working as we collect resources and defeat enemies, as well as the game over message if we complete the objectives before the time is up, or if we failed to complete them, as we can see in the following screenshot:

Figure 15.4 – The GAME OVER popup and the objective progress in the top-right corner

We have finished developing the last feature of our RTS game, and it is now ready to play! Add more enemies, buildings, elements on the map, objectives, and even new levels to explore all the features we have and make it your own as well.

Summary

Well done for reaching the end of this chapter! We have finished developing all the major features of an RTS game and our *Dragoncraft* game is ready to play, as well as having more levels.

In this chapter, we added the last feature of the book, which was the game progression using objectives. We learned how to set up objectives for time, resource collection, and enemies defeated, and added them into a ScriptableObject that can be used to create different objectives for each game level.

We also implemented a component for tracking the player's progression in each objective, as well as displaying the status of each objective in the UI to make it easier for the player to track the objectives.

We also learned how to pause the game using the timescale, and added this to our fully functional pause popup.

Finally, we learned how to evaluate whether the player won or lost the game and how to trigger a Game Over popup with visual feedback for the player, with a Restart button that allows you to retry the level.

In *Chapter 16, Exporting and Expanding Your Game*, we are going to finish our game development journey with a few suggestions on how to expand the *Dragoncraft* game with more features. We will also learn how to export our game from the Unity Engine to run on desktop platforms, and a few ways to share the game you just made.

Further reading

In this chapter, we were introduced to a few concepts and APIs that are worth finding out more about:

- *StringBuilder Class*: `https://learn.microsoft.com/en-us/dotnet/api/system.text.stringbuilder`

- *Time.timeScale*: `https://docs.unity3d.com/2023.1/Documentation/ScriptReference/Time-timeScale.html`

- *Application.Quit*: `https://docs.unity3d.com/2023.1/Documentation/ScriptReference/Application.Quit.html`

16
Exporting and Expanding Your Game

It is fun to develop our own games and play them in the Unity Editor, but it is even more fun to share what we have built with other people so they can also play the games we have developed.

In this chapter, we are going to see how to manually export the game for desktop platforms, such as Windows and macOS, with custom build options, and also how to automate this process using a custom build script that will export different builds depending on our build options.

You will also learn how to expand the game by adding background music to our levels and sound effects to give the players feedback on their actions, as well as how to add more content, such as new units, enemies, objectives, and levels – everything you need to make *Dragoncraft* your own game, using the tools and features we developed through the book.

So, in this chapter, we will cover the following topics:

- Exporting Dragoncraft for Desktop
- Creating a build system
- Expanding Dragoncraft

Technical requirements

The project setup in this chapter with the imported assets can be found on GitHub at `https://github.com/PacktPublishing/Creating-an-RTS-game-in-Unity-2023` in the `Chapter16` folder inside the project.

Exporting Dragoncraft for desktop

We have finished developing our *Dragoncraft* game with lots of cool features, and now we are ready to share it so other people can play the game you developed. The Unity Engine has a lot of platform options for exporting the game; however, some of them, such as gaming consoles, require licenses and extra setup. We are going to focus on the desktop platform, which is the most common and easy to use when sharing your game.

In this section, we are going to prepare our code and our project settings to export the game for desktop, and then how to manually build and run the game from the Unity Editor using the **Build Settings** window.

Preparing the Editor scripts

When we export our game from the Unity Engine to run on any platform, some pieces of code might not be available to run on the given platform, and that will cause an error when we try to build the game.

In our project, we used some APIs from Unity that are available only when we are running the game in the Editor, such as `UnityEditor`, or using the custom menu options we created there, such as the **Dragoncraft** menu. To build and run the game on desktop platforms such as Windows or macOS, we need to make sure the APIs that exist only on the Editor are not included when exporting the game to desktop platforms.

Before configuring the project to export a build for desktop platforms, we need to adjust a few scripts to make sure their code will run only on the Editor. We are going to add a **conditional compilation** directive that will allow us to selectively include a piece of code in the compilation, using **scripting symbols**.

In the following code block, `#if` and `#endif` are conditional compilation directives and UNITY_ EDITOR is a scripting symbol defined in the Unity Engine:

```
#if UNITY_EDITOR
...
#endif
```

Any piece of code that is between the conditional compilation directives `#if` and `#endif` will only be included in the build if the scripting symbol is defined by our code or the Unity Engine in this case.

When we are using the Unity Editor, or running the game there, the UNITY_EDITOR scripting symbol is defined by the Unity Engine so any code between the conditional compilation directives will be executed. However, when we are building and running the game on the desktop platform, for example, the piece of code between the conditional compilation directives with the UNITY_EDITOR scripting symbol is not included or executed.

Now, let's add the `#if` and `#endif` conditional compilation directives with the `UNITY_EDITOR` scripting symbol in a few scripts that should only be available in the Editor:

1. Open the `LevelDataEditor` script, located inside the **Scripts | Editor** folder.

2. Add the `#if UNITY_EDITOR` line as the first line of the script, at the very beginning of the file.

3. Add the `#endif` line as the last line of the script at the very end of the file.

4. Open the `EnemyDebugger` script, located inside the **Scripts | Editor | Debug** folder, and repeat *steps 2* and *3*.

5. Open the `ResourceDebugger` script, located inside the **Scripts | Editor | Debug** folder, and repeat *steps 2* and *3*.

6. Open the `UnitDebugger` script, located inside the **Scripts | Editor | Debug** folder, and repeat *steps 2* and *3*.

All scripts that use features from the Editor are now prepared to only work in the Unity Editor and, the most important thing, not cause errors when we export a build to any other platform. Now, we are ready to export our game using the **Build Settings** window.

Exporting the desktop build manually

The Unity Engine has a build system that can be accessed using the **Build Settings** window, where we can select what scenes to include in the build, what platform we want to export the build to, and many other configurations that are specific to each platform.

Let's export our first build for the desktop platform:

1. Open the **Build Settings** window using **File | Build Settings…**.

2. Drag the **Level01** scene from the **Inspector** view and drop it in the **Build Settings** window under **Scenes In Build**.

3. Drag the **GameUI** scene from the Inspector view and drop it in the **Build Settings** window under **Scenes In Build** and below the **Level01** scene.

4. In the left-hand side panel named **Platform**, click **Windows, Mac, Linux**.

5. Then, on the right-hand side, select **Windows** in the **Target Platform** property (if you are using macOS, select the **macOS** option instead).

6. Click on the checkbox on the right side of the **Development Build** property.

7. Click on the **Build** button at the bottom of the **Build Settings** window and select a folder where Unity will export the build. A good practice is creating a folder named `Builds` in the root of the project directory (not inside the `Assets` folder).

8. Once Unity finishes building, a new window from your OS will be opened with all the files and executables exported for the game to run. You can double-click on the executable file and run the game to test it on your computer.

We included two scenes in the build, **Level01** and **GameUI**. As soon as the player runs the game, the first scene in the list will be loaded as the first level of the game. It is important that, in our project, the first scene is **Level01** and the second is **GameUI**, which will be loaded additively by the LevelManager script. The following screenshot shows the two scenes in the **Build Settings** window, as well as **Windows** as the **Target Platform** and the **Development Build** option enabled:

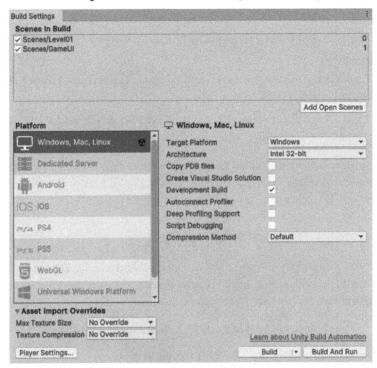

Figure 16.1 – The Build Settings window

We enabled the **Development Build** option in the **Build Settings** window to tell Unity to include scripting debug symbols in our build – these are the recommended settings for testing our game because if an error happens, it will be easy to identify the root cause. In this case, the type of build is **Debug**, and should be used only for tests or analysis. The other type of build is **Release**, where the game does not have any test or debug code and is intended for the final users – the players.

The **Development Build** option also adds the **Profiler** to the build, which lets us connect the build to the Unity Editor and debug in real time. The Profiler is an amazing and essential tool for debugging the game, as well as for analyzing and identifying performance issues and optimizing the game. We are not going to see it in this book, but there are a few links in the *Further reading* section of this chapter with more information on how to use it.

Now that we have learned how to manually export a build from Unity, let's create a script to automate the export process and add custom options, such as different scenes, depending on the build configuration.

Creating a build system

We have learned how to export a build in the Editor using the **Build Settings** window. Now, we are going to see how to automate this process by creating a new `BuildSystem` script, which will use similar options to export the build through the `UnityEditor` API in the code instead of the **Build Settings** window.

Create a new script in the **Scripts | Editor** folder, name it `BuildSystem`, and replace the content with the following code block:

```
#if UNITY_EDITOR
using UnityEditor.Build.Reporting;
using UnityEditor;
using UnityEngine;

namespace Dragoncraft
{
  public static class BuildSystem
  {
    private static string[] _releaseScenes = new[] {
      "Assets/Scenes/Level01.unity",
      "Assets/Scenes/GameUI.unity"
    };

    private static string[] _debugScenes = new[] {
      "Assets/Scenes/Playground.unity",
      "Assets/Scenes/GameUI.unity"
    };
  }
}
#endif
```

The `BuildSystem` class uses features from the Editor platform, so we need to include the `#if UNITY_EDITOR` and `#endif` conditional compilation directives as the first and last lines of the script, respectively. Inside the `BuildSystem` class, we have a couple of properties:

- The `releaseScenes` property is a list of scenes to be included in a release build, which in this case will include the `Level01` and `GameUI` scenes in the build, in that order

- The `debugScenes` property is a list of scenes to be included in a debug build, which includes the `Playground` and `GameUI` scenes in the build

The `releaseScenes` and `debugScenes` properties will be used as parameters in the following `Build` method:

```
private static void Build(BuildTarget buildTarget,
    BuildOptions buildOptions, string[] scenes, string outputPath)
{
    BuildPlayerOptions options = new BuildPlayerOptions();
    options.scenes = scenes;
    options.locationPathName = outputPath;
    options.target = buildTarget;
    options.options = buildOptions;

    BuildReport report = BuildPipeline.BuildPlayer(options);
    BuildSummary summary = report.summary;

    if (summary.result == BuildResult.Succeeded)
    {
        Debug.Log($"Build succeeded: {summary.outputPath}");
    }
    else
    {
        Debug.LogError($"Build failed: {summary.totalErrors}");
    }
}
```

The `Build` method has a few parameters that we are going to use as building options, and it is flexible enough to handle different settings if needed:

- The `buildTarget` parameter is where we set what platform we want to build, for example, Windows or macOS.

- The `buildOptions` parameter has many different values that can be used together to create very specific builds. The **Development Build** option that we saw in the **Build Settings** window will be used in this parameter.

- The `scenes` parameter will be either `debugScenes` or `releaseScenes`, depending on the type of build we want to export.

- The `outputPath` parameter is a path, relative to the root of the project, to the folder we are going to export the build to, including the build name.

Now that we have seen what each parameter is for, let's see how they are used inside the Build method. In the first line of the Build method, we are creating an object of the BuildPlayerOptions type, which we will use to set up our build before exporting it:

- The scenes property is set using the value of the scenes parameter, which is the list of the scenes that will be included in the build

- Next, we set the locationPathName property using the outputPath parameter, which has the relative path and name of the build

- Using the buildTarget parameter, we set the value of the target property, indicating what platform we want to build for

- The last property, options, uses the buildOptions parameter to set the options we want to include in the build

Once we have configured the buildPlayerOptions variable, we can export our game using this variable in the BuildPlayer method, from the BuildPipeline API. This line will make the Editor export the game, and the Build method will be on hold until the build is finished. After the build is exported, the BuildPlayer method will return the build process information in the report variable, and then we can get the summary property from the report variable to check whether the build was successful or not.

If BuildResult is Succeeded, we use Debug.Log to print a message in the **Console** view with the output path of the exported build. Otherwise, if the build export failed, we use Debug.LogError to print an error message in the **Console** view with the totalErrors property value.

Now that we have finished the Build method, we can use it to export builds. Let's start creating the new BuildWindowsDebug method in the following code snippet:

```
[MenuItem("Dragoncraft/Build/Build Windows (Debug)")]
private static void BuildWindowsDebug()
{
    string outputPath =
        "Builds/Windows/Debug/Dragoncraft.exe";

    Build(BuildTarget.StandaloneWindows,
        BuildOptions.Development, _debugScenes, outputPath);
}
```

The BuildWindowsDebug method, as the name suggests, uses the Build method to export a build for Windows, using the StandaloneWindows property, and the Development build option to make it a debug build. We are also using the debugScenes property here to include the scenes for the debug build and the output with Debug in the path we want to export the game executable to. We also included the MenuItem attribute in the BuildWindowsDebug method to add a new option in the **Dragoncraft** menu that will list all the build options.

Now that we have a method that exports a debug build for the Windows platform, let's also include a new method, BuildWindowsRelease, to export another build for the Windows platform, but this time as a release build type. Add the following code block to the BuildSystem class:

```
[MenuItem("Dragoncraft/Build/Build Windows (Release)")]
private static void BuildWindowsRelease()
{
  string outputPath =
    "Builds/Windows/Release/Dragoncraft.exe";

  Build(BuildTarget.StandaloneWindows,
    BuildOptions.None, _releaseScenes, outputPath);
}
```

The BuildWindowsRelease method has most of the same things we used in the previous BuildWindowsDebug method, but this time we changed the output to have Release in the path, changed the build options to None, and changed the scenes to the releaseScenes property. The menu option also has a different name to reflect that the BuildWindowsRelease method will export the release build.

If you are using Windows, these methods are ready to be used. However, if you are using macOS you might need to add a new module to the Unity Editor. We will see how to add support to export a Windows build to macOS, and vice versa, in the next section, where we are going to include methods to export the build for macOS and to export both Windows and macOS executables at the same time.

Adding more platforms

When we install the Unity Engine in a Windows OS, the support for macOS builds is not added automatically. We need to install an additional module to include the macOS platform support, and that is the same when using macOS – we need to install an additional module to support the Windows platform.

Let's include the additional module so we can export builds for both Windows and macOS platforms:

1. Close the Unity Editor, open the Unity Hub application and, on the left side, select **Installs** to see the list of Unity Engines installed.

2. Click on the configuration button with a cog icon in the **Unity 2023.1** (or later) install that you are using and, in the new window, select the option based on your OS:

    A. Select **Mac Build Support (Mono)** if you are using Windows OS.

    B. Select **Windows Build Support (Mono)** if you are using macOS.

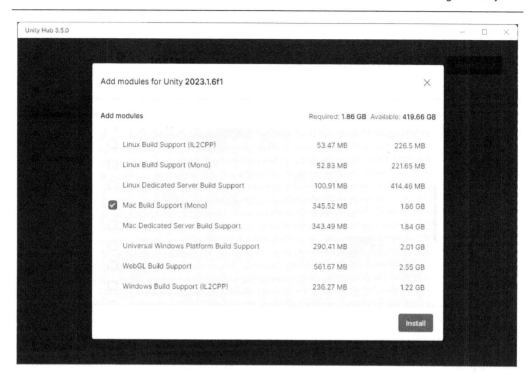

Figure 16.2 – Adding a new module to Unity

3. Click on the **Install** button at the bottom of the window and wait for the Unity Hub to download and install the additional module to your Unity installation.

4. Once the installation process is done, you can open the **Dragoncraft** project again.

Now that we have added the macOS platform support on Windows (or, depending on your OS, Windows platform support on macOS), we can add a couple of new methods to also export builds for macOS. Add the following `BuildOSXDebug` method to the `BuildSystem` class:

```
[MenuItem("Dragoncraft/Build/Build OSX (Debug)")]
private static void BuildOSXDebug()
{
   string outputPath = "Builds/OSX/Debug/Dragoncraft";

   Build(BuildTarget.StandaloneOSX,
      BuildOptions.Development, _debugScenes, outputPath);
}
```

The `BuildOSXDebug` method is almost the same as the `BuildWindowsDebug` method, with the differences being the build target as `StandaloneOSX`, the output having the `OSX` in the path, and the name of the menu option to export this build.

> **Note**
>
> OSX and macOS refer to the same platform – nowadays, the name macOS is most popular; however, Unity has used the name OSX since the release of its first engine.

After we have finished adding the `BuildOSXDebug` method to export debug builds on macOS, we can add the following `BuildOSXRelease` method to export the release builds for the same platform:

```
[MenuItem("Dragoncraft/Build/Build OSX (Release)")]
private static void BuildOSXRelease()
{

  string outputPath = "Builds/OSX/Release/Dragoncraft";

  Build(BuildTarget.StandaloneOSX,
    BuildOptions.None, _releaseScenes, outputPath);
}
```

The `BuildOSXRelease` method is also like the `BuildWindowsRelease` method, and the differences are the same as we just saw between the `BuildOSXDebug` method and the `BuildWindowsDebug` method.

The last method we are going to add to the `BuildSystem` class is the following `BuildAll` method, which we will use to export all the build targets and types we added to this class:

```
[MenuItem("Dragoncraft/Build/Build All (Debug and Release)")]
private static void BuildAll()
{
  BuildWindowsDebug();
  BuildWindowsRelease();
  BuildOSXDebug();
  BuildOSXRelease();
}
```

In the `BuildAll` method, we are executing in sequence the method to export the debug build for Windows, then the release build for Windows, followed by the method to export the debug build for macOS, and the last method to export the release build for macOS.

We added a `MenuItem` attribute to the `BuildAll` method to make it visible in the **Dragoncraft** menu, as we can see in the following screenshot:

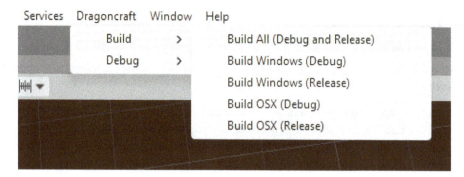

Figure 16.3 – The new Build menu with all options

Now, when we click **Build All (Debug and Release)**, Unity will export the four builds in sequence for us. The following screenshot shows the logs in the **Console** view with the output path of each build:

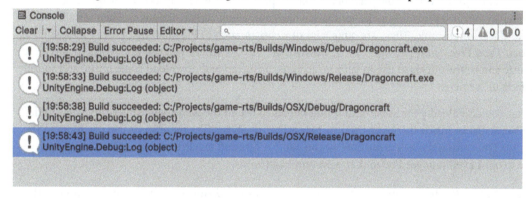

Figure 16.4 – Logs showing all generated builds

We have just finished creating our simple but flexible build script and automated all four builds to be generated using a single menu option. Keep in mind that Windows builds will not work on macOS, and that macOS builds will not work on Windows. Although Unity can export builds, it is not possible to run targets that are different from your OS.

You can zip the release folders with the Windows and macOS builds and share them with anyone to try your game, but remember to include all the contents in the folder since only the executable file is not enough to run the game.

Now, let's look at how we can expand our game by adding music and sound effects, as well as a few things you can do after finishing the book to add more content to the *Dragoncraft* game.

Expanding Dragoncraft

We have a complete RTS game with many features implemented. Now, it is time to see what we can do to expand it by adding more content, such as new enemies and units, a few variations, new upgrades, and new levels.

The first content we are going to see is how to add music and sound effects to the game, which is simple enough that after a couple of examples, you will be able to add as much as you want. After the music and sound effects, we will discuss how to add more content to the game, such as a main menu, new levels, units, enemies, and objectives, using the existing tools and features that we have developed in the book.

Adding music and sound effects

Music and sound effects are big parts of the game experience, helping players become immersed in the gameplay, but also providing audio feedback when completing certain actions or when certain events take place, for example. There are quite a lot of websites, such as Pixabay (`https://pixabay.com/`) and Freesound (`https://freesound.org/`), where you can download royalty-free music and sound effects to use in your projects, so go ahead and download some music and sound effects to use here.

After you download your music and sound effects to use in the game, you can create a new folder called `Sounds` inside the `Assets` directory and copy all the files there so we can use them in the Editor. Once this is done, let's see how we can add background music to our levels by adding a sound file to our **LevelManager** Prefab that is present in all levels:

1. In the **Project** view, double-click on the **LevelManager** Prefab located inside the **Prefabs | Level** folder.

2. In the **Hierarchy** view, right-click on the **LevelManager** GameObject, select **Create Empty**, and name the new GameObject `Music`.

3. Left-click on the newly added **Music** GameObject and, in the **Inspector** view, click on the **Add Component** button. Search for **Audio Source** and double-click on the component to add it.

4. In the **Audio Source** component, click on the circle button on the right side of the **AudioClip** property and, in the new window, select a music file that you imported into the project. In this example, we are using a file named `background_music`.

5. Click on the checkbox on the right side of the **Loop** property to enable it, so the music will continue to play even after it reaches the end.

6. Change the **Volume** property to **0.5** to make the music an ambient sound and to avoid overlapping any sound effects.

7. Save the changes by clicking **File | Save**.

Now, this music will play on every level that is using the **LevelManager** Prefab. If you want to use different music for each level, you can open the scene you want to have the new music and change **AudioClip** to the new file in the **LevelManager** GameObject that is in the scene, so the change is not applied to the Prefab. The following screenshot shows the **Audio Source** configured in the **LevelManager** Prefab with the `background_music` file:

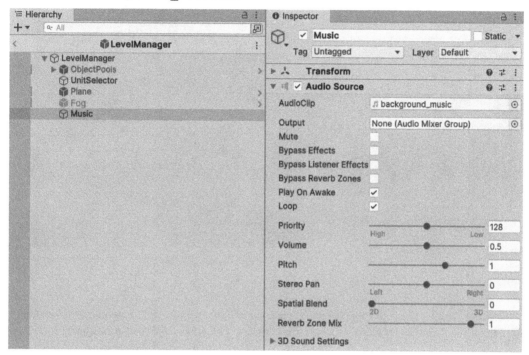

Figure 16.5 – The level music

After adding the background music to the level, we can add a sound effect to one action in the game. The sound effects use the same **Audio Source** component to play the **AudioClip** file, but the difference here is in the configuration since we want our sound effect to play only once (no **Loop** property is enabled) and the **Volume** property has the max value of **1**.

Let's add a new sound effect to the damage feedback that is displayed in the UI when either the enemy or the unit takes damage in the game. The following screenshot shows an **Audio Source** component added to the **DamageFeedback** Prefab, located inside the **Prefabs | UI** folder. Every time a unit or enemy takes damage and the **DamageFeedback** GameObject is created, the `sfx_attack` file is played once to give sound feedback:

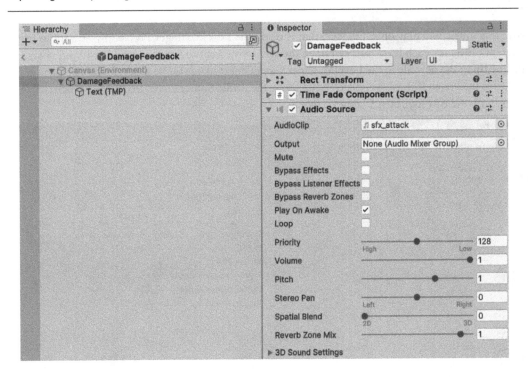

Figure 16.6 – The attack sound effect

If you cannot hear the music or sound effects while testing the game in the Editor, remember to enable the audio in the **Game** view (use the sound icon in the top-right corner of the **Game** view).

As you can see, it is quite simple to add music and sound effects to our game. Now, look at the project and find more places where a sound effect can be added to give feedback. It could be when a pop-up window opens, for example, or when an upgrade is purchased. Sound effects are good feedback for actions in the game, but you should limit music to only one track per level, otherwise it will not be pleasant for the player.

Now, let's move on to the last section of this book and see how to add more content to the *Dragoncraft* game.

Adding more content

The **Playground** and **Level01** scenes are great examples of what a level can contain: dynamic objectives and enemies to interact with. They are good starting points for you to create your own games, but in actuality, they are not challenging enough to play. The following sections are suggestions of content you can add to your game to expand it, as well as what chapters you can refer to when in doubt.

Adding new levels

Adding a new level is as simple as duplicating the **Level01** scene to use it as a template, and then modifying the configuration in the **LevelManager** GameObject. You can create a new `LevelData` ScriptableObject with a different level configuration and add a new `ObjectiveData` object with different objects for the player in the new level.

You can also manually add new elements to the map, such as more trees, houses, and props – there are plenty of Prefabs ready to use inside the `ThirdParty/RPGPP_LT/Prefabs` folder for you to add to your new levels. Do not forget to add the **Food** and **Wood** Prefabs that we created as resources as well.

Look at *Chapter 3, Getting Started with Our Level Design*, to review how to configure the level, and *Chapter 15, Tracking Progression and Objects*, to review how to create objectives for the game. You can also use *Chapter 11, Adding Enemies to the Map*, as a reference to set up the enemies on the map with spawn points and enemy groups.

Adding a main menu

Since you are adding more content to the game, it might be a good idea to create a main menu with options to select what level to play. We covered a lot of UI configuration in *Chapter 4, Creating the User Interface and HUD*, and it should help you to set up a new UI scene for the main menu. Remember to look at the `ThirdParty/GUI_parts` folder to explore other UI elements you can use to customize the interface.

Another change that might help with the game progression is adding a new button to **GameOverPopup**, in the **GameUI** scene, with an option to play the next level. As a suggestion, you could create a simple script to control what the next level is.

Adding new enemies and units

Adding new enemies and units is also simple but requires a few steps since you might want to test the attributes as well to make sure the new enemy or unit is not overpowered or too weak for the game. *Chapter 5, Spawning an Army of Units*, will help you to configure new variations of the Warrior and the Mage units, with different attributes, and *Chapter 9, Adding Enemies*, has everything you need to configure new variations of the Orc and Golem enemies.

You can also add new Dragons to the game using the other Prefabs located in the `ThirdParty/FreeDragons/Prefabs` folder. To test and balance the unit and enemy attributes, you can simulate battles using the unit tests we set up in *Chapter 12, Balancing the Game's Difficulty*.

Chapter 14, Crafting Buildings and Defense Towers, can be used to expand the unit store to set up new configurations and even add new upgrades for the units. It is also a good example of how to add a new type of unit, since we added the Defense Tower as a new unit.

You have plenty of information in this book to help you make *Dragoncraft* your own game, or even create a totally new RTS game with the features we learned about in this book.

Summary

In this final chapter, we learned how to export our game as a build for the Windows and macOS platforms, and how to set up different build options and scenes for debug and release builds. We also learned how to automate the build process by creating a custom build script, as well as how to add new modules to the Unity installation. Finally, we learned how to add music and sound effects to the game, and a list of different content that can be added to the project using the tools and features we developed.

And with that, congratulations – you have reached the end of the book and created an RTS game from scratch! Take a moment to appreciate the game you have created and the new game development skills you learned through this book. Although the focus of this book was to create an RTS game, most of the content can be used to develop different game genres, and now you have the foundation to create many other projects. Congratulations one more time, and good luck with your future projects!

Further reading

Here are a few links from the Unity documentation to learn more about the topics we have covered in this chapter as well as the Profiler, which we did not cover:

- *Conditional Compilation*: `https://docs.unity3d.com/2023.1/Documentation/Manual/PlatformDependentCompilation.html`

- *Build Settings*: `https://docs.unity3d.com/2023.1/Documentation/Manual/BuildSettings.html`

- *Profiler overview*: `https://docs.unity3d.com/2023.1/Documentation/Manual/Profiler.html`

- *Analysis*: `https://docs.unity3d.com/2023.1/Documentation/Manual/analysis.html`

- *BuildPipeline.BuildPlayer*: `https://docs.unity3d.com/2023.1/Documentation/ScriptReference/BuildPipeline.BuildPlayer.html`

Index

A

A* (A star) algorithm 209, 210
animation state connections
  removing 236
Animator 138
Assembly Definition file 346
assets
  importing, from Unity Asset Store 20
  organizing, from Unity Asset Store 21, 22

B

baking 213
base scene
  camera 26, 27
  Directional Light settings 26
  ground foundation 27, 28
  level scene 28-30
  Light settings 25
  scenes 24
  setting up 24
BaseSpawner class
  creating 107, 108
bitwise 164
buildings
  Barracks 10
  Blacksmith 10
  crafting 412
  Defense Tower 10
  dragging 412-420
  dropping 412-420
  Farm 10
  Resource Store, configuring 420
  Town Hall 10
  upgrading 412
build system
  creating 503-506
  platforms, adding 506-509

C

camera controller
  creating 55-60
camera, RTS game 5
Canon Tower package 458
Canvas
  setting up 66
  using, for responsive UI 64-70
Canvas Scaler 66
character life cycle
  implementing 304-308
characters, RTS game 5
code refactoring 274

collision component

creating 290-296

Command pattern 166

example 166

conditional compilation directive 500

custom editor, for level configuration

adding 42, 43

buttons, adding 47-49

level details, adding 44, 45

level slots, adding 45, 46

custom group, Prefabs

creating 35, 36

D

damage calculation 274

base class, adding for characters 277-281

base class, adding for data 274-277

damage, adding to ranged attack 288, 289

EnemyComponent, updating 281-284

UnitComponent, updating 284-287

damage feedback, in UI 255

managing 256-258

Prefabs, setting up 260-263

testing 263-267

text over time, fading 258-260

debugging

NavMesh 226-230

debugging tool

creating, for Editor 118

Defense Tower 457

adding, as unit 460

layer, adding 459, 460

model, adding 458, 459

prefab, creating 466, 467

scripts, modifying 460-466

testing 468, 469

directional light 24

Dragoncraft game 6

bosses 7, 8

characters 7

content, adding 512

desktop build, exporting manually 501, 502

Editor scripts, preparing 500, 501

enemies 7, 8

enemies and units, adding 513

expanding 510

experience 9

exporting, for desktop 500

gameplay mechanics 10

gameplay modes 10

gameplay overview 8

levels, adding 513

main menu, adding 513

modes 10

music and sound effects, adding 510-512

outline 6

world 6

Dynamic Link Library (DLL) 346

E

Editor

debugging tool, creating for 118

debug script, creating 123-125

enemies

configuring 236-242

enemy component, creating 244-247

enemy messages, creating 243, 244

enemy spawner, configuring 250-252

enemy spawner, creating 247-249

enemy spawner, testing 253-255

patrol behavior, adding to 336-340

spawning 242, 243

EnemyComponent

updating 281, 282

Enemy NavMesh
creating 297-301

Event Queue pattern. *See* **Message Queue pattern**

Explore, Expand, Exploit, Exterminate (4X) 4

F

fog, adding on map 321
cameras, updating 322, 323
fog component, creating 323-329
Level Manager, updating 332-336
material, creating 329, 330
new layer, adding 322
Prefab, creating 330, 331
Fog of War 321

G

game
condition, to win or lose 489-492
pausing 488, 489
game configuration
ScriptableObjects, creating for 37-40
game design document (GDD)
creating 6
game design, RTS game 4
Game Over popup
adding 492-496
GameUI
updating 181-183
Global Unique Identifier (GUID) 114
graphical user interface (GUI) 43
Graphic Raycaster 66, 182
Greedy Best-First algorithm 208

H

Health Points (HP) 85, 255
high-definition (HD) 68

I

inheritance 276
integrated development environment (IDE) 13

K

KeyCode 135

L

level configuration
custom editor, adding for 42-45
ScriptableObjects, creating for 40-42
level design, RTS game 4
LevelManager Prefab 130
level scene, preparing 136

M

Mage character
adding 191
data, creating 191, 192
object pool, creating 194, 195
ranged attacks, setting up with fireballs 196-203
spawner, creating 193, 194
map configuration
ScriptableObjects, using for 36, 37
map editor
creating, for speeding up map creation 49-54

map layout
creating, with Prefabs 30-34
Mask component 93
MenuButton GameObject
creating 72-74
Mesh Renderer 27
Message Queue pattern 101, 108
implementing 109-112
minimap 91
rendering 91-93
multiplayer online battle arena (MOBA) 4

N

NavMesh 207, 210-213
adding, to RTS game 213-217
debugging 226-230
updating 301-303
NavMesh Agent component 217
adding, to RTS game 217-220
NavMesh Obstacle component 220
adding, to RTS game 220-226
Nested Prefabs 331
NUnit unit testing framework 350

O

Objective component
creating 478-486
objectives 472
setting up 472-476
tracking 476-478
Objectives UI
creating 486-488
Object-Oriented Programming (OOP) 276
Object Pool
creating, for Warrior unit 118-122

Object Pooling pattern 101
implementing 105-107
used, for spawning units 104

P

pathfinder 208
A* (A star) 209, 210
Greedy Best-First 208
implementing, with NavMesh 213
NavMesh 210, 211
Unity AI Navigation package 211-213
path nodes 210
patrol behavior
adding, to enemies 336-340
PausePopup GameObject
creating 75-79
physics settings
updating 270-273
plane 27
player control, RTS game 5, 6
Points of Visibility (POVs) 210
pop-up window background
adding 434
close button, adding 435, 436
error message, adding 436, 437
Prefabs 23
custom group, creating 35, 36
first map layout, creating with 30-34
Prefabs and UI preparation 128
debug options, adding 128-130
level scene, preparing 136-138
performing 130, 131
selected area, drawing in UI 132-136
Prototype pattern 249

R

Raycast 145

real-time strategy (RTS) games 3

battle difficulty, balancing 374-379

camera 5

characters 5

classic examples 4

Dune II 4

game design 4

level design 4

modern games 4

NavMesh Agent component, adding 217-220

NavMesh component, adding 213-217

NavMesh Obstacle component,
adding 220-226

player control 5, 6

StarCraft 4

Warcraft III 4

real-time tactics (RTT) 4

Render Texture

creating 92, 93

using 87

resource gathering 398

Resource layer, adding 399, 400

resources, producing manually 401-405

resource generation 382-385

LevelManager prefab, updating 397, 398

LevelManager script, updating 394-396

player's inventory, managing 385-390

resources, producing automatically 390-394

resources

testing 406-409

Resource Store

configuring 420-426

Resource Store UI

creating 426-433

responsive UI

creating, with Canvas 64-70

S

ScriptableObjects 23, 36

creating, for game configuration 37-40

creating, for level configuration 40-42

using, for map configuration 36, 37

scripting symbols

using 500

selected units

actions, setting up 163

bitwise operation, for actions 164, 165

Command pattern 166-168

details, setting up 154, 155

dynamic buttons, setting up 168-176

GameObject class, extending 155-158

portrait model, updating 161-163

UI, updating with 154

updating 158-161

Singleton pattern 109, 110

spawn points

creating 312-321

Sprites 69

State design pattern 282

state pattern 188

Store UI

button, adding 433, 434

container, adding for items 438, 439

creating 433

pop-up window background, adding 434

scripts, adding 441-445

scripts, configuring 441-445

store items, adding 440, 441

T

ternary conditional operator 188

Test-Driven Development (TDD) 350

TextMeshPro (TMP) 70

TowerComponent class
attackCooldown 461
attack property 461
cooldown property 461
targets variable 461

tuple 364

turn-based strategy (TBS) game 4

U

UI scene
loading 94-97

UI, setting up with Prefabs 70, 71
action buttons, adding 82-85
details panel, creating with
custom camera 85-91
MenuButton GameObject, creating 72-74
PausePopup GameObject 75-79
resources counter, displaying 79-82

UnitComponent class
updating 284-287

UnitComponent script 184

units
attacking animations 184-190
collisions, setting up 179-181
custom color, selecting 139-143
for attacking 176-178
for defending 176-178
layers, setting up 179-181
moving 147-150
selecting 138, 139
spawning, with Object Pooling pattern 104

unit ScriptableObject
configuring 102-104

unit selector component
defining 143-147

unit spawning system
BaseSpawner class, creating 107, 108
message interface, creating 112
Message Queue pattern,
implementing 109-112
Object Pooling pattern,
implementing 105-107
resource type, creating 115
UI, updating 116-118
warrior unit spawner,
implementing 113, 114

Unit Store
configuring 446-449
creating 445
testing 456

Unit Store UI
button, adding 453
creating 449
listener, adding 455, 456
listener, removing 455, 456
pop-up window, adding 453, 454
supporting scripts, updating 450-452
supporting UI elements, updating 453
upgrade message, adding 456

unit tests
battles, simulating against multiple
enemies 369-374
battles, simulating against
single enemy 362-368
test script, adding 356-359
test script, creating 350-356
used, for simulating battles 359-362
writing 346

Unity AI Navigation package 211-213
NavMesh 213
NavMesh Agent 213
NavMesh Obstacle 213
Off-Mesh Link 213
Unity Asset Store 13
assets, importing from 20-22
assets, organizing 20-22
URL 20
Unity Editor 14
downloading 14
installing 15, 16
URL 14
Unity engine 13
Unity Hub 14
Unity Test Framework (UTF) 346
setting up 346-349

V

Visual Studio Code 13, 16
URL 16
using, as default IDE 16-19

W

Warrior unit
Object Pool, creating for 118-122
warrior unit spawner
implementing 113, 114

www.packtpub.com

Subscribe to our online digital library for full access to over 7,000 books and videos, as well as industry leading tools to help you plan your personal development and advance your career. For more information, please visit our website.

Why subscribe?

- Spend less time learning and more time coding with practical eBooks and Videos from over 4,000 industry professionals

- Improve your learning with Skill Plans built especially for you

- Get a free eBook or video every month

- Fully searchable for easy access to vital information

- Copy and paste, print, and bookmark content

Did you know that Packt offers eBook versions of every book published, with PDF and ePub files available? You can upgrade to the eBook version at packtpub.com and as a print book customer, you are entitled to a discount on the eBook copy. Get in touch with us at customercare@packtpub.com for more details.

At www.packtpub.com, you can also read a collection of free technical articles, sign up for a range of free newsletters, and receive exclusive discounts and offers on Packt books and eBooks.

Other Books You May Enjoy

If you enjoyed this book, you may be interested in these other books by Packt:

Unity 2022 Mobile Game Development

John P. Doran

ISBN: 978-1-80461-372-6

- Design responsive UIs for your mobile games
- Detect collisions, receive user input, and create player movements
- Create interesting gameplay elements using mobile device input
- Add custom icons and presentation options
- Keep players engaged by using Unity s mobile notification package
- Integrate social media into your projects
- Add augmented reality features to your game for real-world appeal
- Make your games juicy with post-processing and particle effects

Become a Unity Shaders Guru

Mina Pêcheux

ISBN: 978-1-83763-674-7

- Understand the main differences between the legacy render pipeline and the SRP
- Create shaders in Unity with HLSL code and the Shader Graph 10 tool
- Implement common game shaders for VFX, animation, procedural generation, and more
- Experiment with offloading work from the CPU to the GPU
- Identify different optimization tools and their uses
- Discover useful URP shaders and re-adapt them in your projects

Packt is searching for authors like you

If you're interested in becoming an author for Packt, please visit `authors.packtpub.com` and apply today. We have worked with thousands of developers and tech professionals, just like you, to help them share their insight with the global tech community. You can make a general application, apply for a specific hot topic that we are recruiting an author for, or submit your own idea.

Share Your Thoughts

Now you've finished *Creating an RTS Game in Unity 2023*, we'd love to hear your thoughts! Scan the QR code below to go straight to the Amazon review page for this book and share your feedback or leave a review on the site that you purchased it from.

`https://packt.link/r/1-804-61324-X`

Your review is important to us and the tech community and will help us make sure we're delivering excellent quality content.

Download a free PDF copy of this book

Thanks for purchasing this book!

Do you like to read on the go but are unable to carry your print books everywhere?

Is your eBook purchase not compatible with the device of your choice?

Don't worry, now with every Packt book you get a DRM-free PDF version of that book at no cost.

Read anywhere, any place, on any device. Search, copy, and paste code from your favorite technical books directly into your application.

The perks don't stop there, you can get exclusive access to discounts, newsletters, and great free content in your inbox daily

Follow these simple steps to get the benefits:

1. Scan the QR code or visit the link below

https://packt.link/free-ebook/9781804613245

2. Submit your proof of purchase
3. That's it! We'll send your free PDF and other benefits to your email directly

www.ingramcontent.com/pod-product-compliance
Lightning Source LLC
Chambersburg PA
CBHW060638060326
40690CB00020B/4436